Lecture Notes in Physics

New Series m: Monographs

Springer

Berlin
Heidelberg
New York
Barcelona
Budapest
Hong Kong
London
Milan
Paris
Tokyo

The Editorial Policy for Monographs

The series Lecture Notes in Physics reports new developments in physical research and teaching - quickly, informally, and at a high level. The type of material considered for publication in the New Series m includes monographs presenting original research or new angles in a classical field. The timeliness of a manuscript is more important than its form, which may be preliminary or tentative. Manuscripts should be reasonably self-contained. They will often present not only results of the author(s) but also related work by other people and will provide sufficient motivation, examples, and applications.

The manuscripts or a detailed description thereof should be submitted either to one of the series editors or to the managing editor. The proposal is then carefully refereed. A final decision concerning publication can often only be made on the basis of the complete manuscript, but otherwise the editors will try to make a preliminary decision as definite as they can on the basis of the available information.

Manuscripts should be no less than 100 and preferably no more than 400 pages in length. Final manuscripts should preferably be in English, or possibly in French or German. They should include a table of contents and an informative introduction accessible also to readers not particularly familiar with the topic treated. Authors are free to use the material in other publications. However, if extensive use is made elsewhere, the publisher should be informed. Authors receive jointly 50 complimentary copies of their book. They are entitled to purchase further copies of their book at a reduced rate. As a rule no reprints of individual contributions can be supplied. No royalty is paid on Lecture Notes in Physics volumes. Commitment to publish is made by letter of interest rather than by signing a formal contract. Springer-Verlag secures the copyright for each volume.

The Production Process

The books are hardbound, and quality paper appropriate to the needs of the author(s) is used. Publication time is about ten weeks. More than twenty years of experience guarantee authors the best possible service. To reach the goal of rapid publication at a low price the technique of photographic reproduction from a camera-ready manuscript was chosen. This process shifts the main responsibility for the technical quality considerably from the publisher to the author. We therefore urge all authors to observe very carefully our guidelines for the preparation of camera-ready manuscripts, which we will supply on request. This applies especially to the quality of figures and halftones submitted for publication. Figures should be submitted as originals or glossy prints, as very often Xerox copies are not suitable for reproduction. For the same reason, any writing within figures should not be smaller than 2.5 mm. It might be useful to look at some of the volumes already published or, especially if some atypical text is planned, to write to the Physics Editorial Department of Springer-Verlag direct. This avoids mistakes and time-consuming correspondence during the production period.

As a special service, we offer free of charge LaTeX and TeX macro packages to format the text according to Springer-Verlag's quality requirements. We strongly recommend authors to make use of this offer, as the result will be a book of considerably improved technical quality.

Manuscripts not meeting the technical standard of the series will have to be returned for improvement.

For further information please contact Springer-Verlag, Physics Editorial Department II, Tiergartenstrasse 17, D-69121 Heidelberg, Germany.

Paul Busch   Marian Grabowski
Pekka J. Lahti

Operational
Quantum Physics

Springer

Authors

Paul Busch
Institute for Theoretical Physics
University of Cologne
D-50937 Cologne, Germany
and
Department of Applied Mathematics
The University of Hull
Kingston upon Hull HU6 7RX, United Kingdom

Marian Grabowski
Institute of Physics and
Institute of Philosophy
Nicolaus Copernicus University
PL-87100 Toruń, Poland

Pekka J. Lahti
Department of Physics
University of Turku
SF-20500 Turku, Finland

ISBN 3-540-59358-6 Springer-Verlag Berlin Heidelberg New York

CIP data applied for.

© Springer-Verlag Berlin Heidelberg 1995
Printed in Germany

Typesetting: Camera-ready by authors using TeX
SPIN: 10127367   55/3142-543210 - Printed on acid-free paper

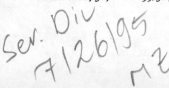

In memory of Eugene P. Wigner

Preface

Operational Quantum Physics offers a systematic presentation of quantum mechanics which makes exhaustive use of the full probabilistic structure of this theory. Accordingly the notion of an observable as a positive operator valued (POV) measure is explained in great detail, and the ensuing quantum measurement theory is developed and applied both to a resolution of long-standing conceptual and interpretational puzzles in the foundations of quantum mechanics, and to an analysis of various recent fundamental experiments. Fundamental to the present approach is the distinction between sharp and unsharp observables: the former correspond to the commonly used self-adjoint operators or their spectral measures, while the latter are given by POV measures that are not built by projection operators.

The book, or different parts of it, will be of interest to advanced students or researchers in quantum physics, to philosophers of physics, as well as to mathematicians working on operator valued measures. The first two chapters provide the motivations behind and a systematic development of the physical concepts and mathematical language. It is here where the measurement theoretic and operational foundations are laid for a realistic interpretation of quantum mechanics as a theory for individual systems. This interpretation, which has been the authors' guide in their research work in quantum physics, seems to come closest to physicists' practice in devising theoretical models and conceiving new experiments. It is illustrated in the phase space measurement model of Chapter VI, where Heisenberg's interpretation of the indeterminacy relation in terms of individual, irreducible quantum inaccuracies is demonstrated. Apart from this instance, most of the other issues treated do not presuppose adherence to such an interpretation of quantum mechanics.

Chapter III illustrates the use of POV measures in carrying out the covariance point of view for the operational definition of an observable. The underlying space–time symmetry is that of the Galilei group. While an analogous programme can be carried out for the Einstein-relativistic case, it has not been included here since the corresponding measurement theoretic foundation does not exist. The aspect of covariance offers a physically satisfactory unifying approach and at the same time opens up a variety of mathematical questions.

The foundations of quantum mechanics are addressed in Chapters IV–VI. For example, the possibility of measuring jointly noncommuting observables is spelled out conceptually and in terms of realisable models, some of them proposed recently in the field of quantum optics from which the majority of the experimental examples provided in Chapter VII are drawn.

Although this book applies quantum measurement theory as a tool in foundational investigations and analyses of experiments, it does not address the perennial measurement problem. It does, however, provide evidence that unsharp observables may be important in developing a coherent quantum picture of classical physical phenomena and thus of the occurrence of definite events. In fact POV measures allow for a novel concept of coarse-graining, as reviewed in Chapter V and applied in Chapter VI in the characterisation of a quasi-classical phase space measurement situation.

This book has grown out of research work that the authors have enjoyed carrying out in collaborations over more than the past decade. Support in the form of Fellowships, grants and exchange programs has been extended to us by the Academy of Finland, the Alexander von Humboldt-Foundation, the BMFT in Bonn, the Polish Academy of Sciences, the Research Institute for Theoretical Physics in Helsinki, and the University of Turku Foundation. In the final stage of this work one of us (PB) enjoyed hospitality and support from the Department of Physics, Harvard University.

Cambridge – Toruń – Turku
February 28, 1995

Paul Busch
Marian Grabowski
Pekka Lahti

Contents

Prologue

The theory of quantum mechanics on Hilbert space has been the basis of fruitful and deep research into virtually all branches of physics for nearly seventy years. There seems to be no instance of a conflict between theoretical predictions and experimental results. In view of this success it is remarkable that a few conceptual problems have resisted any attempted resolution even until now. Some of them became tractable once the probabilistic structure of quantum mechanics was properly appreciated in its full generality. Gleason's theorem and the introduction of the notion of an observable as a *positive operator valued* (POV) *measure* were the crucial steps in this development. Interestingly, the latter discovery was made independently in a variety of rather disparate areas of quantum physics, motivations ranging from foundational interests to fairly practical needs. This wide scope of the concept of POV measure already demonstrates its status as an integral part of the basic structure of quantum mechanics. The traditional notion of observables as self-adjoint operators constitutes a special case, represented by projection operator valued (PV, or spectral) measures.

The incorporation of POV measures into the quantum vocabulary has not only opened up new ways of approaching longstanding theoretical puzzles, it also gave rise to an elaboration of quantum measurement theory into a conceptually sound and mathematically rigorous, powerful tool for analysing physical experiments. The ensuing research activities have led to a variety of reviews and monographs dealing with such diverse topics as probabilistic and statistical aspects of quantum theory, quantum estimation and detection, quantum theory of open systems, photon counting theory, or quantum mechanics on phase space. Much of this work is done on a high technical level and has thus contributed to setting new standards of rigour for investigations in the foundations of quantum physics. At the same time the POV measure approach to quantum observables has gradually induced a probabilistic reformulation of quantum theory that is conceptually simpler and closer to experimental practice than the traditional approach.

The present book is a result of two intimately related lines of research efforts that took place in the past decade. On the one hand quantum measurement theory has found manifold successful applications leading to new insights in the analysis of fundamental experiments. On the other hand considerable progress has been made in working out the operational conditions needed for associating POV measures with properties of physical systems. Both developments have contributed to fully appreciating the relevance of quantum mechanics as a theory of *individual* objects. The need for such a *realistic* interpretation of quantum theory is strikingly evident in these days where one is witnessing worldwide activities in carrying out exciting experiments with single microsystems such as atoms, ions, neutrons, or photons – experiments which formerly could only be conceived as thought experiments.

Our primary concern is twofold. First we wish to demonstrate the amazing capabilities of the quantum formalism if applied in its full-fledged probabilistic form. The advantages of POV measures and of measurement theory, taken as tools of investigation into the quantum world, will be illustrated in several steps and on different levels of sophistication. Yet the notion of a POV measure must itself be subjected to a measurement theoretical analysis in order to elucidate its physical meaning. Reference to this double role of the measurement theory is the ultimate purpose of the term 'operational' appearing in the book title.

The understanding of POV measures as representing observable properties of a physical system will be based on a realistic, individual interpretation of quantum theory. According to this interpretation quantum mechanics describes physical systems existing independently once they have been prepared or identified by observation. Evidence for the presence of a system may be ascertained by means of determining its real (or actual) properties; and using ideal measurements this can be achieved without thereby changing the system in any way. In general, however, a property will be nonobjective (or potential) and may be actualised through measurement. Such repeatable measurements can thus be used to prepare systems with actual properties. It is an important result of quantum measurement theory that the formalism is rich enough to ensure the existence of measurement operations serving these purposes demanded by the realistic interpretation. As a consequence the famous reality criterion by Einstein, Podolsky and Rosen, which states certain predictability on the basis of non-disturbing observations as a sufficient condition for ascertaining actual properties, is naturally incorporated into quantum mechanics, along with establishing the lattice of Hilbert space projection operators as representing the totality of properties of a physical system. Ordinary observables, described as PV measures, refer thus to collections of properties associated with the values of the measured quantities.

This interpretation can be extended to observables represented as POV measures provided that the criterion of reality is relaxed appropriately so as to be applicable in situations to be characterised in terms of approximately real properties. The positive operators in the range of a POV measure, also called *effects*, represent the occurrence of particular outcomes of measurements. The expectation values of the effects are interpreted as the probabilities for these events. Instead of probabilities 1 or 0 one should in general require only probabilities close to these values in order to be able to ascertain that some property is *approximately* real or absent. This leads to a generalised notion of *properties*, which comprises both, the projection operators referred to as *sharp properties*, and certain genuine effects, the so-called *unsharp properties*. Again the non-disturbing and repeatable measurement operations required by such an interpretation will be shown to exist in the quantum formalism.

Unsharp observables arise naturally in the theoretical analysis of experimental procedures. What kind of properties they represent can often be determined by making reference to some known sharp observable. Indeed many POV measures derive from some PV measures by a *coarse-graining* procedure. For example, a function of the position observable Q can be considered as a coarse-grained version of Q; this gives rise, among others, to the discretised versions of Q which are still represented as PV measures. A POV measure associated with Q arises if instead one performs a convolution of the spectral measure with some confidence function. This will be one of our prominent and prototopical examples. We shall refer to such an unsharp observable as a smeared position observable.

In order to avoid misunderstandings regarding our terminology, it is important to be aware of the new, non-classical meaning of the terms 'inaccuracy' and 'unsharpness' in the context of quantum measurements. An inaccurate measurement refers to a situation where instead of a given observable a coarse-grained version of it is measured. By contrast, the term 'unsharp measurement' shall refer to the measurement of an unsharp observable. One should bear in mind the conceptual difference between the *relation* of coarse-graining that can exist between pairs of observables, and the *property* of being an unsharp observable that pertains to some POV measures. Some, but not all, unsharp observables arise as smeared versions of a sharp observable. The unsharpness in question should not in general be taken as an imperfect perception of an underlying more sharply determined property. On the contrary, this term is intended to describe possible elements of reality whose preparation and determination are subject to inherent limitations.

The unsharpness brought about by coarse-graining may or may not admit the kind of ignorance interpretation familiar from classical physical experimentation. In general, however, the unsharpness is a reflection of a genuine quantum indeterminacy. This turns out to be the case, e.g., with the individual measurement interpretation of the uncertainty relation. The question of how to interpret the source of unsharpness and inaccuracy will be investigated in numerous examples.

The realistic interpretation of quantum theory outlined here must be regarded as tentative in one important respect. According to this interpretation physical reality is described as it emerges when investigated by measuring processes, which are themselves physical processes. Accordingly the self-consistency of the realistic interpretation would require a solution of the so-called measurement problem, which has not been achieved yet. We shall have to leave open whether a new theory of macrosystems is needed in order to explain the occurrence of objective pointer readings, or whether the universality of quantum mechanics can be held up in this respect. A discussion of these problems is offered in a monograph on the quantum theory of measurement coauthored by two of the present writers [1.1]. That book may be viewed as providing the general theory, the present enterprise being devoted to elaborating various kinds of applications.

The first chapter is intended to serve as a guide to our programme. It offers an informal survey showing how POV measures arise in quantum physics and which new possibilities they bring about. The examples and problems touched upon here are taken up and explained in appropriate detail later in the book. Chapter I closes with a brief historical account of those reviews and monographs that gave direction to our own work in the field.

Chapter II provides a systematic presentation of the concept of an observable as a POV measure. Mathematical as well as operational issues are discussed here. Quantum measurement theory is developed to the extent needed for providing the operational foundations of the interpretation of POV measures sketched out above. In addition, measurement theory sets the frame not only for a rigorous treatment of conceptual and interpretational problems but also for the analysis of experiments. The remaining chapters are concerned with various kinds of examples (Chapter III) and applications of POV measures. The extended quantum language provides new insights on topics like incommensurability, complementarity and other limitations on measurability; the indeterminacy relations will be seen to emerge as conditions for *unsharp* joint measurability (Chapter IV). Various aspects of quantum uncertainty, coarse-graining, state inference, and informational completeness, appear in a new light if unsharpness is taken into account (Chapter V). Phase space representations of states open up new perspectives upon the quantum/classical relation, and an operationally founded classical limit procedure is presented on the basis of a phase space measurement model, along with an operational justification of the individual interpretation of the indeterminacy relations (Chapter VI). Finally, detailed descriptions will be given, in terms of POV measures, of some fundamental experiments, ranging from the classic Stern-Gerlach experiment and other polarisation measurements towards advanced present-day examples such as photon number and phase measurements, quantum nondemolition devices and demonstrations of unsharp wave-particle duality (Chapter VII).

I. Introduction

I.1 Observables as POV measures

One may identify at least four different ways in which POV measures enter the arena of quantum physical investigations. First, they are an integral part of the probabilistic structure of quantum mechanics. Second, they arise in a natural way in the theoretical description of many experiments. Furthermore, some of the long-standing conceptual problems of quantum mechanics appear in an entirely new light if reformulated in terms of POV measures. Finally, thinking of observables as POV measures leads one to envisage new perspectives that could not have been seen in a frame based on ordinary observables only.

I.1.1 Statistical analysis of an experiment.
In analysing the general features of any physical experiment one is able to specify those mathematical structures that are relevant to the theoretical description of an experiment (Figure 1.1). Any type of physical system is characterised by means of a collection of *preparation* procedures, the application of which prepare the system in a *state T*. The set of states is taken to be convex, thus accounting for the fact that different preparation procedures can be combined to produce mixtures of states. Given a system prepared in a state T, then a measurement can be applied, leading to the *registration* of some outcome ω_i. For illustrative purposes, we assume a finite set of pointer readings $\Omega = \{\omega_1, \ldots, \omega_n\}$. The very existence of physical experience is due to the fact that one is able to observe regularities in the event sequences occurring in nature. In particular, physical experimentation as sketched above would lose its meaning, were there not a probabilistic connection between the occurrence of a registration and the preceding preparation. Hence, any pair (T, ω_i) of a state T and an outcome ω_i should determine a probability $p(\omega_i \,|\, T)$,

$$(T, \omega_i) \;\mapsto\; p(\omega_i \,|\, T) \tag{1.1}$$

which in a long run of repeated measurements (N trials) is approximated by the relative frequency $N(\omega_i)/N$ of the occurence of the outcome ω_i. It should be noted that different preparation procedures may be statistically equivalent in that they yield the same statistics for all possible measurements. Therefore the states T correspond, strictly speaking, to equivalence classes of preparation procedures. Similarly, different registration procedures may be statistically equivalent in the sense of yielding the same probabilities in every state. This gives rise to the definition of an observable as an equivalence class of measurements. In fact, the map (1.1) can be viewed in two ways. First, any outcome ω_i induces a state functional E_i,

$$E_i : T \;\mapsto\; E_i(T) := p(\omega_i \,|\, T) \tag{1.2}$$

called an *effect*. Now the measured *observable* may be defined as the map assigning to each outcome ω_i its associated effect:

$$E : \omega_i \;\mapsto\; E_i \tag{1.3}$$

According to the second reading of (1.1), any state T fixes a probability distribution

$$p_T : \omega_i \mapsto p_T(\omega_i) := p(\omega_i \,|\, T) \tag{1.4}$$

In the simple case of a discrete experiment the properties of such a probability measure are summarised in the positivity $(p_T(\omega_i) \geq 0)$ and normalisation $(\sum_i p_T(\omega_i) = 1)$ conditions. In view of (1.3) the mapping $T \mapsto p_T$ is defined by the observable E. Since the properties of p_T are naturally transferred to E, an observable will appropriately be called an *effect valued measure*.

It is natural to assume that any state functional E_i preserves the convex structure of the set of states, that is, it associates with any mixture of states the corresponding convex combination of probabilities. This is taken as a reflection of the statistical independence of a long run of identical measurements performed on an ensemble of mutually independent systems. Effects are thus represented as linear functionals on the space of states.

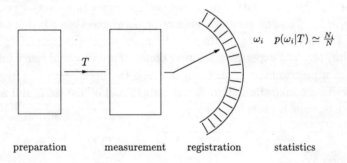

preparation   measurement   registration   statistics

Figure 1.1. Scheme of a physical experiment

In Hilbert space quantum mechanics, where states T are represented by state operators, positive trace one operators $\hat{T}$, the (linear) probability functionals $T \mapsto p_T(\omega_i)$ can be shown to be of the form

$$T \mapsto p_T(\omega_i) = \mathrm{tr}\big[\hat{T}\hat{E}_i\big] \tag{1.5}$$

The $\hat{E}_i$ are positive linear operators adding up to the unit operator. Consequently, effects E_i are represented by positive operators $\hat{E}_i$ and observables by *positive operator valued measures*, respectively. From now on we will identify states and effects with the respective operators, thus dropping the hats from the letters.

In summary, a quantum mechanical experiment is represented by a pair (T, E) where T is the prepared state and E is a POV measure. E corresponds to the measured observable in the sense of the *minimal interpretation* of the numbers $\text{tr}[TE_i]$ as the probabilities for an outcome ω_i to occur if the measurement in question is performed on a system in state T. For vector states associated with unit vectors φ of the Hilbert space $\mathcal{H}$ the probabilities (1.5) can be written in the inner product form $\langle \varphi \,|\, E_i\varphi \rangle$.

I.1.2 POV measures arising from actual measurements. The following simple examples demonstrate how POV measures appear in a natural way in quantum physical investigations, especially when one aims at realistic descriptions.

Example 1. The following 'laboratory report' of the historic Stern-Gerlach experiment [1.2] stands quite in contrast to the usual textbook 'caricatures'. A beam of silver atoms, produced in a furnace, is directed through an inhomogeneous magnetic field, eventually impinging on a glass plate (Figure 1.2). The run time in the original experiment was 8 hours. Comparison was made with a similar experiment with the magnet turned off, run time 4.5 hours. The result of the magnet-off case was a single bar of silver on the glass approximately 1.1 mm long, 0.06-0.1 mm wide. In the magnet-on case, a pair-of-lips shape appeared on the glass 1.1 mm long, one lip 0.11 mm wide, the other 0.20 mm wide, the maximum gap between the upper and lower lips being approximately the order of magnitude of the width of the lips. Both lips appeared deflected relative to the position of the bar. Only visual measurements through a microscope were made. No statistics on the distributions were made, nor did one obtain 'two spots' as is stated in some texts. The beam was clearly split into distinguishable but not disjoint beams; yet this was considered to be enough to justify the conclusion that some property had been demonstrated. Gerlach and Stern viewed this property as 'space quantisation in a magnetic field.'

The Stern-Gerlach experiment is simple enough to admit a detailed account of the idealisations leading to its interpretation as a sharp spin measurement. Such an analysis will be deferred to Chapter VII, while for the moment a simplified description shall suffice to show that, strictly speaking, only an unsharp spin observable, hence a POV measure, is obtained.

Let an atom carrying spin-$\frac{1}{2}$ be prepared in a spin state $\varphi = c_+\varphi_+ + c_-\varphi_-$. Its center of mass will be represented by a wave packet ϕ, so that the initial state of the atom is $\phi \otimes \varphi$. Upon passage through a Stern-Gerlach magnet (oriented so as to measure the spin orientation associated with the eigenstates $\varphi_\pm$) the atom undergoes a unitary evolution which couples the spin degrees of freedom with its translational motion:

$$\Psi \equiv U(\phi \otimes \varphi) = c_+ \phi_+ \otimes \varphi_+ + c_- \phi_- \otimes \varphi_- \tag{1.6}$$

The states $\phi_\pm$ represent the wave packets deflected up or down due to the action of the (inhomogeneous) magnetic field.

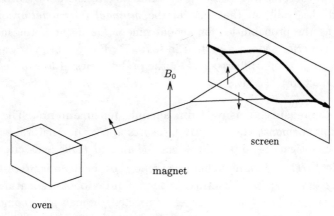

Figure 1.2. Scheme of the Stern-Gerlach experiment

The next step should be to describe the registration of spots on the screen. To this end one would need to take into account the interaction between the silver atom and the molecules of the plate, which should establish the proper coupling between the positions of the atom and the plate molecules. The observable which corresponds to the measurement of the spots (called the screen observable) shall be modelled by means of projection operators P_+ and P_- corresponding to the localisation in the upper or lower half planes of the screen. The corresponding probabilities can then be expressed with respect to the incoming spin state φ as follows:

$$\langle \Psi | P_\pm \otimes I \Psi \rangle = |c_+|^2 \langle \phi_+ | P_\pm \phi_+ \rangle + |c_-|^2 \langle \phi_- | P_\pm \phi_- \rangle =: \langle \varphi | F_\pm \varphi \rangle \tag{1.7}$$

where the effects

$$F_\pm = \langle \phi_+ | P_\pm \phi_+ \rangle P[\varphi_+] + \langle \phi_- | P_\pm \phi_- \rangle P[\varphi_-] \tag{1.8}$$

constitute the unsharp spin observable actually measured in this experiment. One may immediately confirm that $F_+ + F_- = I$; however, the effects (1.8) are no projections, i.e., $F_\pm^2 \neq F_\pm$, unless their eigenvalues $\langle \phi_+ | P_\pm \phi_+ \rangle$ and $\langle \phi_- | P_\pm \phi_- \rangle$ are 0 or 1. If the center of mass wave packets $\phi_\pm$ were well separated and localised in the appropriate half planes, i.e., if $\langle \phi_\pm | P_\pm \phi_\pm \rangle = 1$ and thus $\langle \phi_\pm | P_\mp \phi_\pm \rangle = 0$, one would have recovered the familiar textbook description with $F_\pm$ coinciding just with the projections $P[\varphi_\pm]$. However due to the spreading of wave packets this could only be achieved approximately and for special initial states ϕ.

This example provides a theoretical picture for possible sources of experimental inaccuracies which are due to the quantum indeterminacy inherent in the center of mass wave function. Even if the spin is prepared sharply prior to measurement, its value can only be ascertained with some uncertainty. If there was no definite spin value initially, there will not be one afterwards. Thus spin remains indeterminate, in general.

Example 2. We investigate next how the photon statistics measured by a photodetector are altered due to the presence of a beam splitter. Let us imagine a set-up as described in Figure 1.3. An input signal described by a state T, corresponding to a single-mode photon field, propagates in a given direction and enters a photodetector D. The detector D is assumed to record with unit efficiency the number of photons. In other words, it serves as an ideal detector for the number observable $N = \sum n|n\rangle\langle n|$ of the mode. Hence, if the field is in a k-photon number state $T = |k\rangle\langle k|$, the statistics collected by D corresponds to the probability $p(n|k) := |\langle n|k\rangle|^2 = \delta_{kn}$. Now a partially transparent mirror is placed in front of the photodetector as a lossless beam splitter BS of transparency ε. The beam splitter induces a coupling between the signal mode in state T and a local field mode in the vacuum state $|0\rangle$.

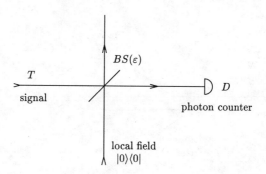

Figure 1.3. Photon beam splitter arrangement

Thinking in accordance with the classical picture of a splitting of wave intensities into fractions, one would expect that given k input photons the detector would count εk photons. By contrast, a quantum mechanical calculation (Section VII.3.1) leads to binomial photon statistics,

$$p(n|k,\varepsilon) = \binom{k}{n}\varepsilon^n (1-\varepsilon)^{k-n} \qquad (1.9)$$

Due to the presence of BS an input state T is changed into a state T' which is then analysed in the detector. In order to interpret the photon statistics $n \mapsto \mathrm{tr}\big[T'|n\rangle\langle n|\big]$ with respect to the input state T it is necessary to find a POV measure $n \mapsto F_n$ such that

$$\mathrm{tr}\big[T F_n\big] = \mathrm{tr}\big[T' |n\rangle\langle n|\big] \qquad (1.10)$$

for all n. It turns out that the F_n are as follows:

$$F_n := \sum_{m=n}^{\infty} \binom{m}{n} \varepsilon^n (1-\varepsilon)^{m-n} |m\rangle\langle m| \qquad (1.11)$$

These are obviously positive operators, $F_n \geq 0$, and they sum up to unity, $\sum F_n = I$. The effects F_n are no projections unless the beam splitter is totally transparent. The source of unsharpness in this experiment is the process of mixing two field modes, which is again of a genuine quantum nature and must not be understood as a classical random process.

Example 3. Yet another type of motivation for POV measures is furnished by the idea of compound measurements. For example, an experimental set-up may consist of a combination of different devices measuring some spin component of a spin-$\frac{1}{2}$ object, say, in the x-, y-, and z-directions (Figure 1.4). Assume that a mechanism is built in which chooses, with apriori probability w_i, $i = x, y, z$, one of the devices and then carries out the measurement of the relevant spin quantity with the spectral decomposition $s_i = \frac{1}{2}E_+^i - \frac{1}{2}E_-^i$, $i = x, y, z$.

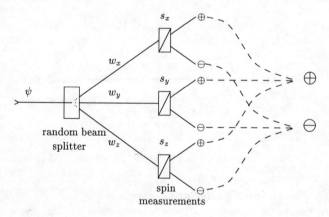

Figure 1.4. A compound spin measurement

There are various possibilities to collect and exploit the measurement statistics of such an experiment. First one could be interested in gaining as much statistical information as possible on the input spin state. It turns out that the present set-up provides an example of an informationally complete measurement in the sense that its statistics determine uniquely the initial state of the measured system. The outcomes of the experiment are represented by the six effects $w_i E^i_\pm$, $i = x, y, z$. These effects constitute a POV measure $(i, \pm) \mapsto w_i E^i_\pm$ on the outcome space $\Omega = \{(i, +), (i, -)| \ i = x, y, z\}$. Clearly, $w_i E^i_\pm \geq O$, and $\sum_i w_i (E^i_+ + E^i_-) = I$. The informational completeness of the above measurement and the resulting unsharp observable correspond to the fact that the statistics of the spin components s_x, s_y, s_z of a spin-$\frac{1}{2}$ system determine its state.

In addition to illustrating the informational completeness of compound observables, this device furnishes an example of the destruction of information by means of randomisation. To see this, imagine that the machine records the outcomes ('spin up' $\hat{=} E^i_+$, 'spin down' $\hat{=} E^i_-$, $i = x, y, z$) only with respect to their sign, up or down, without noting the spin component which was actually measured. The resulting statistics of 'ups' and 'downs' are then given by the sum $N_\pm = N^x_\pm + N^y_\pm + N^z_\pm$ of the three measurements. If N is the total number of measurements carried out, then the relative frequencies of the measurement outcomes can be written as

$$\frac{N_\pm}{N} = \sum w_i \frac{N^i_\pm}{N^i} \tag{1.12}$$

where $N^i = w_i N$. In the limit of large N these frequencies should approach the quantum mechanical probabilities

$$p_\pm = \langle \psi | \left(w_x E^x_\pm + w_y E^y_\pm + w_z E^z_\pm \right) \psi \rangle =: \langle \psi | F_\pm \psi \rangle \tag{1.13}$$

with ψ denoting the initial state of the spin object.

The interpretation of randomness in this experiment depends on the applied splitting mechanism. If a random switch forces each individual particle to take one route, then the weights w_i represent subjective ignorance as to which measurement was actually carried out. On the other hand, if the beam is split according to some diffraction technique, as is done in neutron interferometry, then each particle is represented by a coherent wave function which consists of three parts localised in the respective beam paths. In this case one is dealing with a genuine quantum indeterminacy. In Chapter VII we shall analyse a test of complementarity with single photons, where such an unsharp observable is involved.

Example 4. Unsharp observables arise in a direct way in the context of sequential measurements. Consider a beam of photons propagating in the z-direction and passing through two polarisers placed one after the other in the beam path (Figure 1.5).

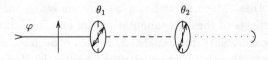

Figure 1.5. Crossed polarisers as a sequential measurement set-up

Assume that the initial photon state φ corresponds to a linear polarisation in some fixed direction and the polarisers are oriented with angles θ_1 and θ_2 relative to that direction and represented by projections P_{θ_1} and P_{θ_2}, respectively. Then the transition probability is given by iterated application of the Malus law,

$$
\begin{aligned}
p_\varphi(\theta_1, \theta_2) &= \|P_{\theta_2} P_{\theta_1} \varphi\|^2 = \langle \varphi \,|\, P_{\theta_1} P_{\theta_2} P_{\theta_1} \varphi \rangle \\
&= \cos^2(\theta_2 - \theta_1) \cos^2(\theta_1)
\end{aligned}
\tag{1.14}
$$

This probability corresponds to the sequential 'yes-yes' outcome, and it can be given as $\langle \varphi \,|\, F_{+,+} \varphi \rangle$ in terms of the effect $F_{+,+} = P_{\theta_1} P_{\theta_2} P_{\theta_1}$ in the initial state φ of the incoming photon. Without going into further details it is obvious that the full measurement statistics collected by a sequential measurement are again conveniently comprised by a POV measure.

I.1.3 Conceptual problems requiring POV measures.
There was a class of conceptual problems in the foundations of quantum theory which could be tackled only after a proper formulation in terms of POV measures was found. Typically these puzzles appeared in the form of a conflict between familiar classical physical ideas and some 'strange' implications of the quantum formalism. In each case the resolution consisted of rephrasing a strict no-go verdict excluding certain sharp measurements into a positive statement expressing the possibility of unsharp measurements subject to some limitations. We shall sketch out three prominent examples: the incommensurability or even complementarity of observables, the problem of repeatable measurements, and limitations of measurability due to conservation laws. A more comprehensive account of these issues requires the tools of quantum measurement theory and will be taken up in later chapters.

Example 1. The noncommutativity of certain pairs of self-adjoint operators is commonly interpreted as the root of the incommensurability of the corresponding observables – a feature that is alien to classical physics. In fact ever since the discovery of the fundamental 'exchange relation'

$$
QP - PQ = i\hbar I
\tag{1.15}
$$

the joint measurability of position and momentum has been a subject of lasting debates in quantum mechanics. It was argued that these observables are complementary; all their measurements are mutually exclusive and thus cannot be performed together. Much effort has gone into analysing various operational and probabilistic aspects of the noncommutativity of observables, aiming at conditions for the joint measurability, or *coexistence*, of such quantities. However, these investigations did not solve the dilemma that arose with the Heisenberg interpretation of the uncertainty relation

$$
\delta q \cdot \delta p \geq \tfrac{1}{2}\hbar
\tag{1.16}
$$

Indeed, being a consequence of the commutation relation (1.15), this inequality describes, in the first instance, the impossibility of preparing states in which position and momentum would have both arbitrarily sharp values. This probabilistic interpretation as a scatter relation is based on the identification of the entities δq, δp with the standard deviations of Q, P in some state T. Accordingly the uncertainty relation refers to measurement series for position and momentum, performed separately on ensembles of systems prepared in one and the same state.

It is plausible that the possibilities of measurement should conform with those of preparation. Thus, rather than accepting the mere incommensurability of position and momentum, the pioneers of quantum theory considered, in a semi-quantitative way, a variety of thought experiments, such as the gamma ray microscope, in order to demonstrate that joint measurements of these quantities should be possible in principle. The crucial idea was indeed that such measurements must not be too accurate, the limits of precision being given by (1.16). While the measurement indeterminacy interpretation of the uncertainty relation is commonly accepted and rephrased in many texts, its tenability was nevertheless long questioned due to the lacking rigorous incorporation of the idea of inaccurate measurements into the quantum formalism. A solution of this problem was achieved relatively late in the history of quantum physics and furnishes one of the first instances of introducing unsharp observables represented as POV measures.

There exist procedures of coarse-graining, such as the convolution of a PV measure with some confidence measure, which transform a given sharp observable into a smeared version, an unsharp observable. It then becomes possible to associate with a pair of incommensurable observables a new pair of coexistent unsharp observables which are smeared versions of the original ones. Whether two such unsharp observables are coexistent or not depends on the degree of smearing involved. In the case of position and momentum it is precisely the uncertainty relation which serves to characterise the amount of smearing required for their joint measurability. In this way Heisenberg's inaccuracy interpretation of the uncertainty relations is justified, after all; these inequalities do express conditions for a positive possibility. It is easy to construct POV measures representing unsharp joint position and momentum measurements. However, we shall be content here with giving only the ensuing quantum mechanical probability distributions. Let $\psi_{qp} := \exp\left(-\frac{i}{\hbar}(qP - pQ)\right)\psi$ denote the phase space translates of a fixed unit vector $\psi \in \mathcal{H}$, then

$$(q,p) \mapsto \rho_\varphi(q,p) := \frac{1}{2\pi\hbar} \left|\langle \varphi \mid \psi_{qp} \rangle\right|^2 \tag{1.17}$$

is a phase space probability density for any vector state φ. The family of vectors ψ_{qp} is a set of coherent states, so that the normalisation condition is just a consequence of their overcompleteness relation. Clearly, the marginal distributions are not the ordinary position and momentum distributions $|\varphi(q)|^2$ and $|\hat{\varphi}(p)|^2$ [where $\hat{\varphi}$ denotes the Fourier transform of the wave function φ], but they are obtained from them by means of convolutions with the distributions $|\psi(q)|^2$ and $|\hat{\psi}(p)|^2$, respectively. Their variances are found to be related to the variances of Q and P in a vector state φ, $\mathrm{Var}(Q, \varphi)$ and $\mathrm{Var}(P, \varphi)$, as

$$
\begin{aligned}
\mathrm{Var}(q, \rho_\varphi) &= \mathrm{Var}(Q, \varphi) + \mathrm{Var}(Q, \psi) \\
\mathrm{Var}(p, \rho_\varphi) &= \mathrm{Var}(P, \varphi) + \mathrm{Var}(P, \psi)
\end{aligned}
\tag{1.18}
$$

so that

$$
\mathrm{Var}(q, \rho_\varphi) \cdot \mathrm{Var}(p, \rho_\varphi) \geq \hbar^2
\tag{1.19}
$$

This already suggests that the measurement inaccuracies must be identified as $\delta q = \mathrm{Var}(Q, \psi)^{\frac{1}{2}}$ and $\delta p = \mathrm{Var}(P, \psi)^{\frac{1}{2}}$.

Example 2. The possibility of interpreting quantum mechanics as a theory of individual systems with definite real properties rests on the existence of measurements which allow one to prepare certain properties. To this end there should be repeatable measurements leaving the system in states which give probability equal to unity for the registered outcome to occur again upon repeated measurements. Interestingly the role of this type of measurement, also known as measurements of the first kind, has always been judged in a very controversial way by different researchers. Still today there is a widespread tendency to just identify 'measurement' with 'repeatable measurement'. On the other hand this notion was often regarded as a completely unrealistic idealisation of no relevance to actual experimentation.

From the point of view of quantum measurement theory it is clear, however, that some, though not all, observables admit repeatable measurements (at least offered by the formalism), while there is a variety of (non-repeatable) measurements associated with any given observable. For example an unsharp observable whose effects do not have the number one as an eigenvalue will never allow a repeatable measurement. There is, however, also a less trivial, large class of observables for which repeatable measurements are excluded for fundamental reasons. A basic result of measurement theory states that observables which admit repeatable measurements are necessarily discrete. Hence continuous observables do not admit repeatable measurements. This means that quantum 'particles' cannot be prepared in principle by specifying their basic observables, position or momentum. This difficulty can be made plausible intuitively by noting that an interval in the value space of a continuous observable can be indefinitely subdivided into finer and finer intervals, to each of which there should correspond a 'localised' eigenstate if a repeatable measurement were available. This would run into having uncountably many mutually orthogonal states, which is excluded in a separable Hilbert space. There are two ways out of this situation. One is to discretise a given continuous observable but this often destroys the defining properties of that observable, for example its covariance. Thus one is forced to consider weaker forms of repeatability, which indeed turn out to be realisable for POV measures representing coarse-grained versions of the given observable.

Example 3. The third type of limitations on measurability to be noted here is the one discovered by Wigner in his study of the implications of symmetry arguments on the description of a physical system. Wigner demonstrated with an example that a physical quantity that does not commute with an additive conserved quantity cannot be measured at all. The argument is simple enough to be reproduced already here. Let ψ_+ and ψ_- denote the eigenstates of the z-component s_z of the spin of a spin-$\frac{1}{2}$ system. Then the eigenvectors of s_x are of the form $\varphi_\pm = \frac{1}{\sqrt{2}}(\psi_+ \pm \psi_-)$. Assume s_x to be measured by coupling it to a measuring system via a unitary map. Then the following state transformation is obtained:

$$\begin{aligned} \varphi_+ \otimes \phi &\rightarrow \varphi_+ \otimes \phi_+ \\ \varphi_- \otimes \phi &\rightarrow \varphi_- \otimes \phi_- \end{aligned} \tag{1.20}$$

Here ϕ, $\phi_\pm$ are unit vectors in the Hilbert space of the measuring system, and the pointer states ϕ_+, ϕ_- are required to be orthogonal to each other. Clearly, s_z does not commute with s_x. If j_z denotes the z-component of the measuring system's angular momentum, then $s_z + j_z$ should be considered as a conserved quantity and therefore commute with any unitary operator leading to (1.20). Using the linearity of this state transformation, one obtains

$$\begin{aligned} \psi_+ \otimes \phi &\rightarrow \psi_+ \otimes \tfrac{1}{2}(\phi_+ + \phi_-) + \psi_- \otimes \tfrac{1}{2}(\phi_+ - \phi_-) \\ \psi_- \otimes \phi &\rightarrow \psi_+ \otimes \tfrac{1}{2}(\phi_+ - \phi_-) + \psi_- \otimes \tfrac{1}{2}(\phi_+ + \phi_-) \end{aligned} \tag{1.21}$$

It is easily seen that the expectation values of $s_z + j_z$ coincide in the two states on the right-hand side of (1.21) while they are different on the left-hand sides. Hence this quantity cannot be conserved.

This argument can be generalised to a large extent. In view of the fundamental conservation laws for linear and angular momenta one has to conclude that quantities like spin and position are not measurable. Again, introducing some measuring errors (which can be made arbitrarily small for sufficiently large measuring devices) helps one to circumvent this obstacle. A straightforward analysis shows that Wigner's original proposal amounts to the introduction of some unsharp observable. Indeed, in the above example it is clear from Equation (1.21) that the violation of the conservation law becomes negligible if the expectation values of j_z are large in the measuring system's states involved. One may therefore expect that only small deviations from the measurement transition (1.20) are required if the conservation law is to be respected. The proposal was to add the terms $\pm \varphi_\mp \otimes \eta$ to the right-hand sides of (1.20). In this way one obtains a measurement with three outcomes represented by a POV measure with the effects $F_+ = (1 - \varepsilon) P[\varphi_+]$, $F_- = (1 - \varepsilon) P[\varphi_-]$ corresponding to the intended properties and $F_o = \varepsilon I$ representing an uncertain outcome. The number $\varepsilon := \|\eta\|^2$ can be made arbitrarily small by suitable choices of the states ϕ, $\phi_\pm$.

I.1.4 New possibilities envisaged with POV measures. The use of POV measures for representing physical quantities has opened up new views upon quantum mechanics and its objects that could not have been seen only on the basis of the notion of observables as PV measures.

First, there are some potential experimental questions which can be conceived theoretically, such as time of occurrence, phase, or photon localisation, for which no representation in terms of spectral measures exists. For instance, following an argument given by Pauli it can be shown that a self-adjoint time operator T, if considered as a quantity conjugate to the Hamiltonian H, would act as a generator of a unitary group $U_h := \exp\left(\frac{i}{\hbar}hT\right)$, $h \in \mathbf{R}$, such that $U_h E^H(X) U_h^* = E^H(X+h)$. This would force the spectrum of H to be all of $\mathbf{R}$, in contradiction to its semiboundedness. Similarly, the difficulties encountered with defining a phase operator for the harmonic oscillator or for spin systems is related to the semiboundedness of the conjugate number or spin observables. In all these cases there do exist consistent descriptions by means of POV measures.

As a second example, the famous complementarity of path and interference observations in two-slit or split-beam experiments is customarily interpreted as a mutual exclusion of the two options. It was the idea of unsharp observables which allowed one to conceive of an experimental realisation of unsharp wave-particle duality. In recent years photon and neutron interferometry experiments were carried out displaying, for instance, 99%-confidence path determination together with a good interference contrast.

There are some features typical of classical physical theories which can be recovered (in some weakened form) in quantum mechanics if the full set of observables is taken into account. First of all, noncommuting pairs of ordinary observables do possess a kind of unsharp counterparts which are coexistent. Coexistence was defined so far only in a probabilistic sense, referring to the existence of a joint observable. The subsequent measurement theoretical models will demonstrate in addition that an interpretation as joint measurability in a proper operational sense is well justified.

Another classical feature of statistical theories is the existence of informationally complete observables, the statistics of which always fix the state uniquely. While in a classical theory there exist sharp observables that are informationally complete, this is not the case in quantum mechanics. Indeed, in the latter theory an informationally complete observable is never a PV measure. As illustrated in Subsection 1.2 for the case of spin-$\frac{1}{2}$ systems, a simple way of producing informationally complete observables consists of mixing measurements of sufficiently many noncommuting quantities. Realisations of informationally complete phase space or polarisation measurements will be given in Chapters VI and VII.

Finally, the general conception of an observable E as a POV measure was seen to give rise to classical embeddings of the quantum mechanical state space into a space of probability measures via the mapping $T \mapsto p_T^E$. This embedding provides the basis for the general method of coarse-graining mentioned earlier which in turn opens up a new approach towards an operationally sound classical limit procedure. In the case of an informationally complete observable, the mapping is injective and affords thus a classical representation of quantum mechanics.

I.2 Historical survey

The systematic application of POV measures has led to important insights in many distinct areas of quantum physics. It suffices here to give an impression of these interesting developments by just indicating the manifold of topics covered. This will enable us to describe the purpose, scope and limits of the present account in a historical perspective.

The programme of Ludwig and his coworkers on the foundations of physics aims at a general probabilistic formulation of quantum mechanics guided by the attempt to reconstruct the Hilbert space structure of the theory. Starting with an abstract convex structure representing preparation procedures, one first derives the so-called statistical duality and then proceeds to axiomatically characterising its Hilbert space realisation. Our Section I.1.1 is a simple short presentation of this line of argument; a more detailed account is found in the books of Ludwig [1.3] and Kraus [1.4]. What goes beyond purely statistical goals is our attempt to formulate the operational foundations for a realistic, individual interpretation of quantum mechanics.

An operational approach to quantum mechanical probability theory was initiated among others by the work of Haag and Kastler [1.5], Mielnik [1.6], and Davies and Lewis [1.7]. Its goal was a mathematically rigorous generalisation of quantum measurement theory in order to deal with the foundations of quantum field theory and quantum stochastic processes as they occur in photon counting theory. An early account of this branch of applications of general observables is provided by the classic of Davies [1.8]. In the monographs of Lindblad [1.9] and Alicki and Lendi [1.10] this theory is applied to a detailed study of quantum dynamical semigroups and the phenomenon of irreversibility. While stochastic processes are not the subject of the present treatise, the operational concepts developed in this approach are employed here in the context of measurement theory (Chapters II, IV, VI, VII).

Statistical applications of quantum mechanics have led to a new quantum estimation and detection theory summarised by Helstrom [1.11]. In a related programme Holevo [1.12] provided mathematically rigorous treatments of quantum estimation techniques and contributed to laying the foundations of quantum communication theory. This subject is currently under investigation in C^*-algebraic quantum mechanics [1.13], where also further examples of concrete POV measures such as time or phase observables are presented. The way in which POV measures arise in the context of state inference is described in Chapter V. In particular coarse-graining will be defined there, and the scope of information theoretic concepts is found to extend in a natural way to the full set of observables.

The problem of constructing quantum objects calls for the implementation of a symmetry group into the theoretical description of physical experiments. Concrete observables should be characterised as covariant POV measures. This view leads to the definition of observables that did not exist as PV measures (Chapter III). In particular the problem of Galilei or Poincaré covariant localisation on phase space gave rise to a formulation of quantum mechanics on phase space [1.14, 1.15] and recently to a general framework for quantum geometry [1.16]. Phase space localisation and its operational foundation will be extensively investigated in Chapters III and VI.

Finally, an entirely new branch of activities arose since the early 1980s when progress in high technology opened up the possibility of performing quantum experiments with single microsystems. We are thus witnessing a situation where direct applications and illustrations of the formerly 'abstract' measurement theory come into sight and invite quantum physicists to enter a promising area of research. We believe that the richness and power of the quantum language developed and reviewed in the subsequent chapters will become fully apparent in future attempts to give detailed realistic accounts of such fundamental experiments. Some examples collected in the concluding Chapter VII may give a flavour of this exciting perspective.

II. Theory

This chapter presents the conceptual and mathematical framework of Hilbert space quantum mechanics. The basic physical notions, states and observables, are defined with reference to the interpretational rules of the theory. Special attention is paid to the representation of observables as POV measures. We shall investigate the structure of the set of observables and introduce some physically motivated methods of constructing POV measures. We also review the Neumark dilation of a POV measure into a PV measure. Quantum measurement theory is developed as an operational basis for the subsequent discussions of interpretational issues as well as physical applications.

II.1 Hilbert space frame

Quantum mechanics on Hilbert space forms the general mathematical setting for our study. We adopt here the formulation of the theory that is based on the dual concepts of states and observables as suggested by the general statistical analysis of an experiment sketched out in Section I.1.1.

There is a great variety of excellent texts on Hilbert space quantum mechanics, any reader being familiar with some of them. However, in order to fix notations and to trim our terminology we wish to recall briefly those items of the theory which are fundamental to carrying out our programme. The standard results quoted below can be found, for instance, in the monographs of Beltrametti and Cassinelli [2.1], Davies [1.8], Holevo [1.12], Jauch [2.2], Kraus [1.4], Ludwig [1.3], and von Neumann [2.3]. We are also using freely the Hilbert space operator theory, as presented, for example, in the work of Reed and Simon [2.4].

II.1.1 Preliminaries. The quantum mechanical description of a physical system S is based on a complex separable Hilbert space $\mathcal{H}$, with the inner product $\langle\cdot|\cdot\rangle$. Quite commonly, *states* and *observables* are represented as unit vectors φ of $\mathcal{H}$ and self-adjoint operators A acting in $\mathcal{H}$, respectively. The number

$$\langle A \rangle_\varphi := \langle \varphi | A\varphi \rangle \tag{1.1}$$

is interpreted as the *expectation value* of the observable A in the state φ. Any two unit vectors φ and ψ which differ only by a phase factor,

$$\varphi = e^{i\alpha}\psi, \quad \alpha \in \mathbf{R} \tag{1.2}$$

give rise to the same expectation values for all observables of the system.

Let A be a discrete observable, that is, a self-adjoint operator with a complete set of eigenvectors. For simplicity we assume that the eigenvalues a_i of A are nondegenerate. Let $\{\varphi_i\} \subset \mathcal{H}$ be a complete orthonormal set of eigenvectors of A, $A\varphi_i = a_i\varphi_i$. Then any $\varphi \in \mathcal{H}$ can be expressed as the Fourier series

$$\varphi = \sum \langle \varphi_i | \varphi \rangle \, \varphi_i \tag{1.3}$$

with $\sum |\langle \varphi_i | \varphi \rangle|^2$ converging to the square of the norm $\|\varphi\| = \langle \varphi | \varphi \rangle^{\frac{1}{2}}$ of φ. One finds

$$\langle A \rangle_\varphi = \sum a_i \, |\langle \varphi_i | \varphi \rangle|^2 \equiv \sum a_i \, p_\varphi^A(a_i) \tag{1.4}$$

where the numbers $p_\varphi^A(a_i)$ are positive and add up to unity. This is to say that $\langle A \rangle_\varphi$ is the expectation value of the probability measure

$$a_i \mapsto p_\varphi^A(a_i) := |\langle \varphi_i | \varphi \rangle|^2 \tag{1.5}$$

defined by the observable A and the state φ. Again one may note that the numbers $p_\varphi^A(a_i)$ do not depend on the phase of φ.

As another familiar example we consider a physical system represented by the Hilbert space $L^2(\mathbf{R}, dx)$, the Lebesgue function space of square integrable complex valued functions on the real line $\mathbf{R}$. The system's position observable is usually identified as the multiplicative operator Q,

$$(Q\psi)(x) = x \, \psi(x) \tag{1.6}$$

so that the expectation value of Q in the state ψ is

$$\langle Q \rangle_\psi = \int_{\mathbf{R}} x \, |\psi(x)|^2 \, dx \tag{1.7}$$

One may quickly verify that the mapping

$$X \mapsto p_\psi^Q(X) := \int_X |\psi(x)|^2 \, dx \tag{1.8}$$

is a probability measure on the real line, that is, $0 = p_\psi^Q(\emptyset) \leq p_\psi^Q(X) \leq p_\psi^Q(\mathbf{R}) = 1$, and $p_\psi^Q(X_1 \cup X_2 \cup \cdots) = p_\psi^Q(X_1) + p_\psi^Q(X_2) + \cdots$ whenever the sets $X_1, X_2, \cdots$ are pairwise disjoint (Borel) subsets of $\mathbf{R}$. The number $\langle Q \rangle_\psi$ is again seen to arise as the expectation value of the probability measure p_ψ^Q defined by the position observable Q and the wave function ψ.

The above examples are simple instances of the spectral theorem for self-adjoint operators which we shall review next. Let $\mathcal{L}(\mathcal{H})$ denote the set of bounded linear operators on $\mathcal{H}$. This is a Banach space with respect to the operator norm, $\|A\| = \sup \{\|A\varphi\| \mid \varphi \in \mathcal{H}, \|\varphi\| = 1\}$. An operator $A \in \mathcal{L}(\mathcal{H})$ is self-adjoint if A equals its adjoint A^*, and it is a projection operator if $A = A^* = A^2$. The notion of adjoint operator can be extended also to unbounded operators defined on a dense domain $D(A)$. Then an operator is self-adjoint whenever $D(A) = D(A^*)$ and $A = A^*$.

With $\mathcal{B}(\mathbf{R})$ we denote the Borel subsets of the real line. A mapping $E : \mathcal{B}(\mathbf{R}) \to \mathcal{L}(\mathcal{H})$ is a *projection valued* (PV) *measure*, or a spectral measure, if

$$E(X) = E(X)^* = E(X)^2 \text{ for all } X \in \mathcal{B}(\mathbf{R}) \tag{1.9a}$$

$$E(\mathbf{R}) = I \tag{1.9b}$$

$$E(\cup X_i) = \sum E(X_i) \text{ for all disjoint sequences } (X_i) \subset \mathcal{B}(\mathbf{R}) \tag{1.9c}$$

where the series converges in the weak operator topology of $\mathcal{L}(\mathcal{H})$.

The defining properties of a PV measure E and the continuity and the additivity properties of the inner product guarantee that for any unit vector φ the mapping

$$X \mapsto p_\varphi^E(X) := \langle \varphi \,|\, E(X)\varphi \rangle \tag{1.10}$$

is a probability measure. If $\varphi \in \mathcal{H}$ is an arbitrary vector, then $X \mapsto \langle \varphi | E(X)\varphi \rangle$ is still a real measure, but with normalisation $\langle \varphi | E(\mathbf{R})\varphi \rangle = \langle \varphi | \varphi \rangle$. These real measures give rise to the definition of a unique self-adjoint operator $A := \int x \, dE(x)$ with its domain of definition $D(A)$ consisting of those vectors $\varphi \in \mathcal{H}$ for which the integral $\int x^2 \, dp_\varphi^E(x)$ is convergent. The converse result is the *spectral theorem* for self-adjoint operators.

THEOREM. *Let A be a self-adjoint operator with the domain $D(A) \subset \mathcal{H}$. There is a unique* PV *measure $E : \mathcal{B}(\mathbf{R}) \to \mathcal{L}(\mathcal{H})$ such that*

$$D(A) = \left\{ \varphi \in \mathcal{H} \,\Big|\, \int_\mathbf{R} x^2 \, d\langle \varphi \,|\, E(x)\varphi \rangle < \infty \right\} \tag{1.11}$$

and for any $\varphi \in D(A)$

$$\langle \varphi \,|\, A\varphi \rangle = \int_\mathbf{R} x \, d\langle \varphi \,|\, E(x)\varphi \rangle \tag{1.12}$$

In order to emphasise the one-to-one correspondence between the self-adjoint operators A and the real PV measures E, we let E^A denote the PV measure associated with A.

On the basis of the spectral theorem, the expectation value $\langle A \rangle_\varphi$ of any observable A in a state φ is just the first moment of the probability measure $p_\varphi^{E^A} \equiv p_\varphi^A$ defined by A and φ. It is these probability measures which must be regarded as the conceptual basis of quantum theory, and the interpretation of the theory starts with spelling out their physical meaning. At the present stage we adopt the *minimal interpretation*, which says that the numbers $\langle \varphi \,|\, E^A(X)\varphi \rangle$ are probabilities for measurement outcomes.

We have already pointed out that the probabilities $p_\varphi^A(X)$ do not depend on the phase of the vector φ. Indeed, all unit vectors $\varphi \in \mathcal{H}$ which define the same one-dimensional projection operator $P[\varphi]$, given as $P[\varphi] \psi = \langle \varphi \mid \psi \rangle \varphi$ for any $\psi \in \mathcal{H}$, give rise to the same probabilities. This justifies the representation of a state not just as a unit vector φ but rather as a one-dimensional projection operator $P[\varphi]$.

There are two important ways to obtain new states from any two states represented by vectors φ_1 and φ_2, namely the procedures of *superposition* and *mixing*. Let $\varphi = c_1\varphi_1 + c_2\varphi_2$ be a superposition, i.e., a linear combination, of the states φ_1 and φ_2, with some coefficients $c_1, c_2 \in \mathbf{C}$ such that $\|\varphi\| = 1$. For any observable A we then have

$$
\begin{aligned}
p_\varphi^A(X) = & \ |c_1|^2 \left\langle \varphi_1 \mid E^A(X)\varphi_1 \right\rangle + |c_2|^2 \left\langle \varphi_2 \mid E^A(X)\varphi_2 \right\rangle \\
& + c_1\bar{c}_2 \left\langle \varphi_2 \mid E^A(X)\varphi_1 \right\rangle + c_2\bar{c}_1 \left\langle \varphi_1 \mid E^A(X)\varphi_2 \right\rangle
\end{aligned}
\tag{1.13}
$$

where the term on the second line accounts for the so-called interference effects. Defining operators of the form $|\varphi_1\rangle\langle\varphi_2|$ as $|\varphi_1\rangle\langle\varphi_2|(\psi) := \langle \varphi_2 \mid \psi \rangle \varphi_1$ for $\psi \in \mathcal{H}$, we may express the state $P[\varphi]$ as

$$
\begin{aligned}
P[\varphi] = & \ P[c_1\varphi_1 + c_2\varphi_2] \\
= & \ |c_1|^2 P[\varphi_1] + |c_2|^2 P[\varphi_2] + c_1\bar{c}_2 |\varphi_1\rangle\langle\varphi_2| + c_2\bar{c}_1 |\varphi_2\rangle\langle\varphi_1|
\end{aligned}
\tag{1.14}
$$

which shows the typical structure of a superposition of states.

The second method of obtaining new states from φ_1 and φ_2 consists of forming their mixture with weights $w, 1 - w$ $(0 \leq w \leq 1)$. For any observable A, the convex combination of the probability measures $p_{\varphi_1}^A$ and $p_{\varphi_2}^A$, with weights $w, 1 - w$ is a probability measure; the number

$$
\begin{aligned}
w\, p_{\varphi_1}^A(X) + (1 - w)\, p_{\varphi_2}^A(X) = & \\
w \left\langle \varphi_1 \mid E^A(X)\varphi_1 \right\rangle + (1 & - w) \left\langle \varphi_2 \mid E^A(X)\varphi_2 \right\rangle
\end{aligned}
\tag{1.15}
$$

being the probability that a measurement of A leads to a result in the set X when the system is in a mixture of the states φ_1 and φ_2 with the weights w and $1 - w$. This state is conveniently expressed as the operator $wP[\varphi_1] + (1 - w)P[\varphi_2]$. A typical situation where mixed states are needed occurs when the preparation of the state of a system contains some ambiguities; it may happen that the system is only known to be in one of the states $\varphi_1, \varphi_2, \cdots$, with the probability $0 \leq w_k \leq 1$ $(\sum w_k = 1)$ for the actual state to be φ_k. Another important instance requiring mixed states is when the system under investigation is a part of a bigger system. The reduced state of a subsystem is then, as a rule, a mixed state even when the state of the whole system is given by a unit vector.

These considerations demonstrate once more that the states of a physical system should not primarily be described as unit vectors φ but rather as operators

$P[\varphi]$. In addition, the set of states should also contain states of the form $\sum_i w_i P[\varphi_i]$. This leads to the representation of a state as a particular operator on $\mathcal{H}$, namely as a positive operator of trace one, also called state operator.

Finally, a closer look at the probability measures (1.10) shows that the idempotency of $E(X)$ is not necessary for p_φ^E being a probability measure. In addition to the normalisation condition (1.9b) and the sigma-additivity property (1.9c), one needs only to require the positivity of $E(X)$. This is to say that for obtaining a probability measure of type (1.10), it is sufficient that E is a normalised positive operator valued measure.

In this way one is led to representing states and observables as positive trace one operators and POV measures, respectively. These descriptions are exhaustive in the following sense. Given the set of observables, the most general way of defining probability measures is the one based on state operators. Conversely, if the set of states is represented by the positive trace-one operators, there is a one-to-one correspondence between the totality of probability measures and the POV measures.

II.1.2 States and observables. In order to spell out the above introduction of states and observables we shall collect here the necessary mathematical ingredients.

An operator $A \in \mathcal{L}(\mathcal{H})$ is positive, $A \geq O$, if $\langle \varphi \,|\, A\varphi \rangle \geq 0$ for all vectors $\varphi \in \mathcal{H}$. A positive operator is always self-adjoint. The relation $A \geq B$ (or $B \leq A$), defined as $A - B \geq O$, is an ordering on the set of self-adjoint bounded operators. Let Ω be a nonempty set and $\mathcal{F}$ a σ-algebra of subsets of Ω so that $(\Omega, \mathcal{F})$ is a measurable space. A *normalised positive operator valued* (POV) *measure* $E : \mathcal{F} \to \mathcal{L}(\mathcal{H})$ on $(\Omega, \mathcal{F})$ is defined through the properties:

$$E(X) \geq O \quad \text{for all } X \in \mathcal{F} \tag{1.16a}$$

$$E(\Omega) = I \tag{1.16b}$$

$$E(\cup X_i) = \sum E(X_i) \quad \text{for all disjoint sequences } (X_i) \subset \mathcal{F} \tag{1.16c}$$

where the series converges in the weak operator topology of $\mathcal{L}(\mathcal{H})$. For any POV measure $E : \mathcal{F} \to \mathcal{L}(\mathcal{H})$ the following two conditions are equivalent:

$$E(X)^2 = E(X) \quad \text{for all } X \in \mathcal{F}$$
$$E(X \cap Y) = E(X)E(Y) \quad \text{for all } X, Y \in \mathcal{F} \tag{1.17}$$

Thus a positive operator valued measure is a projection valued measure exactly when it is multiplicative. If the measurable space $(\Omega, \mathcal{F})$ underlying a PV measure E is the real Borel space $(\mathbf{R}, \mathcal{B}(\mathbf{R}))$, then E determines a unique self-adjoint operator A, and conversely, any real PV measure is determined by a unique self-adjoint operator as already discussed in the preceding subsection.

On the set of positive bounded operators one may define a functional

$$T \,\mapsto\, \mathrm{tr}[T] \,:=\, \sum_i \langle \varphi_i \,|\, T\varphi_i \rangle, \tag{1.18}$$

where $\{\varphi_i\}$ is an orthonormal basis of $\mathcal{H}$. The number $\mathrm{tr}[T]$ is independent of the choice of the basis and is called the *trace* of T. The positive operators of finite trace span an important vector subspace $\mathcal{T}(\mathcal{H})$ of $\mathcal{L}(\mathcal{H})$, the trace class. The trace extends to a positive (bounded) linear functional on $\mathcal{T}(\mathcal{H})$. The formula

$$\|T\|_1 := \mathrm{tr}[|T|] \tag{1.19}$$

with $|T| = (T^*T)^{1/2}$, defines a norm, the trace norm, on $\mathcal{T}(\mathcal{H})$ with respect to which $\mathcal{T}(\mathcal{H})$ is a Banach space.

With $\mathcal{S}(\mathcal{H})$ we denote the set of positive trace one operators,

$$\mathcal{S}(\mathcal{H}) := \{T \in \mathcal{T}(\mathcal{H}) \,|\, T \geq O, \mathrm{tr}[T] = 1\} \tag{1.20}$$

It is a convex set having the one-dimensional projections $P[\varphi]$ as its extreme elements. Any $T \in \mathcal{S}(\mathcal{H})$ can be expressed as a σ-convex combination of some extreme elements $(P[\varphi_i])$: $T = \sum_i w_i P[\varphi_i]$, where (w_i) is a sequence of weights $[0 \leq w_i \leq 1, \sum_i w_i = 1]$ and the series converges in trace norm.

For any POV measure $E : \mathcal{F} \to \mathcal{L}(\mathcal{H})$ and any $T \in \mathcal{S}(\mathcal{H})$ the mapping

$$p_T^E : \mathcal{F} \to [0,1], \; X \mapsto p_T^E(X) := \mathrm{tr}[TE(X)] \tag{1.21}$$

is a probability measure. This follows from the defining properties of E and the continuity and linearity of the trace. Finally, the decomposition of states, $T = \sum w_i P[\varphi_i]$, induces the corresponding decomposition of the probability measures.

We are now ready to fix our general framework. The *states* of a system $\mathcal{S}$ are represented by – and identified with – the elements of $\mathcal{S}(\mathcal{H})$. The notion of a state as a unit vector of $\mathcal{H}$ is subsumed under this general definition in the form of the extremal elements of $\mathcal{S}(\mathcal{H})$. These states $P[\varphi]$ and the corresponding unit vectors $\varphi \in \mathcal{H}$, are referred to as the *vector states*. They are also often called *pure states*. In the absence of superselection rules, all vector states are pure states. Due to the linear structure of $\mathcal{H}$, superpositions of vector states form new vector states; and any vector state can be represented as a superposition of some other vector states. The convexity of the set of states reflects the possibility of preparing new states as *mixtures* of other states.

The notion of an observable provides a representation of the possible events occurring as outcomes of a measurement. In this sense an *observable* is defined as – and identified with – a POV measure $E : \mathcal{F} \to \mathcal{L}(\mathcal{H})$, $X \mapsto E(X)$ on a measurable space $(\Omega, \mathcal{F})$. Since the space $(\Omega, \mathcal{F})$ describes the possible measurement outcomes, or measured values of the observable E, we call it the *value space* of E. The traditional concept of an observable as a self-adjoint operator in $\mathcal{H}$ refers to the PV measures on the real line $\mathbf{R}$. If not stated otherwise, we shall assume that $(\Omega, \mathcal{F})$ is of the form $(\mathbf{R}^n, \mathcal{B}(\mathbf{R}^n))$.

According to (1.21), any pair (E, T) of an observable E and a state T induces a probability measure p_T^E on the value space $(\Omega, \mathcal{F})$ of E. The minimal interpretation of the probability measures p_T^E establishes their relation to measurements: the number $p_T^E(X)$ is the probability that a measurement of the observable E performed on the system $\mathcal{S}$ in the state T leads to a result in the set X.

II.2 Physical and mathematical features of POV measures

II.2.1 Unsharp properties.
Quantum mechanical probabilities are given as expectation values of certain operators, the effects. Any state T induces an expectation functional $B \mapsto \mathrm{tr}[TB]$ on the set $\mathcal{L}(\mathcal{H})$ of bounded operators. The requirement that the numbers $\mathrm{tr}[TB]$ represent probabilities entails that the operator B is positive and bounded by the unit operator: $O \leq B \leq I$. Equivalently the spectrum of any effect is in the interval $[0, 1]$. The set of effects, denoted $\mathcal{E}(\mathcal{H})$, is ordered by the relation $B \leq C$, $B, C \in \mathcal{L}(\mathcal{H})$, and has the least and the greatest elements O and I, respectively. However, this ordering does not make the set of effects a lattice, unless $\mathcal{H} = \mathbf{C}$. There is also a kind of complementation in $\mathcal{E}(\mathcal{H})$. Indeed the map $B \mapsto B^\perp := I - B$, when applied twice, yields the identity $(B^{\perp\perp} = B)$ and reverses the order (if $B \leq C$, then $C^\perp \leq B^\perp$). These two properties guarantee that the de Morgan laws hold true in $\mathcal{E}(\mathcal{H})$ in the sense that if, for instance, the infimum of $B, C \in \mathcal{E}(\mathcal{H})$ exists in $\mathcal{E}(\mathcal{H})$, then also the supremum of their complements $B^\perp$ and $C^\perp$ exists in $\mathcal{E}(\mathcal{H})$ and $I - B \wedge C = (I - B) \vee (I - C)$. The mapping $B \mapsto B^\perp$ is not an orthocomplementation: the infimum of B and $B^\perp$ need not exist at all, and even if it does, it need not be the null effect. Thus the *tertium non datur* is not fulfilled in $\mathcal{E}(\mathcal{H})$. For example, if λ is a number satisfying $0 < \lambda < \frac{1}{2}$ then $B = \lambda I \leq B^\perp$, and thus $B = B \wedge B^\perp$, which is not the null effect.

The set $\mathcal{E}(\mathcal{H})$ of effects is a convex subset of the linear space $\mathcal{L}(\mathcal{H})$. The physical meaning of forming a convex combination $B = \lambda_1 B_1 + \lambda_2 B_2$ of two effects $(0 \leq \lambda_1 = 1 - \lambda_2 \leq 1)$ was illustrated in the random measurement example of subsection I.1.2. The extremal elements of $\mathcal{E}(\mathcal{H})$ are the projection operators which form an orthocomplemented lattice $\mathcal{P}(\mathcal{H})$ with respect to the order and the complementation it inherits from $\mathcal{E}(\mathcal{H})$ [1.8, 2.1].

The Hilbert space language of quantum mechanics admits an interpretation as referring to individual systems and their properties. This claim will be substantiated on the basis of the quantum theory of measurement in the next sections. For an experimental event B (an effect) to correspond to a property of a system it is essential that (1) there is also the counter property $B^\perp$ and that (2) both of them may be realised in some states. In addition, it seems natural to require that (3) a property and its counter property are mutually exclusive in a strict sense, meaning that their greatest lower bound must exist and be equal to the null effect. The product of B and $B^\perp$ is an effect which is a common lower bound to both of these effects and therefore, as a consequence of (3), equals the null effect. This is to say that both B and $B^\perp$ are projections. It is convenient to consider also the projections O and I as (trivial) properties. Then the set of *properties*, as characterised by (1), (2) and (3), coincides with the lattice of projections, that is, the *sharp properties*. If for a sharp property P one has $\text{tr}[TP] = 1$ in a state T, then P is a *real* property in that state.

This notion of reality can be relaxed so as to apply to a wider class of effects, the *regular* effects, whose spectrum extends both below as well as above the value $\frac{1}{2}$. For any regular effect the following weak form (2′) of (2) is satisfied: there exist states T, T' such that $\text{tr}[TB] > \frac{1}{2}$ and $\text{tr}[T'B^\perp] > \frac{1}{2}$. In such states B and $B^\perp$ can be considered as *approximately real* properties, respectively. Any effect C below both a regular effect B and its complement $B^\perp$ satisfies $2C \leq B + B^\perp = I$ so that $C \leq \frac{1}{2}I$. Such a C is not a regular effect itself. Therefore the map $B \mapsto B^\perp$ is an orthocomplementation in the set of regular effects. We take this as a weakened form (3′) of condition (3). The set of *properties*, defined by (1), (2′), (3′), is then

$$\mathcal{E}_p(\mathcal{H}) = \{B \in \mathcal{E}(\mathcal{H}) \,|\, B \not\leq \tfrac{1}{2}I, \, B \not\geq \tfrac{1}{2}I\} \cup \{O, I\} \qquad (2.1)$$

Separating from $\mathcal{E}_p(\mathcal{H})$ the set of *sharp* properties yields the remaining set of *unsharp* properties $\mathcal{E}_u(\mathcal{H}) := \mathcal{E}_p(\mathcal{H}) \setminus \mathcal{P}(\mathcal{H})$.

The question of interpreting effects as a kind of properties was an important issue in the operational approach to quantum mechanics. It has been taken up recently in the course of investigations into the algebraic structure of the set $\mathcal{E}_p(\mathcal{H})$ with the aim to develop a quantum language for unsharp properties [2.5].

On the basis of the notion of an unsharp property it might be straightforward to call an observable unsharp if the nontrivial effects in its range are unsharp properties. In that case the range of the observable would be a Boolean lattice in the ordering of $\mathcal{E}(\mathcal{H})$. However, such a definition turns out unnecessarily restrictive. We say that a POV measure E is an *unsharp observable* if there is an unsharp property in its range.

II.2.2 Coexistence. One of the key notions in the quantum description of a physical system is the *coexistence*, describing the possibility of measuring together several effects or observables. Its rudimentary formal expression is the commutativity of self-adjoint operators. We formalise here the idea of joint measurability in terms of its probabilistic meaning while its measurement theoretical content will be analysed in Section IV.2.

An observable $E : \mathcal{F} \to \mathcal{E}(\mathcal{H})$ is a representation of a class of measurement procedures in the sense that it associates with any state T the probability $p_T^E(X)$ for the occurrence of an outcome $X \in \mathcal{F}$. For a pair of observables E_1, E_2 the question may be raised as to whether their outcome distributions $p_T^{E_1}$ and $p_T^{E_2}$ can be collected within one common measurement procedure for arbitrary states T.

Thus one is asking for the existence of a third observable whose statistics contains those of E_1 and E_2. We say that observables E_1 and E_2 are coexistent whenever this is the case. More explicitly, and with a slight generalisation, a collection of observables $E_i, i \in \mathbf{I}$, is *coexistent* if there is an observable E such that for each $i \in \mathbf{I}$ and for each $X \in \mathcal{F}_i$ there is a $Z \in \mathcal{F}$ such that

$$p_T^{E_i}(X) = p_T^E(Z) \tag{2.2}$$

for all states T. In other words, a set of observables $E_i, i \in \mathbf{I}$, is coexistent if there is an observable E such that the ranges $\mathcal{R}(E_i) := \{E_i(X) \mid X \in \mathcal{F}_i\}$ of all E_i are contained in that of E, that is, $\bigcup_{i \in \mathbf{I}} \mathcal{R}(E_i) \subseteq \mathcal{R}(E)$. Observable E is called a *joint observable* for the E_i.

The notion of coexistence has two obvious but important relaxations. It may happen that for some observables the defining condition is met only for certain value sets and/or for certain states. In particular, we say that observables E_1 and E_2 are partially coexistent if some of their coarse-grainings, induced by some partitions of their value sets, are coexistent. In the extreme case this entails the coexistence of pairs of effects: any two effects $F_1, F_2 \in \mathcal{E}(\mathcal{H})$ are coexistent if there is an observable E such that $F_1 = E(Z_1)$ and $F_2 = E(Z_2)$ for some value sets Z_1 and Z_2. It is an easy exercise to show that F_1 and F_2 are coexistent if and only if there is an effect F_{12} such that $F_{12} \leq F_1$, $F_{12} \leq F_2$ and $F_1 + F_2 - F_{12} \in \mathcal{E}(\mathcal{H})$. In this case a joint observable for F_1, F_2 is given by the four effects F_{12}, $F_{1\bar{2}} := F_1 - F_{12}$, $F_{\bar{1}2} := F_2 - F_{12}$, $F_{\bar{1}\bar{2}} := I - (F_1 + F_2 - F_{12})$. One finds that $F_1 = F_{12} + F_{1\bar{2}}$, $F_{\bar{1}} := I - F_1 = F_{\bar{1}2} + F_{\bar{1}\bar{2}}$, $F_2 = F_{12} + F_{\bar{1}2}$, $F_{\bar{2}} := I - F_2 = F_{1\bar{2}} + F_{\bar{1}\bar{2}}$. The effects F_{ik} represent the joint occurrence of the events specified by the effects F_i and F_k. Note that the existence of positive joint lower bounds is a necessary condition for the coexistence of a pair of effects or simple observables: some of the effects F_{ik} must be nonzero in order to guarantee the normalisation. Identifying F_1, F_2 as $E_1(X), E_2(Y)$, respectively, it follows that the effect F_{12} can be interpreted as representing the joint occurrence of outcomes for a coexistent pair of observables E_1 and E_2 in the sets X, Y. This important aspect of the joint measurability will be illustrated in several examples throughout this book. In general we shall say that a set of effects $\mathcal{A} \subseteq \mathcal{E}(\mathcal{H})$ is coexistent whenever $\mathcal{A}$ is contained in the range of an observable. It is interesting to observe that pairwise coexistence in a set of effects does not guarantee the coexistence of this set.

In concrete applications a joint observable E for a coexistent pair E_1, E_2 will usually be constructed on the product space $\Omega := \Omega_1 \times \Omega_2$ of the two outcome spaces, with $\mathcal{F}$ being some σ-algebra on Ω such that $X \times \Omega_2 \in \mathcal{F}$, $\Omega_2 \times Y \in \mathcal{F}$ for all $X \in \mathcal{F}_1$, $Y \in \mathcal{F}_2$. Thus, if $\mathcal{F}_i = \mathcal{B}(\mathbf{R})$, then $\mathcal{F}$ will be conveniently chosen as $\mathcal{B}(\mathbf{R}^2)$. Then the observables E_1, E_2 are recovered from E as *marginals*: $E_1(X) = E(X \times \Omega_2)$, $E_2(Y) = E(\Omega_1 \times Y)$.

Unlike the case of PV measures, there is no general theory of coexistent sets of effects or observables in quantum mechanics. Some partial results are known but exhaustive characterisations of such observables are still lacking. Therefore we restrict ourselves to reviewing some of the main results on coexistent sets of observables where at least some are sharp observables.

Two observables E_1 and E_2 represented as PV measures on $\big(\mathbf{R}, \mathcal{B}(\mathbf{R})\big)$ are coexistent exactly when they commute [2.6], meaning that for any X and Y

$$E_1(X)\, E_2(Y) \;=\; E_2(Y)\, E_1(X) \tag{2.3}$$

For pairs of POV measures it is still true that they are coexistent if they commute; if one of them is a PV measure then the commutativity is still necessary for the coexistence [2.7]. In general coexistence does not require commutativity. This fact constitutes one of the virtues of POV measures as it allows one to circumvent the incommensurability verdicts for noncommuting sharp observables.

The equivalence of commutativity and coexistence in the case of PV measures is essentially due to a theorem of von Neumann, which says that two self-adjoint operators A and B commute if and ony if they can be expressed as functions of a third self-adjoint operator C, that is, $A = f(C)$ and $B = g(C)$ for some C and for some (real Borel) functions f and g. Condition (2.3) is equivalent to the requirement that the projection valued mapping

$$X \times Y \mapsto E(X \times Y) := E_1(X) \wedge E_2(Y) \tag{2.4}$$

extends to a PV measure E on the product space $(\mathbf{R}^2, \mathcal{B}(\mathbf{R}^2))$, which then constitutes a joint observable for E_1 and E_2. This is to say that the set function

$$X \times Y \mapsto \mathrm{tr}\big[T\big(E_1(X) \wedge E_2(Y)\big)\big] \tag{2.5}$$

extends to a probability measure on $\mathbf{R}^2$ for each state T. Hence coexistent sharp observables E_1 and E_2 have a joint probability $p_T^{E_1,E_2}$ for all states T, with

$$\begin{aligned}
p_T^{E_1,E_2}(X \times Y) &= \mathrm{tr}\big[T\big(E_1(X)\, E_2(Y)\big)\big]\\
p_T^{E_1,E_2}(X \times \mathbf{R}) &= p_T^{E_1}(X), \quad p_T^{E_1,E_2}(\mathbf{R} \times Y) = p_T^{E_2}(Y)
\end{aligned} \tag{2.6}$$

These classic commutativity results, which all go back to von Neumann's theorem, admit an important generalisation to the case of partial commutativity with respect to states. In order to describe these results, let $\mathrm{com}(E_1, E_2)$ denote the set of all vectors $\varphi \in \mathcal{H}$ for which $E_1(X)E_2(Y)\varphi = E_2(Y)E_1(X)\varphi$ for all $X, Y \in \mathcal{B}(\mathbf{R})$. This set is called the *commutativity domain* of the PV measures E_1 and E_2. It is a closed subspace of $\mathcal{H}$ and it is invariant under all the spectral projections $E_1(X)$ and $E_2(Y)$. Therefore, the restrictions of the PV measures E_1 and E_2 on $\mathrm{com}(E_1, E_2)$ define two mutually commuting PV measures $\mathcal{B}(\mathbf{R}) \to \mathcal{L}(\mathrm{com}(E_1, E_2))$. This observation allows one to prove [2.8] that the set function

$$X \times Y \mapsto \langle\varphi\,|\,E_1(X) \wedge E_2(Y)\varphi\rangle \tag{2.7}$$

extends to a probability measure on $\mathbf{R}^2$ exactly for those vector states φ which are in the commutativity domain of E_1 and E_2. For observables represented as PV measures, the commutativity domain provides therefore an appropriate probabilistic characterisation of their degree of coexistence (with respect to states).

II.2.3 Constructing POV measures from PV measures

a) **Smearing PV measures.** Let $(\Omega_1, \mathcal{F}_1)$ and $(\Omega_2, \mathcal{F}_2)$ be two Borel spaces and consider a map $p: \mathcal{F}_2 \times \Omega_1 \mapsto [0, 1]$ such that for each $X \in \mathcal{F}_2$, $p(X, \cdot)$ is a measurable function on Ω_1, and for every $\omega \in \Omega_1$, $p(\cdot, \omega)$ is a probability measure on $\mathcal{F}_2$. If $E : \mathcal{F}_1 \to \mathcal{L}(\mathcal{H})$ is a PV measure, we define a POV measure F on $\mathcal{F}_2$ by the formula

$$F(X) := \int_{\Omega_1} p(X, \omega) \, dE(\omega) \tag{2.8}$$

Physically, p reflects the existence of some source of uncertainty which must be taken into account in interpreting the readings of a measurement. In fact assume that E has an eigenvalue ω such that for an associated eigenstate φ one has $E(\{\omega\})\varphi = \varphi$; then the expectation of $F(X)$ is $\langle \varphi | F(X)\varphi \rangle = p(X, \omega)$. This is to say that under optimal conditions of a calibration situation one can determine the noise inherent in the device measuring the observable F. We shall therefore refer to p as a (conditional) confidence measure. In general a confidence measure p on $\mathcal{F}_2 \times \Omega_1$ induces a map from the POV measures on $\mathcal{F}_1$ to those on $\mathcal{F}_2$. A POV measure F derives from some PV measure according to (2.8) if and only if it is commutative [1.12]. We shall meet a variety of cases where the kind of uncertainty represented by p can be specified in detail.

To spell out the case of a discrete confidence measure, let $\Omega_1 = \{1, 2, \cdots, N\}$, $\Omega_2 = \{1, 2, \cdots, M\}$ be finite or infinite sets, and consider a PV measure $n \mapsto E_n$ on Ω_1. Let (λ_{mn}) be a stochastic matrix, that is, an $M \times N$-matrix with the properties $\lambda_{mn} \geq 0$ and $\sum_m \lambda_{mn} = 1$. Define the confidence measure $p(X, n) := \sum_{m \in X} \lambda_{mn}$. Then the POV measure $m \mapsto F_m$ on Ω_2 induced by (2.8) is given by

$$F_m = \sum_n \lambda_{mn} E_n \tag{2.9}$$

A POV measure of this structure arose in the photocounting example of Eq. (I.1.11).

Another familiar realisation of the confidence measure p is obtained when $\Omega_1 = \Omega_2 = \mathbf{R}$ and the measure is absolutely continuous with a translation invariant density function (confidence distribution) $\rho(\omega', \omega) := e(\omega' - \omega)$. In that case (2.8) corresponds to a convolution of measures, which is commonly used as a description of white noise. To illustrate this interpretation of a smeared PV measure, let us consider the position observable E^Q of a particle with one space degree of freedom. Let e be a confidence distribution, then the following POV measure E^e is an unsharp position observable:

$$E^e(X) := \chi_X * e(Q) = \int_{\mathbf{R} \times \mathbf{R}} \chi_X(x) \, e(y - x) \, dx \, dE^Q(y) \tag{2.10}$$

b) POV measures for open systems. A state transformation is a linear mapping V on $\mathcal{T}(\mathcal{H})$ that sends states to states. Its dual map V^* on $\mathcal{L}(\mathcal{H})$, defined via $\text{tr}\big[(VT)(A)\big] = \text{tr}\big[TV^*(A)\big]$ for $T \in \mathcal{T}(\mathcal{H})$, $A \in \mathcal{L}(\mathcal{H})$, maps positive operators to positive operators and fulfills $V^*(I) = I$. The mapping V^* induces an action Ψ_{V^*} on POV measures: $\Psi_{V^*}(E) = V^* \circ E =: F$. In this way any PV measure E gives rise to a POV measure F, which will not in general be a PV measure unless V^* is of the form $V^*(A) = UAU^*$ for some unitary or antiunitary operator U.

This kind of mapping is applied in formulating the dynamics of open systems. A system interacting with some environment cannot evolve according to the unitary dynamics of closed systems but its time development has to be described by means of nonunitary (trace-preserving) state transformations V [1.9,1.10]. The dual maps V^* furnish the Heisenberg picture of the dynamics. In this situation unsharp observables arise as the effectively measured observables when one tries to measure a sharp observable on a system subjected to external noise. Imagine a system (with Hilbert space $\mathcal{H}$), prepared in some state T, that interacts with another system, its environment (with Hilbert space $\mathcal{H}_e$), in state T_e. If U is the corresponding unitary evolution, then the final state of the compound system is $U(T \otimes T_e)U^*$. Tracing out the environment degrees of freedom yields the system's reduced state, and one obtains the state transformation

$$V : T \mapsto V(T) := \text{tr}_{\mathcal{H}_e}\Big[U(T \otimes T_e)U^*\Big] \qquad (2.11)$$

Assume that a measurement of a sharp observable E is performed on the system in state $V(T)$; then the associated probability distribution is

$$\text{tr}\Big[U(T \otimes T_o)U^* E(X) \otimes I\Big] \;=\; \text{tr}\big[V(T)\,E(X)\big] \;=\; \text{tr}\big[T\,V^*\big(E(X)\big)\big] \qquad (2.12)$$

The mapping $X \mapsto F(X) := V^*\big(E(X)\big)$ is a (normalised) POV measure. This modification of E into F is due to the noise added by the system's environment. Hence in the measurement context the induced map Ψ_{V^*} on the set of POV measures can serve to describe external noise. An example is furnished by the beam splitter experiment of Section I.1.2.

It will be shown in Section 2.4 that any POV measure can be obtained in this way from some PV measure. This general fact can easily be illustrated by means of the two smearing procedures given in a) above. Consider a discrete observable $E_n = |n\rangle\langle n|$, $n = 1, 2, \cdots, N$, then any stochastic $N \times N$-matrix (λ_{mn}) determines a state transformation V via

$$V(T) := \sum_{kn} \lambda_{kn} |k\rangle\langle k| \langle n\,|\,T|n\rangle \qquad (2.13)$$

The dual map applied to the projection E_m yields a POV measure $F : m \mapsto F_m$ of the form (2.9):

$$V^*(E_m) \;=\; \sum_{kn} \lambda_{kn} |n\rangle\langle n| \langle k\,|\,E_m|k\rangle \;=\; \sum_n \lambda_{mn} |n\rangle\langle n| \;=\; F_m \qquad (2.14)$$

For the unsharp position observable (2.10), let $U_y = \exp\left(-\frac{i}{\hbar}yP\right)$ denote the unitary representation of the translation group and define the state transformation

$$V : T \mapsto \int_{\mathbf{R}} e(y)\, U_y^*\, T\, U_y\, dy \qquad (2.15)$$

One can readily verify that

$$
\begin{aligned}
V^*\big(E^Q(X)\big) &= \int_{\mathbf{R}} e(y)\, U_y\, E^Q(X)\, U_y^*\, dy \\
&= \int_{\mathbf{R}} e(y)\, E^Q(X+y)\, dy \;=\; E^e(X)
\end{aligned}
\qquad (2.16)
$$

c) POV measures induced by measurements. Assume that a system in state φ is coupled to a second system (with Hilbert space $\mathcal{H}_a$), initially in state ϕ_a, by means of a unitary interaction operator $U : \varphi \otimes \phi_a \mapsto U(\varphi \otimes \phi_a)$. Consider then a subsequent measurement of an observable E_a of the second system. The corresponding probabilities can be expressed in terms of the first system alone:

$$\big\langle U(\varphi \otimes \phi_a)\,|\, I \otimes E_a(X)\,|\, U(\varphi \otimes \phi_a)\big\rangle \;=:\; \langle \varphi\,|\, E(X)\varphi\rangle \qquad (2.17)$$

Interpreting the second system as a measuring apparatus and E_a as a pointer observable, then the whole process constitutes a measurement performed on the first system in its initial state. The above relation defines the measured observable E which in general will not be a PV measure even if the pointer E_a is one. E can be viewed as a projection of some PV measure in the following sense:

$$E(X) \otimes I\Big|_{\mathcal{H} \otimes [\phi_a]} \;=\; I \otimes P[\phi_a]\, U^*\big(I \otimes E_a(X)\big)U \qquad (2.18)$$

II.2.4 Neumark's theorem. The constructions of POV measures described in parts b) and c) of the preceding subsection start out with PV measures acting in a larger Hilbert space, which are then projected down to the original Hilbert space. These schemes are generic in the sense of the following theorem.

NEUMARK'S THEOREM. *Let* $F : \mathcal{F} \mapsto \mathcal{L}(\mathcal{H})$ *be a POV measure. There exists a Hilbert space* $\widetilde{\mathcal{H}} \supset \mathcal{H}$ *and a PV measure* $E : \mathcal{F} \mapsto \mathcal{L}(\widetilde{\mathcal{H}})$ *such that the equality*

$$F(X)\varphi \;=\; PE(X)\varphi \qquad (2.19)$$

holds for all $\varphi \in \mathcal{H}$ *and for every* $X \in \mathcal{F}$. *The operator* P *is the orthogonal projection of* $\widetilde{\mathcal{H}}$ *onto* $\mathcal{H}$.

The PV measure E is called the spectral dilation of F. There exists a minimal dilation which is unique up to a unitary isomorphism, minimality being defined in the sense that $\widetilde{\mathcal{H}}$ is the smallest Hilbert space containing the union of the closed subspaces $E(X)\mathcal{H}$, $X \in \mathcal{F}$ [2.9–11].

In general the minimal dilation $(\widetilde{\mathcal{H}}, E)$ of a POV measure F representing an observable will not have a direct physical interpretation. It is, however, possible to construct dilations by identifying $\widetilde{\mathcal{H}}$ as a tensor product $\mathcal{H} \otimes \mathcal{H}_o$ of $\mathcal{H}$ with some Hilbert space $\mathcal{H}_o$, which may represent an environment system or a measuring apparatus. Neumark's theorem can then be restated in the following form [1.12].

COROLLARY. *For every* POV *measure* F *acting in* $\mathcal{H}$ *there exists a Hilbert space* $\mathcal{H}_o$, *a state* $P[\phi_o] \in \mathcal{S}(\mathcal{H}_o)$, *and a* PV *measure* E *in* $\mathcal{H} \otimes \mathcal{H}_o$, *such that*

$$\mathrm{tr}_{\mathcal{H}\otimes\mathcal{H}_o}\big[T \otimes P[\phi_o]\,E(X)\big] \; = \; \mathrm{tr}_{\mathcal{H}}\big[T\,F(X)\big] \qquad (2.20)$$

for any $T \in \mathcal{S}(\mathcal{H})$ *and* $X \in \mathcal{F}$. *Moreover,* E *can always be chosen to be of either of the forms* $U^*E(\cdot) \otimes IU$ *or* $U^*I \otimes E(\cdot)U$ *with some unitary map* U.

The isometric embedding of $\mathcal{H}$ into $\widetilde{\mathcal{H}}$ is given by the canonical map onto the subspace $\mathcal{H} \otimes P[\phi_o]$, and the Neumark projection P is $P = I \otimes P[\phi_o]$.

The physical importance of Neumark's theorem derives from the fact that it ensures the existence of a measurement for any observable. This fundamental result of the quantum theory of measurement will be explained in greater detail in Section 3. While it is true that any POV measure can be formally reduced to a PV measure acting on a larger Hilbert space, this does not diminish the need for POV measures in the description of physical systems. If one does not want to stick to an account of experiments solely in terms of pointer observables, thus dealing with phenomena on the level of measuring devices, one has got to perform the Neumark projection: it is this step that enables one to speak of the object under investigation and its measured observable. In Chapters VI and VII we shall encounter various measurement schemes where the Neumark extension of a POV measure is realised by means of coupling the object system to some probe system. Also, using several probes simultaneously allows one to measure independently their respective – mutually commuting – pointer observables, which may be chosen to be PV measures. The resulting process constitutes a Neumark dilation of some observable of the measured system that turns out to be a joint observable of several, possibly noncommuting, unsharp observables.

II.2.5 Symmetric operators and contractions. Neumark's theorem sheds some light on the relation between the traditional operator description and the POV measure representation of observables. According to the spectral theorem the self-adjoint operators are in one-to-one correspondence with the real PV measures. In a similar way certain symmetric operators determine a unique POV measure on $\big(\mathbf{R}, \mathcal{B}(\mathbf{R})\big)$.

A densely defined operator A is *symmetric* if $\langle \varphi \,|\, A\psi \rangle = \langle A\varphi \,|\, \psi \rangle$ for all $\varphi, \psi \in D(A)$. In other words, an operator A is symmetric if $D(A)$ is contained in

$D(A^*)$ and A coincides with A^* on $D(A)$. Not all symmetric operators are self-adjoint. This subtle mathematical difference poses the following physical problem. In the theoretical evaluation of an experiment one usually associates a statistical quantity with an expectation of some operator A. That operator is symmetric in accordance with the fact that measurement outcomes are described as real numbers. In addition, one generally assumes that A is self-adjoint; but there are cases, such as the time delay in scattering theory, where the self-adjointness cannot be achieved. For a symmetric but non-self-adjoint operator A one cannot invoke the spectral theorem to ensure the existence of probability distributions, which would yield the expectation values of A as their first moments. The demand for such a probabilistic interpretation raises the question whether symmetric operators admit self-adjoint extensions. An operator B is an extension of A if $D(A) \subset D(B)$ and $A\varphi = B\varphi$ for $\varphi \in D(A)$. Now there exist symmetric, non-self-adjoint operators which do not possess any symmetric extension *within* the Hilbert space $\mathcal{H}$. These are called *maximal symmetric operators*.

Neumark's theorem guarantees that any maximal symmetric operator A can be extended to a self-adjoint operator $\tilde{A}$ acting in a Hilbert space $\widetilde{\mathcal{H}}$ containing $\mathcal{H}$. Using the spectral decomposition of $\tilde{A}$ in $\widetilde{\mathcal{H}}$ one obtains an analogue of the spectral theorem for maximal symmetric operators. For any $\varphi \in \mathcal{H}$ and $\psi \in D(A)$ one has

$$\langle\varphi|A\psi\rangle = \langle\varphi|\tilde{A}\psi\rangle = \int x\,d\,\langle\varphi|E^{\tilde{A}}(x)\psi\rangle = \int x\,d\,\langle\varphi|PE^{\tilde{A}}(x)\psi\rangle \qquad (2.21)$$

The operator P denotes the projection of $\widetilde{\mathcal{H}}$ onto $\mathcal{H}$. Defining $F(X)\psi = PE^{\tilde{A}}(X)\psi$ for every $\psi \in \mathcal{H}$ and $X \in \mathcal{B}(\mathbf{R})$, one obtains a POV measure F in $\mathcal{H}$ satisfying

$$\langle\varphi\,|\,A\psi\rangle = \int_{-\infty}^{+\infty} x\,d\,\langle\varphi\,|\,F(x)\psi\rangle \qquad (2.22)$$

$$\langle A\psi\,|\,A\psi\rangle = \int_{-\infty}^{+\infty} x^2\,d\,\langle\psi\,|\,F(x)\psi\rangle \qquad (2.23)$$

where $\varphi \in \mathcal{H}$, $\psi \in D(A)$.

The strategy for obtaining a probability description for a symmetric operator consists thus of extending that operator as far as possible, which yields either a self-adjoint or a maximal symmetric operator. Since the Neumark extension of the latter is not unique, the resolution (2.22) of a symmetric operator cannot be unique, either. In the case of a self-adjoint operator A, the formulas (2.22) and (2.23) are satisfied exactly for the spectral measure of A, and the domain of A consists of those vectors for which the integral (2.23) is convergent. For a symmetric operator A, the vectors φ in $D(A)$ satisfy the inequality $\int x^2\,d\,\langle\varphi\,|\,F(x)\varphi\rangle < \infty$. However, it is only for maximal symmetric operators A that the domain $D(A)$ is characterised by this condition; in that case there exists exactly one POV measure F that fulfills

(2.22) and (2.23). For further details of this subject matter the reader may wish to refer to [2.9–10].

This consideration demonstrates clearly that in general a single operator is insufficient for an exhaustive account of all statistical information available in a given experiment. The operator gives only the first moments of the probability distributions, while the latter can be determined completely from the underlying observable represented as a POV measure.

As is well known from probability theory, a probability measure is uniquely specified by the totality of its nth moments. For a POV measure F one may define the *moment operators* $F^{(n)}$, $n = 1, 2, \cdots$, as follows:

$$\langle \varphi \, | \, F^{(n)} \varphi \rangle \; := \; \int_{-\infty}^{+\infty} x^n \, d \, \langle \varphi \, | \, F(x) \varphi \rangle \qquad (2.24)$$

which is to hold for all vectors φ for which the right hand side converges. The spectral calculus for a self-adjoint operator A is based on the fact that the moments of its spectral measure E^A are just A^n. This fundamental feature is lost in the case of a POV measure F since in general $F^{(n)} \neq (F^{(1)})^n$. Thus it is only for self-adjoint operators that their n^{th} powers determine the whole observable. By contrast a maximal symmetric operator does not allow an equally simple reconstruction of its associated observable, represented by the corresponding POV measure.

Besides the maximal symmetric operators there is another important class of operators possessing a spectral representation with respect to a unique POV measure. A bounded linear operator C is a *contraction* if its norm is at most 1. For every contraction C acting on a Hilbert space there exists exactly one POV measure F such that

$$C^n \; = \; \int_0^{2\pi} e^{inx} \, dF(x) \qquad (2.25)$$

for $n = 0, 1, 2, \cdots$. If C is unitary then F is a PV measure. Note that the family of operators C^n is closed under multiplication so that here a kind of functional calculus does apply.

II.3 Quantum measurement theory

To elucidate the various aspects of POV measures as representing quantum physical observables, one needs to take frequent recourse to measurement theory. This theory has two branches, one dealing with the changes experienced by the measured system, the other one considering measurements as physical processes. For our purposes it is convenient to develop both parts independently although we point out their interrelations. We also elaborate the operational basis for the individual interpretation of quantum mechanics.

II.3.1 Operations. Operations and state transformers (Sec. 3.2) constitute the basic tools for the description of changes experienced by a physical system. A change in the system may be due to its temporal evolution as an isolated system, or it may be caused by a coupling to some environment, as it occurs in a measurement. Such a change may well be a violent one, the system may even be destroyed or lose its identity as an individual object; a position measurement with an absorbing plate furnishes a typical example. However important such destructive measurements are in practice, it is nevertheless important to be able to account for the behaviour of a physical system under more general circumstances. To this end one needs to leave open the possibility that the system preserves its identity in the course of events, be it its free temporal evolution or a measurement interaction. Nondestructive measurements on individual atomic objects are no longer only a fiction referred to in thought experiments; for example, the so-called continuous Stern-Gerlach effect allows one to 'perform and *repeat* a quantum measurement on the same individual atomic particle, as often as one pleases' [2.12].

In the formulation of quantum mechanics employed in the present approach any change occurring in a system is represented, in the first instance, as a change of the probability measures: $p_T^E \mapsto m(p_T^E)$. Due to the structure of these measures there is an associated transition of states $[T \mapsto \widehat{m}(T)]$, or observables $[E \mapsto \widetilde{m}(E)]$, or both. Here we use primarily the Schrödinger picture, describing physical changes as state transformations. The equivalent Heisenberg picture follows then in view of the duality between states and observables. Occasionally, the combination of the two, the interaction picture, is also useful.

Let us imagine a measurement performed on a physical system $\mathcal{S}$. If T is the initial state of $\mathcal{S}$, then after the measurement the system will be found, in general, in another state T'. We assume that $\mathcal{S}$ can be subjected to the same measurement in any state. Moreover, the possible changes experienced by the system should be completely determined by the measurement device applied, and in each individual run of the experiment the state change should depend on the respective outcome only. This amounts to saying that the measurement induces a *state transformation* $T \mapsto \widehat{m}(T)$. It turns out convenient to leave open the possibility that the 'states' $\widehat{m}(T)$ are not normalised. Thus one would identify T' as $\widehat{m}(T)/\mathrm{tr}\big[\widehat{m}(T)\big]$. Indeed we shall restrict our attention to those maps $\widehat{m}$ which yield $\mathrm{tr}\big[\widehat{m}(T)\big] \leq 1$ for any state $T \in \mathcal{S}(\mathcal{H})$. Then $\mathrm{tr}\big[\widehat{m}(T)\big]$ admits an interpretation as a relative frequency and this will allow us to base the definitions of state transformers and measurements on the so-called probability reproducibility condition (Sec. 3.3).

If a system $\mathcal{S}$ is prepared in a mixture of states $T = wT_1 + (1-w)T_2$ (so that $\mathcal{S}$ is known to be either in the state T_1 or in T_2), with the respective weights w and $1-w$, then after a measurement with a given outcome, represented by the map $\widehat{m}$, the state of $\mathcal{S}$ is either $\widehat{m}(T_1)$ or $\widehat{m}(T_2)$. Considering a long run of identical measurements, with a number N of occurrences of the same outcome, it is natural to assume that all devices operate independently and in the same way. This is partly reflected by the idea of describing the state change as a map $\hat{m}$. In addition this amounts to requiring the stability of the apparatus in the sense that no back action on the beam of incoming systems occurs. Alternatively one may think of an ensemble of equally prepared systems, each being subjected to the same measurement; then the statistical independence is obviously ensured. It then follows that the number of systems ending up in the states $\hat{m}(T_1)$ and $\hat{m}(T_2)$ is (approximately) given by wN and $(1 - w)N$, respectively. Accordingly, each system is described by the state $\hat{m}\big(T_1 + (1 - w)T_2\big) = w\,\hat{m}(T_1) + (1 - w)\,\hat{m}(T_2)$, with the appropriate ignorance interpretation. Hence under stable measurement conditions the map $\hat{m}$ turns out to be convex from $\mathcal{S}(\mathcal{H})$ to a convex subset of positive operators from $\mathcal{T}(\mathcal{H})$. As such it has a unique linear extension to the space $\mathcal{T}(\mathcal{H})$.

The above assumptions on the nature of the state changes $T \mapsto \hat{m}(T)$ constitute the defining properties of an operation. An *operation* is a positive linear mapping $\Phi : \mathcal{T}(\mathcal{H}) \to \mathcal{T}(\mathcal{H})$, which satisfies

$$0 \leq \operatorname{tr}\big[\Phi T\big] \leq 1 \tag{3.1}$$

for all $T \in \mathcal{S}(\mathcal{H})$ [1.4, 1.8]. (A mapping on a set of operators is called positive if it sends positive operators to positive operators.) We use the terms operation and state transformation as synonyms.

When combined with the trace, any operation Φ defines a unique effect B via the relation

$$\operatorname{tr}\big[\Phi T\big] = \operatorname{tr}\big[TB\big] \tag{3.2}$$

for all $T \in \mathcal{T}(\mathcal{H})$. The correspondence between operations and effects is many-to-one, and the relation $\operatorname{tr}\big[\Phi_1 T\big] = \operatorname{tr}\big[\Phi_2 T\big]$, $T \in \mathcal{S}(\mathcal{H})$, is an equivalence relation among the operations, the equivalence classes being in one-to-one correspondence with the effects.

The dual mapping $\Phi^* : \mathcal{L}(\mathcal{H}) \to \mathcal{L}(\mathcal{H})$ of an operation Φ, defined by the relation $\operatorname{tr}\big[T\Phi^*(A)\big] = \operatorname{tr}\big[\Phi(T)A\big]$, $A \in \mathcal{L}(\mathcal{H})$, $T \in \mathcal{T}(\mathcal{H})$, is a (normal) positive linear mapping [1.8]; the condition (3.1) being equivalent to

$$O \leq \Phi^*(I) \leq I \tag{3.3}$$

The effect B defined by an operation Φ can be expressed in terms of Φ^* as

$$B = \Phi^*(I) \tag{3.4}$$

As an illustration we note that any unitary mapping U on $\mathcal{H}$ defines an operation $\Phi_U : T \mapsto UTU^*$. The corresponding dual map is $(\Phi_U)^* : A \mapsto U^*AU$, and the ensuing effect is the identity operator $(\Phi_U)^*(I) = U^*U = I$. Time-parametrised families of such operations arise as the natural representatives of causal, reversible temporal evolutions. Another important class of operations is induced by the projection operators. With any projection P on $\mathcal{H}$ is associated a so-called Lüders operation $\Phi_L^P : T \mapsto PTP$, with its dual $\left(\Phi_L^P\right)^*(A) = PAP$. The corresponding effect is simply $\left(\Phi_L^P\right)^*(I) = P$. These operations arise in the context of ideal measurements. In general any effect $B \in \mathcal{E}(\mathcal{H})$ gives rise to a Lüders operation $\Phi_L^B : T \mapsto B^{1/2}TB^{1/2}$ in the sense that B is recovered as $\left(\Phi_L^B\right)^*(I) = B$.

For any two operations Φ_1, Φ_2 the composition $\Phi_2 \circ \Phi_1$ is an operation, called a sequential operation as it is obtained by performing first Φ_1 and then, in immediate succession, Φ_2. Exchanging the order in this procedure yields the sequential operation $\Phi_1 \circ \Phi_2$ which in general will not coincide with the former. The associated effects satisfy $(\Phi_2 \circ \Phi_1)^*(I) \leq \Phi_1^*(I)$, as well as $(\Phi_1 \circ \Phi_2)^*(I) \leq \Phi_2^*(I)$. It may happen that these effects coincide so that they would constitute a joint lower bound for $\Phi_1^*(I)$, $\Phi_2^*(I)$. As an example we note that for any two effects $B, C \in \mathcal{E}(\mathcal{H})$ we have $(\Phi_L^C \circ \Phi_L^B)^*(I) = B^{1/2}CB^{1/2}$, whereas $(\Phi_L^B \circ \Phi_L^C)^*(I) = C^{1/2}BC^{1/2}$. For sharp properties P and R this reads $(\Phi_L^P \circ \Phi_L^R)^*(I) = RPR$ and $(\Phi_L^R \circ \Phi_L^P)^*(I) = PRP$ so that the commutativity of the Lüders operations, $\Phi_L^P \circ \Phi_L^R = \Phi_L^R \circ \Phi_L^P$, would amount to having $PR = RP$.

II.3.2 State transformers. The state change experienced by an open system will not in general be deterministic so that each time step of the evolution could be represented by a single operation; it will rather be a stochastic process reflecting the random external influences. Sometimes it is possible to introduce a collection of operations Φ_X determining for each initial state T the possible final states $\Phi_X(T)$ which may occur with probability $\mathrm{tr}\left[\Phi_X(T)\right]$. The labels X indicate the conditions for a particular state transformation Φ_X to be the actual one. We shall refer to such a family of operations as a state transformer. If the system $\mathcal{S}$ is subjected to a manipulation represented by a state transformer $X \mapsto \Phi_X$, then the state of the system changes according to $T \mapsto \Phi_X T$ provided that the conditions described by X are met. This is what happens in a measurement of some observable of $\mathcal{S}$, with an outcome X. The state of the system after the measurement depends not only on its initial state but also on the type of measurement performed as well as the actual outcome. A state transformer $X \mapsto \Phi_X$ is a convenient description of this aspect of a measurement, the outcome X conditioning the state change that has actually taken place. We give the formal definition.

A *state transformer* is a state transformation valued (STV) measure $\mathcal{I} : \mathcal{F} \to \mathcal{L}\left(\mathcal{T}(\mathcal{H})\right)$ on a measurable space $(\Omega, \mathcal{F})$ defined through the properties

$$\mathcal{I}_X(T) \geq O \qquad \text{for all } X \in \mathcal{F}, \ T \in \mathcal{S}(\mathcal{H}) \tag{3.5a}$$

$$\mathrm{tr}\left[\mathcal{I}_\Omega(T)\right] = \mathrm{tr}\left[T\right] \qquad \text{for all } T \in \mathcal{S}(\mathcal{H}) \tag{3.5b}$$

$$\mathcal{I}_{\cup X_i}(T) = \sum \mathcal{I}_{X_i}(T) \quad \text{for all disjoint sequences } (X_i) \subset \mathcal{F}, \text{ all } T \in \mathcal{S}(\mathcal{H}) \tag{3.5c}$$

where the series converge in trace norm. The properties (3.5) guarantee that the mapping $X \mapsto E(X)$, defined as

$$\text{tr}\big[TE(X)\big] := \text{tr}\big[\mathcal{I}_X(T)\big] \quad X \in \mathcal{F},\ T \in \mathcal{S}(\mathcal{H}) \tag{3.6}$$

is a (normalised) POV measure on $(\Omega, \mathcal{F})$. Thus any state transformer $\mathcal{I}$ induces a unique observable E, the *associate observable* of $\mathcal{I}$. Conversely, every observable E has infinitely many *E-compatible* state transformers.

As an example, let $(X_i)_{i \in \mathbf{I}}$ be any countable partition of Ω into disjoint ($\mathcal{F}$-measurable) sets and $(T_i)_{i \in \mathbf{I}}$ a collection of states. Then the following defines an E-compatible state transformer $\mathcal{I}$:

$$\mathcal{I}_X(T) := \sum_{i \in \mathbf{I}} \text{tr}\big[TE(X \cap X_i)\big]T_i \quad X \in \mathcal{F},\ T \in \mathcal{S}(\mathcal{H}) \tag{3.7}$$

There are state transformers that do not change the states (modulo normalisation): for any probability measure λ, let

$$\mathcal{I}_X(T) := \lambda(X)\,T \tag{3.8}$$

The associate observable is given by

$$E(X) = \lambda(X)\,I \tag{3.9}$$

This is to say that a measurement that does not alter the states of the system gives no information at all. In other words any measurement of a nontrivial observable must induce some state changes. A state transformer that does yield only minimal (in a certain sense) disturbances is the Lüders transformer associated with a discrete sharp observable $A = \sum a_i P_i$, defined via

$$\mathcal{I}_{L,X}^{A}(T) := \sum_{a_i \in X} P_i\,T\,P_i = \sum_{a_i \in X} \Phi_L^{P_i}(T) \tag{3.10}$$

where $\Phi_L^{P_i}$ is the Lüders operation for P_i (Sec. 3.1). It is obvious that the operation $\mathcal{I}_{L,\mathbf{R}}^{A}$ leaves unchanged any vector state φ which is an eigenvector of some P_i. This property, known as the *ideality* of a state transformer, is characteristic of the Lüders transformer. On the other hand for all other states φ arising as superpositions of different eigenvectors of A, the state $\mathcal{I}_{L,\{a_i\}}^{A}\big(P[\varphi]\big)$ is an eigenstate of A corresponding to the eigenvalue a_i. This is to say that the Lüders transformer is *repeatable*, acting thus rather invasively on non-eigenstates.

Two state transformers $\mathcal{I}_1$ and $\mathcal{I}_2$ may be combined into a composite state transformer $\mathcal{I}_{12}$ on the product space $(\Omega_1 \times \Omega_2, \mathcal{F}_1 \times \mathcal{F}_2)$. This new state transformer is constructed as the extension to a measure of the following family of sequential operations [1.8]:

$$T \mapsto \mathcal{I}_{12,X \times Y}(T) := \mathcal{I}_{2,Y} \circ \mathcal{I}_{1,X}(T) \tag{3.11}$$

The interpretation of $\mathcal{I}_{12}$ as a sequential state transformer is obvious. It corresponds to a sequential measurement of the observable in question, in the given order. The observable associated with a sequential state transformer $\mathcal{I}_{12}$ is determined by the effects $(\mathcal{I}_{12,X \times Y})^*(I)$, $X \in \mathcal{F}_1$, $Y \in \mathcal{F}_2$. In general one has $\mathcal{I}_{12} \neq \mathcal{I}_{21}$, which is to say that a result of a sequential measurement of any two observables will depend on the order in which the measurements in question are performed on the system. It may happen that for two observables E_1 and E_2 there are commuting state transformers $\mathcal{I}_1$ and $\mathcal{I}_2$ such that

$$\mathcal{I}_{12} = \mathcal{I}_{21} \tag{3.12}$$

This is then sufficient to ensure that these observables are coexistent. Indeed if (3.12) holds true, then $X \times Y \mapsto (\mathcal{I}_{12,X \times Y})^*(I) = (\mathcal{I}_{21,X \times Y})^*(I)$ constitutes a joint observable for E_1 and E_2.

II.3.3 Measurements. Von Neumann's formulation of the measurement process within the quantum theory of compound systems has become the paradigm for quantum measurement theory [2.3]. In recent years this theory was extended in various respects so as to be applicable to measurements of arbitrary observables. In the ensuing general framework the von Neumann model appears as an important special case, characterised as an ideal measurement of a discrete sharp observable. We start with a brief review of this prototypical model and proceed then to summarise the main features of general measurement theory.

In order to construct a measurement for a discrete observable with the spectral decomposition $A = \sum a_i P_i$, first fix an orthonormal basis $\{\varphi_{ij}\}$ for the object system's Hilbert space $\mathcal{H}$ consisting of eigenvectors of A, where the second index accounts for the (possible) degeneracy of the eigenvalue a_i. Consider an apparatus $\mathcal{A}$, with the Hilbert space $\mathcal{H}_{\mathcal{A}}$ of dimension equal to the number of distinct eigenvalues a_i of A. Introduce a 'pointer observable' $Z = \sum_i z_i P[\phi_i]$ [where $\{\phi_i\}$ is any orthonormal basis of $\mathcal{H}_{\mathcal{A}}$] and a unitary mapping U_{vNL} satisfying the relation

$$U_{vNL}(\varphi \otimes \phi) = \sum \langle \varphi_{ij} \,|\, \varphi \rangle \, \varphi_{ij} \otimes \phi_i \tag{3.13}$$

where ϕ denotes a fixed initial state of the apparatus. The mapping U_{vNL} is meant to represent a measurement interaction between $\mathcal{S}$ and $\mathcal{A}$ correlating the values of the measured observable A with those of the pointer observable Z. Thus for any initial state φ of $\mathcal{S}$ the final state of the compound system $\mathcal{S} + \mathcal{A}$ is given by (3.13). In particular, $U_{vNL}(\varphi \otimes \phi)$ determines the state of the apparatus after the measurement as the reduced state $W = \sum p_\varphi^A(a_k) P[\phi_k]$. It is now immediate to observe that the measurement model specified by the items $\langle \mathcal{H}_{\mathcal{A}}, Z, \phi, U_{vNL} \rangle$ satisfies the calibration condition:

$$\text{if} \quad p_\varphi^A(a_i) = 1 \quad \text{then} \quad p_W^Z(z_i) = 1 \tag{3.14}$$

This is equivalent to the seemingly stronger probability reproducibility condition, which serves to characterise A as the measured observable:

$$p_\varphi^A(a_i) = p_W^Z(z_i) \tag{3.15}$$

for all $i = 1, 2 \cdots$, and for any state φ of $\mathcal{S}$. We have thus established the von Neumann-Lüders measurement model. Such a measurement has the peculiar property of being repeatable, that is, it leaves the system always in an eigenstate associated with the registered outcome z_i.

In the construction of a measurement theory for a general observable E one follows the line of reasoning outlined above. Thus one starts with specifying a *measuring apparatus* $\mathcal{A}$ [with Hilbert space $\mathcal{H}_\mathcal{A}$], its *initial state* $T_\mathcal{A} \in \mathcal{S}(\mathcal{H}_\mathcal{A})$, a *pointer observable* $Z : \mathcal{F}_\mathcal{A} \to \mathcal{L}(\mathcal{H}_\mathcal{A})$ [with value space $(\Omega_\mathcal{A}, \mathcal{F}_\mathcal{A})$] and a (measurable) *pointer function* $f : \Omega_\mathcal{A} \to \Omega$ which serves to correlate the pointer values with the values of the measured observable. Next one must find a suitable *measurement coupling*, a positive linear trace preserving state transformation $V : \mathcal{T}(\mathcal{H} \otimes \mathcal{H}_\mathcal{A}) \to \mathcal{T}(\mathcal{H} \otimes \mathcal{H}_\mathcal{A})$ for the compound system $\mathcal{S} + \mathcal{A}$. This map associates with any initial state T of $\mathcal{S}$ a final state of $\mathcal{S} + \mathcal{A}$ given as $V(T \otimes T_\mathcal{A})$. From this one obtains, via partial tracing, the reduced states $\mathcal{R}_\mathcal{S}\big(V(T \otimes T_\mathcal{A})\big)$ and $\mathcal{R}_\mathcal{A}\big(V(T \otimes T_\mathcal{A})\big)$ of $\mathcal{S}$ and $\mathcal{A}$, respectively. The decisive requirement for the quintuple $\langle \mathcal{H}_\mathcal{A}, Z, T_\mathcal{A}, V, f \rangle$ to constitute a measurement of E derives from the interpretation of the numbers $p_T^E(X)$ as probabilities for measurement outcomes. Indeed if $p_T^E(X) = 1$, then one would expect that the pointer observable would show this value with certainty after the measurement, that is, $p_{\mathcal{R}_\mathcal{A}(V(T \otimes T_\mathcal{A}))}^Z \big(f^{-1}(X)\big) = 1$. This is the content of the *calibration condition*. When dealing with arbitrary observables this requirement will not be applicable in all cases. As a matter of fact there are POV measures E admitting $p_T^E(X) = 1$ in no state whenever $E(X) \neq I$. In order to specify sufficient conditions under which a process described by $\langle \mathcal{H}_\mathcal{A}, Z, T_\mathcal{A}, V, f \rangle$ may be claimed to serve as a measurement of an observable E, one therefore needs to stipulate the whole *probability reproducibility condition*: that is, one should require that the full measurement outcome distribution $X \mapsto p_T^E(X)$ be recovered from the pointer statistics: for all $X \in \mathcal{F}$, and $T \in \mathcal{S}(\mathcal{H})$,

$$p_T^E(X) = p_{\mathcal{R}_\mathcal{A}(V(T \otimes T_\mathcal{A}))}^Z \big(f^{-1}(X)\big) \tag{3.16}$$

It should be noted that the term apparatus must not always be understood in a literal sense. In most measurement theoretic models one includes only some microscopic part of the macroscopic device into the description, which could be referred to as a probe system. This is the case throughout this book so that we shall use 'probe' and 'apparatus' as synonyms.

It is obvious that specifying a quintuple $\langle \mathcal{H}_\mathcal{A}, Z, T_\mathcal{A}, V, f \rangle$ which fulfills the condition (3.16) does not yet provide an exhaustive account of the physics underlying a measurement process for the observable E. This goal will require further assumptions and conditions on $\langle \mathcal{H}_A, Z, T_A, V, f \rangle$. In particular one should like to understand how a measurement leads to a definite result. Various kinds of correlation conditions and the so-called objectification requirements have been investigated in this context [1.1]. For the purposes of the present book the probability reproducibility condition will be taken as the crucial (and minimal) condition for $\mathcal{M} := \langle \mathcal{H}_A, Z, T_A, V, f \rangle$ to constitute an E-measurement.

The measurement coupling V is often given by a unitary operator U on $\mathcal{H} \otimes \mathcal{H}_A$ so that $V(T \otimes T_A) = U(T \otimes T_A)U^*$. Furthermore the pointer observable Z is in many cases conveniently chosen to be a PV measure possessing the same value space and scale as E. If in addition the initial apparatus state is a vector state ϕ, then for the resulting *unitary measurement* $\mathcal{M}_U := \langle \mathcal{H}_A, Z, \phi, U \rangle$ of E the condition (3.16) assumes the simpler form

$$p_\varphi^E(X) = \langle U(\varphi \otimes \phi) \mid I \otimes Z(X)U(\varphi \otimes \phi) \rangle \qquad (3.17)$$

for all $X \in \mathcal{F}$ and for all states $\varphi \in \mathcal{H}$. Clearly the measurement $\mathcal{M}_{vNL} := \langle \mathcal{H}_A, Z, \phi, U_{vNL} \rangle$ of a discrete observable $A = \sum a_i P_i$ is a unitary measurement; but it is important to note that one can construct other unitary A-measurements which do not always share the property of being repeatable [2.13].

The relation between processes $\mathcal{M} = \langle \mathcal{H}_A, Z, T_A, V, f \rangle$ and observables E established by the probability reproducibility condition (3.16) allows also a reading opposite to the one just described. In concrete applications one usually starts not with fixing the observable to be measured but rather with analysing an experimental arrangement, considered to yield a measurement $\mathcal{M}$. Equation (3.16) then determines the observable actually measured with this arrangement.

Any measurement $\mathcal{M}$ of an observable E induces a state transformer $\mathcal{I}$ by virtue of the relation

$$\mathcal{I}_X(T) := \mathcal{R}_S\left[V(T \otimes T_A) \cdot I \otimes Z(f^{-1}(X))\right] \qquad (3.18)$$

This state transformer is E-compatible since one has

$$p_T^E(X) = \operatorname{tr}\left[\mathcal{I}_X(T)\right] \quad X \in \mathcal{F},\ T \in \mathcal{S}(\mathcal{H}) \qquad (3.19)$$

This is a confirmation of the fact that the measurement determines uniquely the measured observable. It may happen that two different measurements $\mathcal{M}$ and $\widetilde{\mathcal{M}}$ of E define the same state transformer. Such measurements are said to be *(operationally) equivalent*.

The elements of a measurement are schematically shown in Figure 2.1. The object and apparatus are prepared, coupled and separated. The apparatus is eventually found in a pointer eigenstate $T_{A,k} := \mathcal{R}_A\left[V(T \otimes T_A) \cdot I \otimes Z(\{k\})\right]$ indicating an outcome k, say, according to a fixed discrete reading scale. This entails that the system has also assumed some final state $T_k := \mathcal{R}_\mathcal{S}\left[V(T \otimes T_\mathcal{A}) \cdot I \otimes Z(\{k\})\right]$ depending on that outcome.

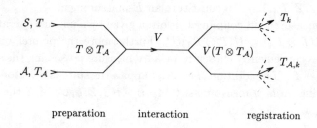

Figure 2.1. Measurement scheme

Any two measurements $\mathcal{M}_1$ and $\mathcal{M}_2$ of observables E_1 and E_2, with induced state transformers $\mathcal{I}_1$ and $\mathcal{I}_2$, can be combined to yield a sequential measurement of the two observables, by performing them on $\mathcal{S}$ one after the other in either order. Without entering into technical details, we let $\mathcal{M}_{12}$ denote the sequential measurement obtained by carrying out first $\mathcal{M}_1$ and then, in immediate succession, $\mathcal{M}_2$. One can show that the resulting state transformer $\mathcal{I}_{12}$ is in fact the sequential state transformer induced by $\mathcal{I}_1$ and $\mathcal{I}_2$ so that $\mathcal{I}_{12} = \mathcal{I}_2 \circ \mathcal{I}_1$ [2.14]. In a similar vein one may construct the sequential measurement $\mathcal{M}_{21}$. Generally $\mathcal{M}_{12}$ and $\mathcal{M}_{21}$ are quite different. However, if these sequential measurements are operationally equivalent, that is, if $\mathcal{I}_{12} = \mathcal{I}_{21}$, then the observables E_1 and E_2 are coexistent. In that case either one of $\mathcal{M}_{12}$ and $\mathcal{M}_{21}$ may serve as a joint measurement for E_1 and E_2.

The construction of the von Neumann-Lüders model $\mathcal{M}_{vNL}$ demonstrates that any discrete sharp observable $A = \sum a_i P_i$ admits a unitary measurement, the enusing state transformer $\mathcal{I}_L^A$ being the Lüders transformer (3.10). This result has been thoroughly generalised: the Hilbert space frame of quantum mechanics permits that any physical quantity, represented as a POV measure, can indeed be measured. In fact according to a fundamental theorem of the quantum theory of measurement any (completely positive) state transformer derives from some unitary measurement [1.4, 2.15]. Therefore for any observable there do exist such measurements.

II.3.4 The standard model. We provide next an example of the general measurement scheme developed in the preceding two subsections. Chapter VII will offer several concrete realisations of this 'standard model'. Suppose that one intends to measure an observable A of the object system by coupling it to some observable B of the measuring apparatus through the interaction

$$U = e^{i\lambda A \otimes B} \tag{3.20}$$

where λ is an appropriate coupling constant. Using the spectral decomposition of A, one may write U in the form

$$U = \int_{\mathbf{R}} E^A(da) \otimes e^{ia\lambda B} \tag{3.21}$$

Then an initial state $\varphi \otimes \phi$ of the compound system is transformed into

$$U(\varphi \otimes \phi) = \int_{\mathbf{R}} E^A(da)\varphi \otimes e^{ia\lambda B}\phi \equiv \int_{\mathbf{R}} E^A(da)\varphi \otimes \phi_{\lambda a} \tag{3.22}$$

where $\phi_{\lambda a} := e^{ia\lambda B}\phi$. If the coupling U is to serve its purpose one needs to choose the initial apparatus state ϕ, the pointer observable Z, and possibly a pointer function f such that $\langle \mathcal{H}_{\mathcal{A}}, Z, \phi, U, f \rangle$ constitutes a measurement of A.

With any specific choice of ϕ, Z, and f, the probability reproducibility condition (3.16) always defines the observable E actually measured by this scheme. To evaluate this condition, one first determines the final state of the apparatus,

$$\mathcal{R}_{\mathcal{A}}\big(P[U(\varphi \otimes \phi)]\big) = \int_{\mathbf{R}} p_{\varphi}^A(da)\, P[\phi_{\lambda a}] \tag{3.23}$$

In view of the coupling constant λ ($\neq 0$) it is convenient to introduce a pointer function $f(x) = \lambda^{-1}x$. Then the measured observable takes the form

$$E(X) = \int_{\mathbf{R}} p_{\phi_{\lambda a}}^Z(\lambda X)\, E^A(da) \tag{3.24}$$

The structure of the effects $E(X)$ show that in general the actually measured observable E is a smeared version of the observable A intended to be measured. The question then is which choices of ϕ, Z, and f would possibly yield $E = E^A$, that is,

$$\int_{\mathbf{R}} p_{\phi_{\lambda a}}^Z(\lambda X)\, E^A(da) = E^A(X) \qquad \text{for all } X \in \mathcal{B}(\mathbf{R}) \tag{3.25}$$

This is the case exactly when $p_{\phi_{\lambda a}}^Z(\lambda X) = \chi_X(a)$. Since $\langle \phi_{\lambda a} \,|\, Z(\lambda X)\phi_{\lambda a} \rangle = \langle \phi \,|\, e^{-ia\lambda B} Z(\lambda X)e^{ia\lambda B}\phi \rangle$, one may also anticipate that an 'optimal' measurement can be obtained by choosing the pointer observable Z to be *conjugate* to B in the sense of covariance (cf. Chapter III):

$$e^{-ia\lambda B} Z(\lambda X)\, e^{ia\lambda B} = Z\big(\lambda(X - a)\big) \tag{3.26}$$

The general form of the state transformer defined by this measurement according to (3.18) can also be given explicitly. Its structure depends on that of the pointer observable Z. Writing $Z(\lambda X) = \int_{\lambda X} |z\rangle\langle z| \, dz$, with a formal resolution of identity, one obtains

$$\mathcal{I}_X(T) = \int_{\lambda X} K_z \, T \, K_z^* \, dz, \qquad K_z = \int_{\mathbf{R}} \langle z | \phi_{\lambda a} \rangle \, E^A(da) \qquad (3.27)$$

Without using this formal expression, one can directly confirm that the outcome probabilities for E, as well as for A, are the same both before and after the measurement. The measurement does not alter these probabilities.

It is instructive to work out two special cases of the standard model. Consider first a discrete observable $A = \sum a_k P_k$. As the probe system, take a particle moving in one degree of freedom so that $\mathcal{H}_A = L^2(\mathbf{R})$, and couple A with the probe's momentum P_A: $U = \exp(-i\lambda A \otimes P_A)$. Since the momentum generates translations of the position, it is indeed natural to choose the position Q_A conjugate to P_A as the pointer observable. An initial state $\varphi \otimes \phi$ of the object-apparatus system is now transformed into $\sum_k P_k \varphi \otimes \phi_{\lambda a_k}$. In the position representation (for $\mathcal{A}$) one has $\phi_{\lambda a_k}(x) = \phi(x - \lambda a_k)$. Assuming that the spacing between the eigenvalues a_k is greater than $\frac{\delta}{\lambda}$ and that ϕ is supported in $\left(-\frac{\delta}{2}, \frac{\delta}{2}\right)$, then the pointer states $\phi_{\lambda a_k}$ are supported in mutually disjoint sets $I_k = \left(a_k - \frac{\delta}{2\lambda}, a_k + \frac{\delta}{2\lambda}\right)$. Therefore

$$E(X) = \sum \langle \phi_{\lambda a_k} | E^{Q_A}(\lambda I_k) \phi_{\lambda a_k} \rangle \, P_k = \sum_{k \in X} P_k = E^A(X) \qquad (3.28)$$

which shows that the actually measured observable is the intended one, A. The state transformer of this measurement is the Lüders state transformer (3.10).

In the case of a continuous observable, such as position, the above model amounts to measuring not the intended observable but a smeared version of it. Indeed taking $A = Q$ and adopting the respective spectral representations, one has $U(\varphi \otimes \phi)(q, x) = \varphi(q)\phi(x - \lambda q)$. Then

$$E(X) = \iint |\phi(q' - \lambda q)|^2 \, \chi_{\lambda X}(q') dq' \, E^Q(dq) = \chi_X * e(Q) \equiv E^e(X) \qquad (3.29)$$

where we have introduced the confidence function e,

$$e(q) := \lambda |\phi(-\lambda q)|^2 \qquad (3.30)$$

Since e cannot be a delta-function, the measured observable E is never the sharp position but an unsharp one. The ensuing state transformer $\mathcal{I}^Q$ is

$$\mathcal{I}_X^Q(T) = \int_X K_q \, T \, K_q^* \, dq \qquad K_q := \sqrt{\lambda}\,\phi(-\lambda(Q - q)) \qquad (3.31)$$

II.3.5 Repeatable measurements. We have seen that the von Neumann-Lüders measurement of an observable $A = \sum a_i P_i$ has a number of remarkable properties.

In particular, it is repeatable, that is, its iterated application leads always to the same result. The probability for obtaining a result a_i upon repetition, under the condition that the preceding measurement just gave this result, is equal to unity. This property can be formulated in various equivalent ways, the most suggestive one being the following:

$$p^A_{T_i}(a_i) \;=\; 1 \qquad \text{whenever} \qquad p^A_\varphi(a_i) \neq 0 \qquad (3.32)$$

Here $T_i := \mathcal{I}^A_L(a_i)P[\varphi]/p^A_\varphi(a_i)$ is the state of $\mathcal{S}$ after the measurement with outcome a_i. An equivalent form of (3.32) is: $\mathrm{tr}\big[\mathcal{I}^A_{L,a_i}\,\mathcal{I}^A_{L,a_i}(T)\big] \;=\; \mathrm{tr}\big[\mathcal{I}^A_{L,a_i}(T)\big]$, and this holds for any state T and all possible values a_i of A.

Generalising these considerations we say that a measurement $\mathcal{M}$ of an observable E is repeatable if its state transformer $\mathcal{I}$ is *repeatable*, that is, if

$$\mathrm{tr}\big[\mathcal{I}_X\mathcal{I}_X(T)\big] \;=\; \mathrm{tr}\big[\mathcal{I}_X(T)\big] \qquad (3.33)$$

for all $T \in \mathcal{S}(\mathcal{H})$ and for all $X \in \mathcal{F}$. Condition (3.33) can be rephrased as $p^E_{T_X}(X) = 1$ [whenever $p^E_T(X) \neq 0$] where $T_X = \mathcal{I}_X T/\mathrm{tr}[\mathcal{I}_X T]$ is the normalised final state of the measured system conditional upon an outcome in X. Since the pioneering work of von Neumann [2.3] it has been an important problem to find out under what circumstances a measurement is repeatable. A necessary condition for this is the following:

THEOREM. *If a measurement is repeatable, then the measured observable is discrete.*

We recall that a POV measure E is discrete if there is a countable subset Ω_0 of Ω such that $E(\Omega_0) = I$. This theorem presumes that the value space $(\Omega, \mathcal{F})$ of the state transformer $\mathcal{I}$ is a standard Borel space. In particular the real Borel space $(\mathbf{R}^n, \mathcal{B}(\mathbf{R}^n))$ is of this kind. The proof of the statement in its present generality is due to Łuczak [2.16]. Ozawa [2.15] established the same result somewhat earlier under the additional assumption that the measurement is equivalent to a unitary measurement.

Not every measurement of a discrete observable, like $A = \sum a_i P_i$, is repeatable. But any discrete sharp observable does possess *some* repeatable measurement such as, for example, a von Neumann-Lüders measurement. Generally a discrete observable $E : \omega_i \mapsto E_i$ admits a repeatable measurement exactly when all the effects $E_i \neq O$ in its range have eigenvalue 1. Indeed if there is a repeatable measurement for E, then to each E_i there belongs some (unit) vector φ_i such that $E_i\varphi_i = \varphi_i$. Conversely if each E_i has such an eigenvector, then the state transformer (3.7) with the choice $X_i = \{\omega_i\}$ and $T_i = P[\varphi_i]$ is repeatable; it is also completely positive so that it admits a repeatable unitary E-measurement.

There are some other measurement theoretical notions which at first sight appear less restrictive than repeatability so that one might agree more easily to consider them as constitutive properties of a measurement. These are the concepts of first kind and value reproducible measurements. A measurement $\mathcal{M}$ of an observable E is of the *first kind* if the probability for a given result is the same both before and after the measurement, that is, if

$$p_T^E(X) = p_{T_\Omega}^E(X) \qquad (3.34)$$

for all outcome sets X and all states T of the system $\mathcal{S}$. One may consider also a weaker formulation, the *value reproducibility condition*, which stipulates that for any X and for all T the following implication should hold:

$$\text{if } p_T^E(X) = 1, \quad \text{then } p_{T_\Omega}^E(X) = 1 \qquad (3.35)$$

It is evident that repeatability implies first-kindness and that the latter property is stronger than value reproducibility. Interestingly, in the case of sharp observables the reverse chain of implications is also valid: a measurement of a sharp observable is repeatable exactly when it is value reproducible or, a fortiori, when it is of the first kind [1.1]. Whenever an E-measurement has one of these properties, and thus all of them, it follows that E is a discrete observable.

To illustrate these notions and their interrelations we refer to the standard model of Section 3.5. We observed already there that a measurement with the coupling $e^{i\lambda A \otimes B}$ never alters the outcome statistics of the actually measured observable E, nor that of A. In other words these measurements are of the first kind. For a discrete sharp observable A we could choose the pointer observable and the initial apparatus state such that the measured observable is A, in which case the measurement is also repeatable. For $A = Q$ the measurement cannot be repeatable since the measured observable $E^{Q,e}$ is not discrete.

II.4 Individual interpretation of quantum mechanics

It is often claimed that quantum mechanics is merely a *statistical* theory, and this statement is accompanied with the remark that only probabilistic predictions of measurement outcomes can be made in general. However, in most practical instances physicists think and speak in terms of *individual* systems, such as elementary particles, considering manipulations performed on them with the aim to prepare and measure certain *properties*. We shall argue next that the Hilbert space language of quantum mechanics is rich enough to provide an operational justification for an *individual* interpretation of this theory. To this end we shall specify those features of the empirical domain of a physical theory which must be regarded as preconditions for the constitution of the objects of that theory. It turns out that quantum mechanics possesses all the structural elements implied by these conditions.

First of all, any probabilistic theory of physics presupposes an event structure associated with the totality of all possible measurement outcomes. The very existence of scientific experience is due to the fact that one is able to find statistical regularities in the observation of certain event sequences. These regularities are represented as law-like probability connections which under certain circumstances (as they are realised, e.g., in the realm of classical physical phenomena) may assume the form of deterministic laws. The possibility of performing reproducible preparations and measurements is reflected in quantum mechanics by the notions of states and effects, the latter being introduced as state functionals assigning probabilities to the possible measurement outcomes. In order to ensure reproducibility one needs to be able to repeat 'the same' measurement under 'the same' conditions (of preparation). Repeated measurements may be done either as a sequence in time or simultaneously at different places. Hence the regularities in question are due to the homogeneity of time and space. The remaining kinematical symmetries come into play by the *objectivity* requirement: different observers should be able to provide consistent descriptions of the same sets of event sequences. These considerations have led to the definition of elementary quantum systems as irreducible projective unitary representations of the Galilei or Poincaré group.

Speaking of systems and their properties presupposes that there exist event sequences which are one by one related to each other so that they can be assigned to some underlying entity persisting in time – an object system. Thus one must be able to observe a system at different times and by means of different measurements. There are two implications of this requirement. First, a system can only be detected as a separate entity if it is to some degree of approximation isolated from its environment. This is to say that at least for some period of times t its state T_t depends only on the original preparation at time t_o and does not reflect any entanglement with the environment, i.e., there must be a deterministic dynamical law $D_{t,t_o} : T_{t_o} \mapsto T_t$ which sends pure states into pure states. The Schrödinger picture of the unitary quantum mechanical dynamics is of that type. Second, objects would never be recognised as such, were there not some *nondestructive* measurement procedures. This is reflected in the operational structure of quantum mechanics. As seen in Section 3.1, to any effect there exists a family of operations describing the state changes due to measurements.

The event space of a single measurement is given by some measurable space $(\Omega, \mathcal{F})$ of outcomes. Any measurement is represented by an observable E, the latter establishing a connection between the outcomes and the object system, $E : \mathcal{F} \to \mathcal{E}(\mathcal{H})$. In this way one is led to consider the set of effects as representing the totality of all possible measurement events. Now the question arises as to what the occurrence of an effect in a measurement allows one to infer about the system under investigation. In Subsection 2.2 of this chapter we have isolated a subset $\mathcal{E}_p(\mathcal{H})$ of effects serving as candidates of what may be called properties. We shall argue that a system must be understood as a carrier of properties. In fact the preparation of a system is generally based on a specification of some of its properties. In this sense there must exist repeatable measurements. A physical quantity can be ascertained as a real property, or an 'element of reality' whenever its value can be predicted with certainty and without changing the system. The decisive question is whether the physical quantities of quantum mechanics can correspond to elements of reality. The answer is in the positive as far as the set of sharp properties is concerned. In the set of operations associated with any sharp property P there is the Lüders operation $\Phi_L^P : T \mapsto PTP$ which is repeatable and ideal. Thus whenever a system is in an eigenstate T of a sharp property P, $PT = T$, then this property corresponds to an element of reality since it can be determined with certainty and without changing the system's state, namely, by application of the corresponding Lüders operation. This is the operational foundation for interpreting projections as properties, or propositions about properties.

Turning now to general (sharp or unsharp) properties and observables, it must be noted that there exist limitations of the measurability which may forbid their interpretation as elements of reality. For example, one may have postulated properties of measurements, such as repeatability, which are not satisfied by any instrument. Such is the case with continuous observables. Furthermore it may happen that measuring systems and interactions available for measurements are limited, e.g., due to conservation laws, to the extent that some of the instruments associated with a POV measure cannot be realised. These problems can be resolved by introducing approximate properties and adopting an accordingly weakened form of the reality criterion. Instead of strict predictability one should only stipulate probability close to one, i.e., greater than $1-\varepsilon$ with some small ε. Also one cannot insist in strict ideality but should only require approximately non-disturbing operations. Now to any effect B there is associated its Lüders operation, $\Phi_L^B : T \mapsto B^{\frac{1}{2}}TB^{\frac{1}{2}}$. This operation is reproducible in the sense that $\text{tr}[TB] \geq 1-\varepsilon$ implies $\text{tr}[\Phi_L^B(T)B]/\text{tr}[\Phi_L^B(T)] \geq 1-\varepsilon$. Φ_L^B is not (approximately) repeatable but only weakly repeatable in the following sense: $\text{tr}[\Phi_L^B\Phi_L^B(T)]/\text{tr}[\Phi_L^B(T)] \geq \text{tr}[\Phi_L^B(T)]$. Most importantly Φ_L^B is approximately ideal as for any state T the state $\Phi_L^B(T)/\text{tr}[\Phi_L^B(T)]$ is close to T (in the trace norm topology) whenever $\text{tr}[TB] \geq 1 - \varepsilon$, ε small:

$$\left\| T - \left[\text{tr}[\Phi_L^B T]\right]^{-1} \Phi_L^B(T) \right\|_1 \leq 2\left(\varepsilon + \sqrt{\varepsilon}\right) \tag{4.1}$$

We have thus shown that quantum mechanics provides a framework in which it is possible to speak of preparing and measuring properties of physical systems. The application of (approximately) repeatable measurements allows one to extend the minimal interpretation of probabilities to a realistic, individual interpretation in the following sense: the number $p_T^E(X)$ is the probability that an (approximately) repeatable measurement of the observable E on a system in state T leads to an outcome in the set X and thus leaves the system in a state in which the effect $E(X)$ is an (approximately) real property.

III. Observables

The representation of the observables of a physical system as self-adjoint operators on Hilbert space is traditionally motivated by making reference to the correspondence principle or by invoking some quantisation scheme, such as the transcription of the Lie algebra structure from phase space to Hilbert space. There exists an alternative approach which refers directly to the operationally relevant features of these observables. This *relativistic approach* has the advantage of being open to the consideration of unsharp observables, which turns out necessary in some cases where sharp 'observables' simply do not exist or are not amenable to measurements.

The idea of symmetry, formalised as invariance under a group of transformations, plays a fundamental role in the quantum, as also in the classical, description of any physical system. The very possibility of scientific experience rests on the fact that the observed objects possess features that ensure their identity under different experimental circumstances. To perform many runs of an experiment under 'the same' conditions requires that the preparation and registration devices behave 'the same way' at different times and locations. Thus it is the invariance of an object under active space-time transformations that first of all enables the collection of scientific experience: reliable experimental facts that can be related to some underlying physical entity. Perhaps even more important is the possibility of communicating such experiences in a unique way. In fact an object would be no 'object', were it not identifiable as 'the same' for different observers, related to each other by means of the various space-time transformations understood in the passive sense. Hence any physical system, as an object of scientific investigations, is necessarily a carrier of properties which are covariant under some kinematical group of space-time transformations.

This consideration shows that the description of a physical system S as existing in space-time should reflect its Galilei or Poincaré invariance: there should exist a group of automorphisms on the algebra of observables, and the corresponding group of convex automorphisms on the set of states, representing the actions of Galilei or Poincaré transformations on the measuring and preparation devices associated with S. This automorphism group gives rise to a projective unitary representation $g \mapsto U_g$ of the symmetry group G on the system's Hilbert space $\mathcal{H}$. Accordingly the Galilei or Einstein relativistic quantum theories of elementary systems are based on irreducible projective unitary representations of the Galilei or Poincaré group on a Hilbert space. These irreducible representations are classified according to the *mass* and *spin* values. The basic quantum observables then appear in a double role:

operationally they arise as covariant entities representing properties of objects; in a more formal sense they come into play as the generators of the one-parameter subgroups. Traditionally the covariance aspect was emphasised primarily for the localisation observable as the property characteristic of a 'particle'. The notion of localisability is formalised as a *system of covariance*, or a *system of imprimitivity*, with respect to the Euclidean group. It turns out that the spectral measures of the ten generators satisfy the covariance requirements characteristic of position, momentum, angular momentum and energy observables, so that the corresponding PV measures can be understood as observables. The formal relationships between these observables obtain thus a natural operational interpretation. In particular, Planck's constant ($\hbar$) arises as a scale parameter fixing the mass unit. Its value can and shall here be chosen to be $\hbar = 1$.

This programme of reconstructing the kinematic structure of Hilbert space quantum mechanics in accordance with the principle of relativity was conceived early by H. Weyl [3.1], tackled in the case of the Poincaré group by Wigner [3.2] and developed systematically by Mackey [3.3]. The problem of covariant localisation was treated by Wightman [3.4], a concise and thorough account being given in the book of Varadarajan [2.6]. As an example we sketch the photon localisation problem. Apart from that we restrict ourselves to the case of the Galilei relativity, reviewing the covariance properties of position, momentum, and angular momentum. Unsharp versions of these observables, which are constructed here, will find applications in later chapters. Conjugate pairs, such as position and momentum, angular momentum and angle, or energy and time, will be seen to admit realisations as coexistent covariant pairs of POV measures. Some of these observables are intrinsically unsharp in that they do not derive from a sharp counterpart.

The covariance point of view can also be applied to other than space-time symmetry groups. As an example we refer to the group of rotations in phase space that corresponds to a canonical symmetry in classical mechanics and gives rise to an angular observable. We shall consider the formally analogous number-phase pair for photons which will find applications in Chapter VII.

III.1 Observables as covariant objects

In the following sections we shall specify those subgroups of the Galilei group $\mathcal{G}$, along with their homogeneous spaces, which are basic for constructing covariant quantum observables. We first recall some basic facts about $\mathcal{G}$ and its representations [2.6, 3.5] and proceed then to formulate the notion of covariance.

Elements of the Galilei group will be parametrised as $g = (b, \mathbf{a}, \mathbf{v}, R)$ according to the action on the space-time variables $(\mathbf{x}, t) \in \mathbf{R}^3 \times \mathbf{R}$:

$$g : (\mathbf{x}, t) \mapsto (\mathbf{x}', t') = (R\mathbf{x} + \mathbf{a} + \mathbf{v}t, \, t + b) \tag{1.1}$$

This corresponds to the following decomposition of $\mathcal{G}$ into subgroups:

$$\mathcal{G} = (\mathcal{Z} \times \mathcal{T}) \times' (\mathcal{V} \times' \mathcal{R}) \tag{1.2}$$

Here $\mathcal{Z}$, $\mathcal{T}$, $\mathcal{V}$, $\mathcal{R}$ are the time and space translation groups and the groups of velocity boosts and rotations, respectively. The symbol $\times'$ denotes the semi-direct product of groups. The action (1.1) is a consequence of the fact that the space-time translation group $\mathcal{Z} \times \mathcal{T}$, being isomorphic to the factor group $\mathcal{G}/(\mathcal{V} \times' \mathcal{R})$, is homomorphic to $\mathcal{G}$ and thus a homogeneous space of $\mathcal{G}$ (see below). The composition law and the form of inverse elements are easily found from (1.1):

$$\begin{aligned}
g'g &= (b', \mathbf{a}', \mathbf{v}', R')(b, \mathbf{a}, \mathbf{v}, R) \\
&= (b' + b, \ \mathbf{a}' + R'\mathbf{a} + b\mathbf{v}', \ \mathbf{v}' + R'\mathbf{v}, \ R'R) \\
g^{-1} &= (b, \mathbf{a}, \mathbf{v}, R)^{-1} = (-b, -R^{-1}(\mathbf{a} - b\mathbf{v}), -R^{-1}\mathbf{v}, R^{-1})
\end{aligned} \tag{1.3}$$

According to the Wigner-Kadison theorem [3.6], the automorphism representation $\mathcal{U}_g$ on the algebra of bounded operators can be expressed by virtue of a projective unitary representation U_g as $\mathcal{U}_g(A) = U_g A U_g^{-1}$ for all $g \in \mathcal{G}$, $A \in \mathcal{L}(\mathcal{H})$. We shall freely use the irreducible unitary projective representations $g \mapsto U_g$ of $\mathcal{G}$ as they appear in the Schrödinger picture. These are based on the Hilbert space

$$\mathcal{H} := L^2(\mathbf{R}^3, \mathbf{C}^{2s+1}) \equiv L^2(\mathbf{R}^3) \otimes \mathbf{C}^{2s+1} \tag{1.4}$$

Here $\mathbf{R}^3$ is meant to be the position space, and elements of $\mathcal{H}$ are conveniently expressed as $(2s + 1)$-component functions $\mathbf{x} \mapsto (\psi_{-s}(\mathbf{x}), \psi_{-s+1}(\mathbf{x}), \cdots, \psi_s(\mathbf{x}))$. Defining functions $\psi_\mu(\mathbf{x}, t)$ square-integrable in $\mathbf{x}$ as

$$\psi_\mu(\mathbf{x}, t) := (\exp(-iHt)\,\psi_\mu)(\mathbf{x}), \qquad H = -\tfrac{1}{2m}\nabla^2 \tag{1.5}$$

one has

$$\begin{aligned}
U_{(b,\mathbf{a},\mathbf{v},R)}\,\psi_\mu(\mathbf{x}, t) = \ &\exp\left[-i\tfrac{1}{2}m\mathbf{v}^2(t - b) + im\mathbf{v}\cdot(\mathbf{x} - \mathbf{a})\right]\cdot \\
&\cdot \sum_{\mu'} D^s_{\mu\mu'}(R)\,\psi_{\mu'}\left(R^{-1}(\mathbf{x} - \mathbf{v}(t - b)) - \mathbf{a}, t - b\right)
\end{aligned} \tag{1.6}$$

or equivalently

$$\begin{aligned}
U_{(b,\mathbf{a},\mathbf{v},R)}\,\psi_\mu(\mathbf{x}', t') &= \exp\left[i\lambda_m(g; \mathbf{x}, t)\right] \sum_{\mu'} D^s_{\mu\mu'}(R)\,\psi_{\mu'}(\mathbf{x}, t) \\
\lambda_m(g; \mathbf{x}, t) &= m\left(\tfrac{1}{2}\mathbf{v}^2 t + \mathbf{v}\cdot R\mathbf{x}\right)
\end{aligned} \tag{1.7}$$

where $(\mathbf{x}', t')$ and $(\mathbf{x}, t)$ are related by (1.1). m, s are the mass and spin values parametrising the representation; $D^s_{\mu\mu'}(R)$ are the matrix elements of the projective irreducible representations of $\mathcal{R}$, which can be expressed as $D^s : \mathcal{R} \to \mathcal{L}(\mathcal{H}_s)$, where $\mathcal{H}_s := \mathbf{C}^{2s+1}$. Integer spin values correspond to the proper representations, while the half-integer values give the projective representations.

A Galilei covariant observable is defined as a POV measure E on a homogeneous space Ω (equipped with a Borel algebra Σ) of $\mathcal{G}$, or some subgroup $\mathcal{K}$, such that the action $g \mapsto \alpha_g$ of $\mathcal{K}$ on Ω commutes with the automorphism group representation of $\mathcal{K}$, $g \mapsto \mathcal{U}_g$, for all effects in the range of E:

$$E\big(\alpha_g(X)\big) \;=\; \mathcal{U}_g\big(E(X)\big) \tag{1.8}$$

for all $X \in \Sigma$, $g \in \mathcal{K}$. We recall the notions entailed in this definition of covariance, without going into topological and measure theoretical details (for these, cf. [2.6]). An action α of a group $\mathcal{K}$ on a space Ω is defined as a map that sends group elements g to maps α_g on Ω with the following properties:

$$\alpha_{gh} \;=\; \alpha_g\,\alpha_h \quad \text{for all } g, h \in \mathcal{K}, \qquad \alpha_e \;=\; \iota_\Omega \tag{1.9}$$

Here e, ι_Ω denote the unit element of $\mathcal{K}$ and the identity function of Ω. Such a space Ω is called a $\mathcal{K}$-*space*. If in addition $\mathcal{K}$ acts transitively on Ω, that is, if for any pair of points $\omega, \omega' \in \Omega$ there exists an element $h \in \mathcal{K}$ such that $\omega' = \alpha_h(\omega)$, then Ω is an *homogeneous* $\mathcal{K}$-space. The homogeneity can be expressed as follows

$$\Omega \;=\; \{\alpha_h(\omega_o)\,|\,h \in \mathcal{K}\} \qquad \text{for some } \omega_o \in \Omega \tag{1.10}$$

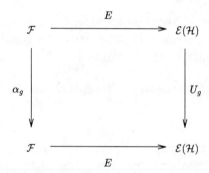

Figure 3.1. Covariance of an observable

We shall be using only automorphism representations which are induced by some unitary projective representations $U : g \mapsto U_g$ as $\mathcal{U}_g(A) = U_g\,A\,U_g^{-1}$. Then the above structures are summarised in the definition of a *system of covariance* (for the Galilei group or some subgroup) as a pair (U, E), where U is a (strongly continuous) unitary projective representation and E is a POV measure on a homogeneous space, equipped with an action of $\mathcal{G}$ (or the subgroup), such that the covariance condition (1.8) is fulfilled. If E is a PV measure, then (U, E) is known as a system of imprimitivity [3.3].

Eq. (1.8) stipulates the commutativity of the diagram shown in Figure 3.1. This figure illustrates how the classical description of the experimental (preparation and registration) equipment precedes the quantum representation of the physical system under investigation. This leads to the important question whether these classical aspects can be ultimately accounted for within quantum theory. Next it becomes evident that an 'object' is constituted as 'something' that persists under a group of symmetry operations, reflected by the existence of a representation. Physically, the covariance requirement means that 'transporting' the registration device associated with the effect $E(X)$ relative to the preparation device by means of the operation $g \in \mathcal{K}$ leads to a new device determining an effect that belongs to the correspondingly transformed value set of the same observable.

III.2 Position and momentum

Position and momentum are perhaps the most familiar observables used in the description of any physical system. Their structures and mutual interrelations carry with them many of the deep questions raised by quantum mechanics, questions addressing the constitution of objects, or the meaning of the uncertainty relations as well as their relevance for the joint measurability of complementary observables. Conceptual and operational aspects of these problems call for the introduction of POV measures towards their solution. We shall investigate in this Section the characteristic properties of the position and momentum observables, considering both their sharp as well as unsharp versions. This finally opens up a way towards introducing phase space observables.

III.2.1 Covariance properties. The very idea of the position of an object is to represent its location in space, to answer the question *where* it *is*. The momentum of an object, given by the product of its mass and velocity, describes its state of motion. Momentum thus accounts for an object's displacements in space, its changes of position. To put these ideas to work, we assume that the system $\mathcal{S}$ under study lives in the Euclidean configuration space $\mathbf{R}^3$. This fact shall be expressed by means of an observable

$$E : \mathcal{B}(\mathbf{R}^3) \to \mathcal{L}(\mathcal{H}), \quad X \mapsto E(X) \tag{2.1}$$

The ensuing probabilities $p_T^E(X) = \mathrm{tr}[TE(X)]$ are interpreted as the probabilities that a position measurement performed on $\mathcal{S}$ in the state T shows this system to be located in the space region $X \subset \mathbf{R}^3$. We call such an observable a *localisation, or position observable*. We do not anticipate that such an observable is unique. It will be so only if it is required to be a sharp observable. This asumption, however, turns out to be problematic from the operational point of view (Sec. IV.1). Similarly, a *momentum observable* shall be described as a POV measure

$$F : \mathcal{B}(\mathbf{R}^3) \to \mathcal{L}(\mathcal{H}), \quad Y \mapsto F(Y) \tag{2.2}$$

with the corresponding measurement outcome probabilities $p_T^F(Y) = \mathrm{tr}[TF(Y)]$.

It is convenient at this point to introduce the component observables according to a Cartesian coordinate system in the respective value spaces $\mathbf{R}^3$. Let $\pi_i : \mathbf{R}^3 \to \mathbf{R}$ denote the ith coordinate projection, $i = 1, 2, 3$, and let the points of the space be represented as the triplets (x_1, x_2, x_3). Then the components of the position along the ith coordinate axis are defined as follows:

$$E^i(X) := E\big(\pi_i^{-1}(X)\big), \quad X \in \mathcal{B}(\mathbf{R}) \tag{2.3}$$

This gives rise to a POV measure E^i on the real line $\mathbf{R}$. If E^i is a PV measure then one obtains a self-adjoint operator

$$Q_i = \int_{\mathbf{R}} x \, dE^i(x) \tag{2.4}$$

In that case E^i as well as the corresponding operator Q_i shall be called the *(sharp) position observables* of the system associated with the coordinate axes $i = 1, 2, 3$. The three position observables Q_1, Q_2, and Q_3 and the PV measure E determine each other uniquely. Indeed E is the unique extension of the mapping

$$X_1 \times X_2 \times X_3 \mapsto E^{Q_1}(X_1) \, E^{Q_2}(X_2) \, E^{Q_3}(X_3) \tag{2.5}$$

defined on the (Borel) cubes $X_1 \times X_2 \times X_3$ of $\mathbf{R}^3$ [2.6]. The spectrum of each Q_i is the whole $\mathbf{R}$ and is obviously continuous since it is assumed that $\mathcal{S}$ can be everywhere in the space $\mathbf{R}^3$.

In a similar way one obtains the component POV measures F^k,

$$F^i(Y) := F\big(\pi_i^{-1}(Y)\big), \quad Y \in \mathcal{B}(\mathbf{R}) \tag{2.6}$$

for momentum and the self-adjoint momentum component operators P_i provided that the F_i are PV measures, thus yielding the *(sharp) momentum observables*:

$$P_i = \int_{\mathbf{R}} y \, dF^i(y) \tag{2.7}$$

In order that the POV measures E and F can be interpreted as the localisation and momentum of an object, it is necessary to specify their operational meaning in terms of the characteristic transformation behaviour under the Galilei group. The entity called 'the position of an object' should be the same up to some shift and rotation irrespective of the location, orientation and state of motion of the position measuring device. Thus position should be covariant under space translations and rotations, and invariant under (velocity) boosts; formally this corresponds to the fact that the position value space $\mathbf{R}^3$ is a homogeneous space of the isochronous Galilei group $\mathcal{G}_o = (\mathcal{T} \times \mathcal{V}) \times' \mathcal{R}$ with respect to the action:

$$\alpha_g = \alpha_{(o,\mathbf{a},\mathbf{v},R)} : \mathbf{x} \mapsto \mathbf{x}' = R\mathbf{x} + \mathbf{a} \tag{2.8}$$

One may thus identify the position space as the subgroup $\mathcal{T}$ of space translations which is itself a homogeneous space of $\mathcal{G}_o$ under this action. The time translation group then serves to label the ensuing family of covariant POV measures as localisations at a given time. Similarly momentum should be covariant under boosts and rotations, and invariant under space-time translations. The corresponding action of $\mathcal{G}$ on the momentum, or velocity space $\mathbf{R}^3$ is given as:

$$\alpha_g = \alpha_{(b,\mathbf{a},\mathbf{v},R)} : \mathbf{u} \mapsto \mathbf{u}' = R\mathbf{u} + \mathbf{v} \tag{2.9}$$

This corresponds to the fact that the boost subgroup $\mathcal{V}$ is a homogeneous space of $\mathcal{G}$. Let $U_{\mathbf{a}}$, $V_{\mathbf{v}}$, W_R denote the subrepresentations of U_g corresponding to the groups of translations, boosts and rotations, respectively. Then the covariance requirements for position and momentum observables E and F are:

$$\begin{aligned}
U_{\mathbf{a}} E(X) U_{\mathbf{a}}^{-1} &= E(X + \mathbf{a}) \\
V_{\mathbf{v}} E(X) V_{\mathbf{v}}^{-1} &= E(X) \\
W_R E(X) W_R^{-1} &= E(RX)
\end{aligned} \tag{2.10}$$

and

$$\begin{aligned}
U_{\mathbf{a}} F(Y) U_{\mathbf{a}}^{-1} &= F(Y) \\
V_{\mathbf{v}} F(Y) V_{\mathbf{v}}^{-1} &= F(Y + m\mathbf{v}) \\
W_R F(Y) W_R^{-1} &= F(RY)
\end{aligned} \tag{2.11}$$

Here m denotes the mass which appears in the multiplier of the given projective representation of $\mathcal{G}$.

It is important to realise that these conditions can be satisfied in many ways. This will be outlined by exploiting the covariance requirements separately and according to whether E and F are assumed to be PV or POV measures. In the first case the solutions are essentially unique.

III.2.2 Sharp position and momentum. Assume first that E, F are PV measures. In view of the Euclidean invariance, the projective subrepresentation of the translation group $\mathcal{T}$ can be chosen to be an ordinary unitary representation [3.3]:

$$\begin{aligned}
U &: \mathbf{a} \mapsto U_{\mathbf{a}} \\
U_{\mathbf{a}+\mathbf{a}'} &= U_{\mathbf{a}} U_{\mathbf{a}'}
\end{aligned} \tag{2.12}$$

$\mathcal{T} = (\mathbf{R}^3, +)$ being a commutative group, any of its continuous unitary representations on $\mathcal{H}$ derives from a PV measure acting on the group of its characters $\mathbf{a} \mapsto e^{i\mathbf{k}\cdot\mathbf{a}}$, $\mathbf{k} \in \mathbf{R}^3$. This is the content of the following theorem [2.4].

STONE'S THEOREM. *For any continuous unitary representation U of $(\mathbf{R}^3, +)$ there is a unique PV measure $F_U : \mathcal{B}(\mathbf{R}^3) \to \mathcal{L}(\mathcal{H})$ such that for all $\mathbf{a} \in \mathbf{R}^3$ one has*

$$U_{\mathbf{a}} = \int_{\mathbf{R}^3} e^{-i\mathbf{a}\cdot\mathbf{k}} \, dF_U(\mathbf{k}) \tag{2.13}$$

Let F_U^i be the ith components of F_U $(i = 1, 2, 3)$, then the induced self-adjoint operators are

$$K_i = \int_{\mathbf{R}} k \, dF_U^i(k) \tag{2.14}$$

This allows one to rewrite $U_{\mathbf{a}}$ in the form

$$U_{\mathbf{a}} = e^{-i\mathbf{a}\cdot\mathbf{K}} \tag{2.15}$$

for a unique triple $\mathbf{K} = (K_1, K_2, K_3)$ of mutually commuting self-adjoint operators.

In a similar way the unitary representation V of the boost group $\mathcal{V}$ can be expressed by virtue of Stone's theorem in terms of a PV measure E_V, or equivalently, in terms of a triple of self-adjoint operators $\mathbf{X} = (X_1, X_2, X_3)$,

$$X_i = \int_{\mathbf{R}} x \, dE_V^i(x) \tag{2.16}$$

such that

$$V_{\mathbf{v}} = e^{i\mathbf{v}\cdot\mathbf{X}} \tag{2.17}$$

In mathematical terms, the covariance of E and F under translations and boosts, respectively, is expressed by saying that the pairs (U, E), (V, F) are transitive systems of imprimitivity with respect to the groups $\mathcal{T}$ and $\mathcal{V}$.

Taking into account the boost invariance of the localisation observable, it follows that the (vector) operators $\mathbf{Q}$ and $\frac{1}{m}\mathbf{X}$ coincide; similarly, the translation invariance of momentum entails that $\mathbf{P}$ can be identified with $\mathbf{K}$:

$$\mathbf{Q} = \frac{1}{m}\mathbf{X}, \quad \mathbf{P} = \mathbf{K} \tag{2.18}$$

So far we have considered observables as specified at a given instant of time. To compare localisation properties (observables) at different times one has to consider in addition the behaviour of the relevant PV measures under time translations. Invoking again Stone's theorem yields a self-adjoint operator H, called the Hamiltonian, generating the representation of the time translation group $\mathcal{Z} = (\mathbf{R}, +)$:

$$U : t \mapsto U_t = e^{-itH} \tag{2.19}$$

For a free elementary system, H is a function of $\mathbf{P}^2$ and is found to be [3.3]

$$H = \frac{1}{2m}\mathbf{P}^2 \tag{2.20}$$

Then it follows that the operators $E_t(X) = e^{-itH} E(X) e^{itH}$ constitute covariant sharp localisation observables associated with each instant of time, giving rise to self-adjoint (vector) operators $\mathbf{Q}_t$. As a consequence, the *velocity operator* $\dot{\mathbf{Q}} := i[H, \mathbf{Q}] = i(H\mathbf{Q} - \mathbf{Q}H)$ turns out to be proportional to momentum:

$$\mathbf{P} = m\dot{\mathbf{Q}} \tag{2.21}$$

The remaining requirements from Equations (2.10) and (2.11), the covariance under rotations, ensure the (Euclidean) vector nature of the position and momentum observables $\mathbf{Q}$ and $\mathbf{P}$. To summarise, the boost and translation generators are found to be the unique solutions of the covariance conditions in terms of PV measures, uniqueness being up to unitary equivalence. They are thus identified as *the* sharp position and momentum observables of a Galilei invariant object. In the literature it is sometimes argued that position and momentum observables (as PV measures) are uniquely given already by the Euclidean covariance of the localisation observable. This conclusion rests, however, on defining the momentum observable as the translation generator. In the present approach momentum is taken as a fundamental observable whose characteristic property is the boost covariance.

It is instructive to spell out the meaning of the covariance conditions (2.10) and (2.11) in relation to the notions of equivalence and identity of observables. As just pointed out, in a given representation of the Galilei group the systems of covariance $(U, \mathcal{T})$, $(V, \mathcal{V})$ are unique modulo unitary equivalence. The covariance $E(X + \mathbf{a}) = U_\mathbf{a} E(X) U_\mathbf{a}^{-1}$ can be rephrased by stating that the family of projections $\tilde{E}(X) := U_\mathbf{a}^{-1} E(X+\mathbf{a}) U_\mathbf{a}$ constitutes a localisation observable that is identical to E. Further identifications among the unitarily equivalent translation covariant POV measures are then obtained, e.g., by exploiting rotation covariance. Indeed the observable E' given by $E'(X) := W_R^{-1} E(RX) W_R$ is again translation covariant and is identified with E by invoking the rotation covariance of localisation observables. On the other hand, unitarily equivalent pairs of localisation observables at different times cannot be identified. This shows to what far-reaching extent the group structure determines properties of the observables.

The translation covariance condition of (2.10) can also be written in terms of the operators Q_1, Q_2, Q_3 and [in view of (2.15), (2.18)] P_1, P_2, P_3:

$$U_\mathbf{a} Q_i U_\mathbf{a}^{-1} = Q_i - a_i I \tag{2.22}$$

for all $i = 1, 2, 3$. Taking a formal derivation with respect to the parameter $\mathbf{a}$ at the point $\mathbf{a} = O$, this yields *the canonical commutation relations* between the position and momentum components Q_i and P_i (to hold on a dense domain):

$$\begin{aligned} Q_k P_k - P_k Q_k &= iI \\ Q_k P_l - P_l Q_k &= O \quad (k \neq l) \end{aligned} \tag{2.23}$$

which is summarised by saying that position and momentum are a *canonically conjugate* pair of observables. Any two self-adjoint operators Q, P satisfying the relation $QP - PQ = iI$ (on a dense domain) are also called a *Heisenberg pair*.

In addition to yielding the canonical commutation relations (2.23), the covariance condition (2.10) implies that the position-momentum pair (E, F), or $(\mathbf{Q}, \mathbf{P})$ is essentially unique. The only irreducible pair (E, F) refers to the case of a spin-0 object, whereas the other possible solutions of (2.10) arise from the spin-s cases. In all of them the spectrum of the position observable has homogeneous degeneracy with multiplicity $2s + 1$ [2.6]. We shall sketch here only the solution for the spinless case, which is sufficient for our purposes.

The *Schrödinger representation* of the position and momentum observables of a spinless elementary system is given by means of the familiar operators in the Hilbert space $L^2(\mathbf{R}^3)$ of square integrable complex valued functions on $\mathbf{R}^3$. For any $\mathbf{a} \in \mathcal{T}$ define $(U_{S,\mathbf{a}}\varphi)(\mathbf{x}) := \varphi(\mathbf{x} - \mathbf{a})$. This constitutes a strongly continuous unitary representation of the translation group $\mathcal{T}$ in $L^2(\mathbf{R}^3)$ [2.6]. Applying Stone's theorem we have $U_{S,\mathbf{a}} = \exp(-i\mathbf{a} \cdot \mathbf{P}_S)$, with

$$(\mathbf{P}_S\varphi)(\mathbf{x}) = i \lim_{\mathbf{a} \to 0} \left[\tfrac{1}{\mathbf{a}} \left(U_{S,\mathbf{a}}\varphi - \varphi \right)(\mathbf{x}) \right], \quad \mathbf{x} \in \mathbf{R}^3 \tag{2.24}$$

and one obtains

$$\mathbf{P}_S = -i\nabla \tag{2.25}$$

Furthermore the mapping $X \mapsto E_S(X)$ with

$$E_S(X)\varphi = \chi_X \cdot \varphi \tag{2.26}$$

(where χ_X denotes the characteristic function of the set X) is a PV measure on $\mathcal{B}(\mathbf{R}^3)$ with the appropriate translation covariance, so that (U_S, E_S) constitutes a system of covariance with respect to the configuration space. Moreover, the pair (U_S, E_S) is irreducible (that is, the only operators commuting with all $U_{S,\mathbf{a}}$ and $E_S(X)$ are multiples of the identity I) and is therefore unique (modulo unitary equivalence) [2.6]. The self-adjoint operators $Q_{S,k}$ associated with E_S^k are simply the multiplication operators

$$(Q_{S,k}\varphi)(\mathbf{x}) = x_k\,\varphi(\mathbf{x}) \tag{2.27}$$

Taken together with $\mathbf{P}_S$ they are easily seen to satisfy the canonical commutation relations (2.23).

Any two (complex, separable infinite-dimensional) Hilbert spaces are unitarily equivalent, that is, isometrically isomorphic. Let $W : L^2(\mathbf{R}^3) \to \mathcal{H}$ be a unitary mapping. With the definition $\widetilde{U}_{\mathbf{a}} = W\,U_{S,\mathbf{a}}\,W^{-1}$ one obtains a representation of $\mathcal{T}$ on $\mathcal{H}$. Moreover $Z \mapsto \widetilde{E}(X) = W\,E_S(X)\,W^{-1}$ is a translation covariant projection valued measure $\mathcal{B}(\mathbf{R}^3) \to \mathcal{L}(\mathcal{H})$, so that we have constructed an irreducible system of covariance $(\widetilde{U}, \widetilde{E})$ for the translation group $\mathcal{T}$, based on $\mathcal{B}(\mathbf{R}^3)$ and acting on $\mathcal{H}$. The question then is, what is its relation to the pair (U, E) considered above. As already noted if (U, E) is irreducible, then it is unitarily equivalent to the *Schrödinger pair* (U_S, E_S), that is, (U, E) is of the type $(\widetilde{U}, \widetilde{E})$ [2.6]. For a spinless object, the position-momentum pair $(\mathbf{Q}, \mathbf{P})$ is irreducible.

The Schrödinger representation of position and momentum can be used to derive the unitary equivalence of $\mathbf{Q}$ and $\mathbf{P}$, or the corresponding PV measures E and F. This equivalence is expressed by means of the Fourier-Plancherel transformation

$$U_F : L^2(\mathbf{R}^3, \mathbf{dq}) \rightarrow L^2(\mathbf{R}^3, \mathbf{dp})$$
$$(U_F\varphi)(\mathbf{p}) = (2\pi)^{-3/2} \int_{\mathbf{R}^3} e^{-i\mathbf{q}\cdot\mathbf{p}} \, \varphi(\mathbf{q}) \, d\mathbf{q} \tag{2.28}$$

The components $Q_{S,k}$ and $P_{S,k}$ are thus related as follows:

$$P_{S,k} = U_F^{-1} Q_{S,k} U_F \tag{2.29}$$

This relation is responsible for the fundamental coupling properties of position and momentum: their support property (complementarity), dispersion property (the uncertainty relations), and their total noncommutativity [see Chapter IV].

The covariance condition (2.10) implies not only the canonical commutation relations (2.23) but in fact somewhat more. If $\mathbf{p} \mapsto V_{\mathbf{p}} := \exp(i\mathbf{p} \cdot \mathbf{Q})$ is the unitary representation of $(\mathbf{R}^3, +)$ on $\mathcal{H}$ given by E, then one may confirm that the covariance condition is equivalent to

$$U_{\mathbf{q}} V_{\mathbf{p}} = e^{-i\mathbf{q}\cdot\mathbf{p}} V_{\mathbf{p}} U_{\mathbf{q}} \tag{2.30}$$

This is the *Weyl form* of the commutation relations between position and momentum. A pair $(\mathbf{Q}, \mathbf{P})$ satisfying (2.30) is called a *Weyl pair*. Hence any Weyl pair giving rise to an irreducible realisation of (2.10) is unitarily equivalent to a Schrödinger pair $(\mathbf{Q}_S, \mathbf{P}_S)$.

III.2.3 Unsharp position and momentum. Position and momentum observables are characterised by their Galilei covariance properties. But these do not single out the PV measures E and F constructed in the preceding subsection as the only solutions. In addition to the sharp observables E and F there are covariant POV measures representing unsharp position and momentum observables.

The general solution of the set of covariance conditions (2.10) and (2.11) is not known. The commutative solutions can be completely characterised. Using the representation (II.2.8) of commutative POV measures, one can show [3.7, 3.8] that the only solutions of (2.10), (2.11) are the convolutions E^μ and F^ν of E and F with some confidence measures μ and ν on $\mathcal{B}(\mathbf{R}^3)$, where, for example,

$$E^\mu(X) := (\mu * E)(X) = \int E(X + \mathbf{q}) \, d\mu(\mathbf{q}) \tag{2.31}$$

The boost invariance of E^μ and translation invariance of F^ν are trivially satisfied while the rotation covariance requires that the confidence measures are rotation invariant. If μ and ν are Dirac measures concentrated at the origin then E^μ and F^ν are the original sharp observables. It will suffice for our purposes to consider the case of absolutely continuous measures associated with the probability densities $e, f \in L^1(\mathbf{R}^3)$, so that, for instance, $\mu(X) = \int_X e(\mathbf{x})\,d\mathbf{x}$. We shall assume e, f to be bounded functions, which will ensure that the subsequent operator-valued integrals are well-defined. We denote the convolution of functions f, h as follows:

$$(h * f)(\mathbf{q}) = \int h(\mathbf{q}')\, f(\mathbf{q} - \mathbf{q}')\, d\mathbf{q}' \tag{2.32}$$

Then the measures E^μ and F^ν obtain the form

$$
\begin{aligned}
X \mapsto E^\mu(X) &\equiv E^e(X) = \int_{\mathbf{R}^3} e(\mathbf{q})\, E(X + \mathbf{q})\, d\mathbf{q} = (\chi_X * e)(\mathbf{Q}) \\
Y \mapsto F^\nu(Y) &\equiv F^f(Y) = \int_{\mathbf{R}^3} f(\mathbf{p})\, F(Y + \mathbf{p})\, d\mathbf{p} = (\chi_Y * f)(\mathbf{P})
\end{aligned}
\tag{2.33}
$$

Accordingly, the effects $E^e(X)$, $F^f(Y)$ can be viewed as resulting from averaging over the families of projections $E(X + \mathbf{q})$, $F(Y + \mathbf{p})$ with weights $e(\mathbf{q}), f(\mathbf{p})$, respectively.

If the confidence distributions e, f have vanishing first and finite second moments, then

$$
\begin{aligned}
\int_{\mathbf{R}^3} \mathbf{q}\, dE^e(\mathbf{q}) &= \int_{\mathbf{R}^3} \mathbf{q}\, dE(\mathbf{q}) = \mathbf{Q} \\
\int_{\mathbf{R}^3} \mathbf{p}\, dF^f(\mathbf{p}) &= \int_{\mathbf{R}^3} \mathbf{p}\, dF(\mathbf{p}) = \mathbf{P}
\end{aligned}
\tag{2.34}
$$

and

$$
\begin{aligned}
\mathrm{Var}(E_i^e, \varphi) &= \mathrm{Var}(E_i, \varphi) + \mathrm{Var}(e_i) \\
\mathrm{Var}(F_i^f, \varphi) &= \mathrm{Var}(F_i, \varphi) + \mathrm{Var}(f_i)
\end{aligned}
\tag{2.35}
$$

for each component $i = 1, 2, 3$. Here, e.g., $\mathrm{Var}(E_i^e, \varphi)$ denotes the variance of the probability measure $p_\varphi^{E_i^e}$.

In this way the covariant unsharp localisation observables are seen to arise as smeared versions of the corresponding sharp observables, via convolutions with some confidence distributions e, f. Choosing sufficiently large but finite intervals X, Y, one can ensure that the effects $E^e(X)$ and $F^f(Y)$ are unsharp properties. Therefore E^e and F^f are indeed unsharp observables.

III.2.4 Phase space observables. Localisation on phase space $\Gamma = \mathbf{R}^3 \times \mathbf{R}^3$ is to be represented by a POV measure G that is characterised via phase space translation and rotation covariance. The phase space Γ can be identified with the space of world lines $\mathcal{T} \times \mathcal{V}$ as a homogeneous space of the isochronous Galilei group with respect to the action:

$$\alpha_g = \alpha_{(0,\mathbf{a},\mathbf{v},R)} : \; (\mathbf{x}, \mathbf{u}) \; \mapsto \; (R\mathbf{x} + \mathbf{a}, R\mathbf{u} + \mathbf{v}) \tag{2.36}$$

With respect to a given irreducible representation we shall denote both the phase space variables and the shift parameters as $(\mathbf{q}, \mathbf{p})$ [where $\mathbf{p} = m\mathbf{v}$]. The projective representation associated with the phase space translation group can be given as the Heisenberg-Weyl algebra,

$$W_{\mathbf{qp}} \; = \; \exp(-i\mathbf{q} \cdot \mathbf{P} + i\mathbf{p} \cdot \mathbf{Q}) \; = \; e^{\frac{i}{2}\mathbf{q}\cdot\mathbf{P}} U_{\mathbf{q}} V_{\mathbf{p}} \quad (\mathbf{q}, \mathbf{p}) \in \Gamma \tag{2.37}$$

satisfying

$$W_{\mathbf{qp}} W_{\mathbf{q'p'}} \; = \; \exp\!\left(\tfrac{i}{2}(\mathbf{p} \cdot \mathbf{q'} - \mathbf{q} \cdot \mathbf{p'})\right) W_{\mathbf{q+q'},\mathbf{p+p'}} \tag{2.38}$$

The translation covariance of a phase space POV measure G reads

$$W_{\mathbf{qp}} \, G(Z) \, W_{\mathbf{qp}}^{-1} \; = \; G\big(Z + (\mathbf{q}, \mathbf{p})\big), \tag{2.39}$$

whereas its rotation covariance is given as

$$W_R \, G(Z) \, W_R^{-1} \; = \; G(RZ) \tag{2.40}$$

for any $Z \in \mathcal{B}(\Gamma)$. Here $RZ := \big\{ (R\mathbf{q}, R\mathbf{p}) \,|\, (\mathbf{q}, \mathbf{p}) \in Z \big\}$.

It is easily seen that there cannot exist any phase space POV measure such that its marginals were the sharp position and momentum observables. In fact if G were such, then for any pair of bounded sets X and Y one would have

$$G(X \times Y) \; \leq \; E^{\mathbf{Q}}(X) \wedge F^{\mathbf{P}}(Y) \; = \; O, \tag{2.41}$$

which is a contradiction; for the crucial equality, see Section IV.2.4.

Instead of looking for the general solution of the covariance conditions (2.39), (2.40), we shall only study phase space POV measures having a positive bounded operator valued density $(\mathbf{q}, \mathbf{p}) \mapsto S_{\mathbf{qp}}$ which is continuous. Then G takes the form

$$Z \; \mapsto \; G(Z) \; := \; (2\pi)^{-3} \int_Z S_{\mathbf{qp}} \, d\mathbf{q} \, d\mathbf{p} \tag{2.42}$$

The translation covariance (2.39) gives

$$S_{\mathbf{qp}} \; = \; W_{\mathbf{qp}} \, S_0 \, W_{\mathbf{qp}}^{-1}, \quad (\mathbf{q}, \mathbf{p}) \in \Gamma \tag{2.43}$$

whereas the rotation covariance (2.40) of G implies that

$$S_0 \; = \; W_R \, S_0 \, W_R^{-1} \tag{2.44}$$

The requirement of the positivity of the operator $G(Z)$ is equivalent to that of S_0. Together with the normalisation condition

$$G(\Gamma) = (2\pi)^{-3} \int_\Gamma S_{\mathbf{q}\mathbf{p}} \, d\mathbf{q} \, d\mathbf{p} = I \qquad (2.45)$$

this implies that S_0 must be a positive trace class operator of trace one [3.8]. The normalisation is then readily verified; the following computation will at once yield also the marginal observables. Let φ be any unit vector and denote the matrix elements of an operator B in the configuration representation as $B(\mathbf{q}', \mathbf{q}'')$,

$$B\varphi(\mathbf{q}') = \int B(\mathbf{q}', \mathbf{q}'') \, \varphi(\mathbf{q}'') \, d\mathbf{q}'' \qquad (2.46)$$

Then one obtains:

$$
\begin{aligned}
\langle \varphi | G(X \times \mathbf{R}^3) \, \varphi \rangle &= (2\pi)^{-3} \int_X d\mathbf{q} \int_{\mathbf{R}^3} d\mathbf{p} \, \langle \varphi \,|\, S_{\mathbf{q}\mathbf{p}} \varphi \rangle \\
&= (2\pi)^{-3} \int_X d\mathbf{q} \int_{\mathbf{R}^3} d\mathbf{p} \int_{\mathbf{R}^3} d\mathbf{q}' \int_{\mathbf{R}^3} d\mathbf{q}'' \, \overline{\varphi(\mathbf{q}' + \mathbf{q})} \, \varphi(\mathbf{q}'' + \mathbf{q}) \cdot \\
&\qquad \cdot S_0(\mathbf{q}', \mathbf{q}'') \exp\big(i\mathbf{p} \cdot (\mathbf{q}' - \mathbf{q}'')\big) \\
&= \int_X d\mathbf{q} \int_{\mathbf{R}^3} d\mathbf{q}' \, |\varphi(\mathbf{q}' + \mathbf{q})|^2 \, S_0(\mathbf{q}', \mathbf{q}') = \langle \varphi \,|\, \chi_X * e(\mathbf{Q}) \, \varphi \rangle
\end{aligned}
\qquad (2.47)
$$

In the third line we have used the representation $\delta(\mathbf{x}) = (2\pi)^{-3} \int_{\mathbf{R}^3} \exp(i\mathbf{p} \cdot \mathbf{x}) \, d\mathbf{p}$ for the Dirac distribution, while the fourth line is obtained with the identification $e(\mathbf{q}) \equiv S_0(\mathbf{q}, \mathbf{q})$. A similar computation yields the second marginal observable, so that we have the following:

$$
\begin{aligned}
G(X \times \mathbf{R}^3) &= E^e(X) \\
G(\mathbf{R}^3 \times Y) &= F^f(Y)
\end{aligned}
\qquad (2.48)
$$

where

$$
\begin{aligned}
e(\mathbf{q}) &= S_0(\mathbf{q}, \mathbf{q}) \\
f(\mathbf{p}) &= \widehat{S}_0(\mathbf{p}, \mathbf{p})
\end{aligned}
\qquad (2.49)
$$

Here $\widehat{S}_0(\mathbf{p}', \mathbf{p}'')$ denotes a matrix element of S_0 in the momentum representation. We shall from now on assume that the system $\mathcal{S}$ has only one spatial degree of freedom so that $\Gamma = \mathbf{R}^2$. To emphasise the dependence of the phase space observable G on S_0 we denote $G = G_{S_0}$.

According to Equations (2.48) the marginal observables of G_{S_0} are unsharp position and momentum observables, respectively. We may thus conclude that any pair of unsharp position and momentum observables, with confidence functions e

and f deriving from one and the same $S_0 \in \mathcal{S}(\mathcal{H})$ as in (2.49) are coexistent. Conversely, this condition on e, f is also necessary for E^e, F^f to possess a continuous phase space observable as their joint observable [3.8, 3.9]. Pairs E^e, F^f with confidence functions (e, f) of the form (2.49) are called *Fourier couples*. Evidently such e, f obey the uncertainty relation

$$\mathrm{Var}(e)\,\mathrm{Var}(f) \geq \tfrac{1}{4} \tag{2.50}$$

which can be directly seen from the following connections with the variances for state operators:

$$\begin{aligned}
\mathrm{Var}(e) &= \mathrm{Var}(Q, S_0) \\
\mathrm{Var}(f) &= \mathrm{Var}(P, S_0)
\end{aligned} \tag{2.51}$$

Thus the uncertainty relation for the confidence functions is a necessary condition for unsharp position and momentum observables to be jointly measurable. We summarise this discussion into the following theorem [3.8].

THEOREM. *A pair E^e, F^f of unsharp position and momentum observables are jointly measurable by means of a continuous phase space observable if they are a Fourier couple. In this case the variances of e, f satisfy the uncertainty relation.*

It is not known whether E^e, F^f must be a Fourier couple in order to be coexistent.

We can now see how the uncertainty relations for position and momentum arise in the statistics of a phase space measurement:

$$\begin{aligned}
\mathrm{Var}(E^e, T)\,\mathrm{Var}(F^f, T) &= \mathrm{Var}(Q, T)\,\mathrm{Var}(P, T) + \mathrm{Var}(e)\,\mathrm{Var}(f) \\
&\quad + \mathrm{Var}(Q, T)\,\mathrm{Var}(f) + \mathrm{Var}(P, T)\,\mathrm{Var}(e)
\end{aligned} \tag{2.52}$$

The first two contributions on the right-hand side satisfy an uncertainty relation of their own, $\mathrm{Var}(Q, T)\,\mathrm{Var}(P, T) \geq \tfrac{1}{4}$ and $\mathrm{Var}(e)\,\mathrm{Var}(f) \geq \tfrac{1}{4}$. The remaining two terms must be estimated jointly:

$$\mathrm{Var}(Q, T)\,\mathrm{Var}(f) + \mathrm{Var}(P, T)\,\mathrm{Var}(e) \geq \frac{1}{4}\left[\frac{\mathrm{Var}(Q, T)}{\mathrm{Var}(e)} + \frac{\mathrm{Var}(e)}{\mathrm{Var}(Q, T)}\right] \geq \frac{1}{2} \tag{2.53}$$

Taken together, this yields

$$\mathrm{Var}(E^e, T)\,\mathrm{Var}(F^f, T) \geq 1 \tag{2.54}$$

Hence the statistical scatter relations for joint position-momentum measurements are more restrictive than the corresponding relations for independent measurements of sharp position and momentum. The latter part accounts for the limitations of preparing states with too well-defined position and momentum values. According to Equation (2.52) there are further contributions reflecting the irreducible measuring inaccuracy as well as the unavoidable mutual influence of the position and momentum measurements when these are carried out 'simultaneously'. The measurement model of Section VI.3 will substantiate this interpretation.

Having thus established the conceptual tools for the consideration of phase space measurements, it is interesting to observe that one can construct phase space representations of the Hilbert space. In other words there do exist representations of the vector states as probability amplitudes on phase space. This offers also a convenient way to realise a Neumark dilation of the phase space observable (2.42), which we sketch for the two dimensional case $\Gamma = \mathbf{R}^2$ and for an observable G_ξ generated by a unit vector ξ. We denote $\xi_{qp} = W_{qp}\xi$, and we introduce the Hilbert space $\widetilde{\mathcal{H}} := L^2(\Gamma, m)$, with $dm(q,p) = dq\, dp/2\pi$. The mapping

$$W_\xi : \mathcal{H} \ \rightarrow \ \widetilde{\mathcal{H}}, \ \ \phi \mapsto \phi_\xi, \ \ \phi_\xi(q,p) := \ \langle \xi_{qp} \,|\, \phi \rangle_{\mathcal{H}} \tag{2.55}$$

is a linear isometry. In fact, by (2.45)

$$\begin{aligned}
\langle \psi_\xi \,|\, \phi_\xi \rangle_{\widetilde{\mathcal{H}}} \ &= \ \int_\Gamma \langle \psi \,|\, \xi_{qp} \rangle_{\mathcal{H}} \ \langle \xi_{qp} \,|\, \phi \rangle_{\mathcal{H}} \ dm(q,p) \\
&= \ \langle \psi \,|\, G_\xi(\Gamma)\phi \rangle_{\mathcal{H}} \ = \ \langle \psi \,|\, \phi \rangle_{\mathcal{H}}
\end{aligned} \tag{2.56}$$

The mapping W_ξ is onto a closed subspace of $\widetilde{\mathcal{H}}$, $W_\xi(\mathcal{H}) \equiv \widetilde{\mathcal{H}}_\xi$. That $\widetilde{\mathcal{H}}_\xi$ is not all of $\widetilde{\mathcal{H}}$ is clear from the outset since the unit vectors of $\widetilde{\mathcal{H}}_\xi$ (which is isomorphic to the Hilbert space $\mathcal{H}$) must obey the uncertainty relation, which is not generally satisfied for the distributions $(q,p) \mapsto |\phi(q,p)|^2$ induced by the 'states' ϕ of $\widetilde{\mathcal{H}}$. The projection $\mathbf{P}_\xi = W_\xi W_\xi^*$, from $\widetilde{\mathcal{H}}$ onto $\widetilde{\mathcal{H}}_\xi$ can be written as follows:

$$\mathbf{P}_\xi \ = \ W_\xi W_\xi^* \ = \ \int_\Gamma |\widetilde{\xi}_{qp}\rangle\langle \widetilde{\xi}_{qp}| \, dm(q,p) \tag{2.57}$$

Here $\widetilde{\xi}_{qp} = W_\xi \xi_{qp}$, and these are unit vectors in $\widetilde{\mathcal{H}}$. The phase space observable G_ξ can be represented on $\widetilde{\mathcal{H}}_\xi$ as follows

$$Z \mapsto \widetilde{G}_\xi(Z) \ = \ W_\xi \, G_\xi(Z) \, W_\xi^* \ = \ \int_Z |\widetilde{\xi}_{qp}\rangle\langle \widetilde{\xi}_{qp}| \, dm(q,p) \tag{2.58}$$

This observable is just the Neumark projection of the canonical PV measure on $\mathcal{B}(\Gamma)$,

$$\begin{aligned}
E : \mathcal{B}(\Gamma) \ &\rightarrow \ \mathcal{L}(\widetilde{\mathcal{H}}), \ \ Z \mapsto E(Z) \\
E(Z)\,\phi(q,p) \ &= \ \chi_Z(q,p)\,\phi(q,p)
\end{aligned} \tag{2.59}$$

Indeed, the projection of E onto $\widetilde{\mathcal{H}}_\xi$ is

$$\begin{aligned}
\mathbf{P}_\xi E(Z)\mathbf{P}_\xi \ &= \ \int_\Gamma dm(q,p) \int_\Gamma dm(q',p')|\widetilde{\xi}_{qp}\rangle\langle \widetilde{\xi}_{qp}|\, E(Z)\widetilde{\xi}_{q'p'}\rangle\langle \widetilde{\xi}_{q',p'}| \\
&= \ \int_Z dm(q,p) \int_\Gamma dm(q',p')\,|\, \widetilde{\xi}_{qp}\rangle\langle \widetilde{\xi}_{qp}\,|\, \widetilde{\xi}_{q',p'}\rangle\langle \widetilde{\xi}_{q',p'}| \\
&= \ \widetilde{G}_\xi(Z)\,\mathbf{P}_\xi
\end{aligned} \tag{2.60}$$

The spaces $\widetilde{\mathcal{H}}_\xi$ are precisely those subspaces of $\widetilde{\mathcal{H}}$ which host an irreducible representation of the Galilei group [1.14, 1.15]. Conversely, all such representations are associated with some phase space observable G_ξ having an operator density. This explains why this type of phase space observables are appropriate for the description of elementary systems.

III.3 Angular momentum and angle

III.3.1 Covariance properties. According to the considerations of Section 1, any irreducible system of covariance (U, E) of the space translation group is unitarily equivalent to the one realised by the Schrödinger representation of position and momentum corresponding to spin-0 elementary systems. There are further localisation systems of covariance, which can be exhaustively classified according to the possible higher spin values. These are given as Euclidean systems of covariance (W, E), where W is a unitary projective representation of the Euclidean group $\mathcal{E} = \mathcal{T} \times' \mathcal{R}$, the semi-direct product of the translation and rotation groups:

$$W_{\mathbf{a},R}\, E(Z)\, W_{\mathbf{a},R}^{-1} \;=\; E\big(R(Z + \mathbf{a})\big) \tag{3.1}$$

for all $Z \in \mathcal{B}(\mathbf{R}^3)$, $(\mathbf{a}, R) \in \mathcal{E}$. For a given spin value s the system (W, E) acts on the Hilbert space $\mathcal{H} = L^2(\mathbf{R}^3) \otimes \mathcal{H}_s$. In this subrepresentation of (1.6) the operators $W_{\mathbf{a},R}$ act as follows:

$$\big(W_{\mathbf{a},R}\psi\big)(\mathbf{x}) \;=\; \big(D^s(R)\psi\big)\big(R^{-1}(\mathbf{x} - \mathbf{a})\big) \tag{3.2}$$

From this formula it is straightforward to determine the Cartesian generators of the translations and rotations. The former are summarised as $\mathbf{P}_S \otimes I_{\mathbf{C}^{2s+1}}$, while the latter are of the form

$$\mathbf{J} \;=\; \mathbf{L} \otimes I_{\mathbf{C}^{2s+1}} + I_{L^2(\mathbf{R}^3)} \otimes \mathbf{s} \;\equiv\; \mathbf{L} + \mathbf{s} \tag{3.3}$$

$\mathbf{L}$ and $\mathbf{s}$ are the vectors of orbital and spin angular momentum, respectively. One obtains $\mathbf{L} = \mathbf{Q} \times \mathbf{P}$ and both, the components of $\mathbf{L}$ and $\mathbf{s}$, satisfy the commutation relations characteristic of angular momentum $\mathbf{J}$:

$$[J_j, J_k] \;=\; i\,\epsilon_{jkl}\, J_l \tag{3.4}$$

The spectra of the L_j are $\mathbf{Z}$, the set of integers, while those of the s_j consist of the $2s + 1$ eigenvalues $-s, -s+1, \cdots, s$. This entails that the spectra of the J_j are either $\mathbf{Z}$ (integer spin) or $\mathbf{Z} + \frac{1}{2}$ (half-integer spin).

We proceed now to characterise the angular momentum as a covariant physical quantity. The usual classical picture is that of a (pseudo-) vectorial entity $\mathbf{J}'$ describing the state of motion with regard to a fixed reference point. This suggests to postulate the following behaviour under the action of the translation, boost, and rotation groups:

$$U_{\mathbf{a}}\, \mathbf{J}'\, U_{\mathbf{a}}^{-1} \;=\; \mathbf{J}' - \mathbf{a} \times \mathbf{P}$$
$$V_{\mathbf{v}}\, \mathbf{J}'\, V_{\mathbf{v}}^{-1} \;=\; \mathbf{J}' - m\mathbf{Q} \times \mathbf{v} \qquad (3.5)$$
$$W_R\, \mathbf{J}'\, W_R^{-1} \;=\; R^{-1}\mathbf{J}'$$

This system is equivalent to the set of commutation relations

$$\left[J_k', P_l \right] \;=\; i\varepsilon_{kln} P_n$$
$$\left[J_k', Q_l \right] \;=\; i\varepsilon_{kln} Q_n \qquad (3.6)$$
$$\left[J_k', J_l \right] \;=\; i\varepsilon_{kln} J_n'$$

With respect to the free particle Hamiltonian (2.20) one has in addition the invariance under time translations:

$$U_b\, \mathbf{J}'\, U_b^{-1} \;=\; \mathbf{J}', \qquad [\mathbf{J}', H] \;=\; O \qquad (3.7)$$

It is well known that the rotation generators themselves satisfy these covariance and commutation relations; that is, putting $\mathbf{J}' = \mathbf{J}$ yields a solution to the systems (3.5). Other solutions $\mathbf{J}'$ differ from this one only in the spin part. Indeed the operator $\Delta\mathbf{J} := \mathbf{J}' - \mathbf{J}$ satisfies $\left[\Delta J_k, P_l\right] = \left[\Delta J_k, Q_l\right] = O$ and is thus of the form $I_{L^2(\mathbf{R}^3)} \otimes \mathbf{A}$. Hence $\mathbf{J}' = \mathbf{L} \otimes I + I \otimes (\mathbf{s} + \mathbf{A})$. From the remaining commutation relations it follows that $\left[A_k, s_l\right] = i\varepsilon_{kln} A_n$, which fixes $\mathbf{A}$ to be of the form $\mathbf{A} = \bigl(f(s_1), f(s_2), f(s_3)\bigr)$ with f being an antisymmetric function on the spectra of the s_k. In addition the operator $\mathbf{A} \cdot \mathbf{A}$ commutes with all s_k and is therefore a multiple of the unit operator. A possible solution is $A_k = f(s_k) = c s_k$, $c \in \mathbf{R}$, showing that covariance alone is not sufficient to fix the relative scales between orbital angular momentum and spin. Exploiting the full rotation covariance of $\mathbf{A}$, $D^s(R)\,\mathbf{A}\,D^s(R)^{-1} = R^{-1}\mathbf{A}$, $R \in \mathcal{R}$, one can prove that $f(x) = cx$ is indeed the only (differentiable) solution [3.3].

The relations (3.5) describe the transformation properties of 'angular momentum' just like in classical mechanics. However, in quantum mechanics one is facing the situation that a set of *noncommuting* operators must be taken for representing *one single* vectorial physical quantity. In addition it should be noted that (3.5) is not based on a proper homogeneous space of the isochronous Galilei group. In order to circumvent these obstacles at a physically appealing characterisation of angular momentum, one has to introduce appropriate POV measures. The appearance of $\mathbf{Q}, \mathbf{P}$ in (3.5) makes it evident that the relevant homogeneous space Ω should take into account the phase space variables. We define

$$\Omega \;:=\; \bigl\{ (\mathbf{q}, \mathbf{p}, \mathbf{q} \times \mathbf{p}, \mathbf{n}) \,\big|\, (\mathbf{q}, \mathbf{p}) \in \Gamma,\ \mathbf{n} \in \mathbf{R}^3,\ \mathbf{n} \cdot \mathbf{n} = 1 \bigr\} \;=\; \Omega_o \times S^2 \qquad (3.8)$$

where $S^2 = \bigl\{ \mathbf{n} \in \mathbf{R}^3 \,\big|\, \mathbf{n} \cdot \mathbf{n} = 1 \bigr\}$. The sphere S^2 is included in Ω in order to reflect the possible presence of an internal degree of freedom compatible with the covariant action of the isochronous Galilei group $\mathcal{G}_o$. The idea is that a system may carry with itself a rotation covariant property that is independent of its location and state of motion. Adopting the action (2.36) on Γ, we get

$$\alpha_{(o,\mathbf{a},\mathbf{v},R)} : \ \omega \ = \ (\mathbf{q},\, \mathbf{p},\, \mathbf{q} \times \mathbf{p},\, \mathbf{n}) \mapsto$$
$$\omega' = \big(R\mathbf{q} + \mathbf{a}, R\mathbf{p} + m\mathbf{v}, (R\mathbf{q} + \mathbf{a}) \times (R\mathbf{p} + m\mathbf{v}), R\mathbf{n}\big) \tag{3.9}$$

Angular momentum should now be represented by means of a POV measure M on Ω with the covariance

$$U_g\, M(Z)\, U_g^{-1} \ = \ M\big(\alpha_g(Z)\big) \tag{3.10}$$

for some suitable σ-algebra $\mathcal{B}_o(\Omega)$ of subsets of Ω. U_g is obtained from the representation (1.6) by putting $b = 0$. Angular momentum being a function on phase space, this POV measure should reflect the information provided by that observable and not more. This will be achieved by taking into account the characteristic symmetries of angular momentum in the definition of $\mathcal{B}_o(\Omega)$. We introduce the projection Π_o of Ω onto the space of angular momentum values,

$$\Pi_o : \ \Omega \ \to \ \mathbf{R}^3 \times S^2, \qquad \omega \mapsto (\mathbf{q} \times \mathbf{p},\, \mathbf{n}) \tag{3.11}$$

If $\mathcal{B}\big(\mathbf{R}^3 \times S^2\big)$ is the σ-algebra generated by the Cartesian product of the Borel algebras $\mathcal{B}\big(\mathbf{R}^3\big)$ and $\mathcal{B}\big(S^2\big)$, the relevant algebra is

$$\mathcal{B}_o(\Omega) \ := \ \big\{\Pi_o^{-1}(Z)\,\big|\, Z \in \mathcal{B}\big(\mathbf{R}^3 \times S^2\big)\big\} \equiv \mathcal{B}_o(\Omega_o) \times \mathcal{B}(S^2) \tag{3.12}$$

In the following subsections we shall construct a class of such unsharp angular momentum observables, which are generated by the mappings

$$Z_1 \times Z_2 \ \mapsto \ M(Z_1 \times Z_2) \ = \ M_o(Z_1) \otimes M_s(Z_2) \tag{3.13}$$

Here $Z_1 \in \mathcal{B}_o(\Omega_o)$, $Z_2 \in \mathcal{B}(S^2)$, and $Z_1 \mapsto M_o(Z_1)$, $Z_2 \mapsto M_s(Z_2)$ are covariant POV measures acting on the Hilbert spaces $L^2\big(\mathbf{R}^3\big)$ and $\mathbf{C}^{2s+1}$, respectively.

III.3.2 Orbital angular momentum. Constructing the orbital part M_o of an angular momentum POV measure amounts to discussing the spinless case. We shall adopt the definitions of Π_o, $\mathcal{B}_o(\Omega)$ of the preceding subsection, but with the understanding that the variable $\mathbf{n}$ is suppressed throughout. We define a map Λ from Γ onto Ω_o,

$$\Lambda : (\mathbf{q}, \mathbf{p}) \mapsto \Lambda(\mathbf{q}, \mathbf{p}) := (\mathbf{q}, \mathbf{p}, \mathbf{q} \times \mathbf{p}) \tag{3.14}$$

This map represents the fact that angular momentum is a function on phase space. Thus for a POV measure M_o to satisfy the covariance (3.10) under the action (3.9), it is sufficient that M_o reflects the covariance characteristic of a phase space observable. The corresponding solutions of (3.10) are of the form

$$Z \ \mapsto \ M_o(Z) := G\big(\Lambda^{-1}(Z)\big) \qquad Z \in \mathcal{B}_o(\Omega_o) \tag{3.15}$$

where G is any phase space observable. It may be noted that the covariance holds even for all $Z \in \mathcal{B}(\Omega)$. The projection Π_o restricts this measure to $\mathcal{B}_o(\Omega_o)$ and hence to its marginal corresponding to the angular momentum. In order to explicate the relation of M_o with the covariant set of noncommuting operators $\mathbf{L}$, we determine the first moment of the measure M_o. Using the covariance properties of M_o, one can show that the self-adjoint operators $L'_k := \int_{\Omega_o} (\mathbf{q} \times \mathbf{p})_k \, dM_o$ satisfy the commutations relations (3.6). Hence one has

$$\int_{\Omega_o} \mathbf{q} \times \mathbf{p} \, dM_o \ = \ \mathbf{L} \ = \ \mathbf{Q} \times \mathbf{P} \tag{3.16}$$

We thus recover the self-adjoint generator of rotations for the spin-zero representation.

Besides satsifying (3.5), the angular momentum operators possess some additional invariance properties: they are left unchanged both by rotations in phase space and by scale transformations. Let us introduce the unitary groups

$$U_\gamma^\Gamma \ := \ e^{i\gamma \frac{1}{2}(\mathbf{P}^2 + \mathbf{Q}^2)}, \qquad \gamma \in [0, 2\pi) \tag{3.17}$$

$$U_\tau^s \ := \ e^{i\tau(\mathbf{Q}\cdot\mathbf{P} + \mathbf{P}\cdot\mathbf{Q})}, \qquad \tau \in \mathbf{R} \tag{3.18}$$

The actions of these groups on $\mathbf{Q}$ and $\mathbf{P}$ are

$$\begin{aligned} U_\gamma^\Gamma \, \mathbf{Q} \left(U_\gamma^\Gamma\right)^{-1} \ &= \ \cos\gamma \, \mathbf{Q} + \sin\gamma \, \mathbf{P} \\ U_\gamma^\Gamma \, \mathbf{P} \left(U_\gamma^\Gamma\right)^{-1} \ &= \ -\sin\gamma \, \mathbf{Q} + \cos\gamma \, \mathbf{P} \end{aligned} \tag{3.19}$$

and

$$U_\tau^s \, \mathbf{Q} \left(U_\tau^s\right)^{-1} \ = \ e^\tau \, \mathbf{Q}, \quad U_\tau^s \, \mathbf{P} \left(U_\tau^s\right)^{-1} \ = \ e^{-\tau} \, \mathbf{P} \tag{3.20}$$

It is also straightforward to confirm the invariance relations for $\mathbf{L}$:

$$U_\gamma^\Gamma \, \mathbf{L} \left(U_\gamma^\Gamma\right)^{-1} \ = \ \mathbf{L} \tag{3.21}$$

$$U_\tau^s \, \mathbf{L} \left(U_\tau^s\right)^{-1} \ = \ \mathbf{L} \tag{3.22}$$

From (3.19), (3.20) one may infer the actions α_γ^Γ and α_τ^s on Ω_o:

$$\alpha_\gamma^\Gamma : (\mathbf{q}, \mathbf{p}, \mathbf{q} \times \mathbf{p}) \ \mapsto \ (\cos\gamma\mathbf{q} + \sin\gamma\mathbf{p}, -\sin\gamma\mathbf{q} + \cos\gamma\mathbf{p}, \mathbf{q} \times \mathbf{p}) \tag{3.23}$$

$$\alpha_\tau^s : (\mathbf{q}, \mathbf{p}, \mathbf{q} \times \mathbf{p}) \ \mapsto \ (e^\tau\mathbf{q}, e^{-\tau}\mathbf{p}, \mathbf{q} \times \mathbf{p}) \tag{3.24}$$

The phase space variables are not themselves invariant under the rotations α_γ^Γ and scale transformations α_τ^s. (Here and in the sequel we use the same notations for the actions on phase space Γ and on Ω_o.) But the subsets of Ω_o collected in $\mathcal{B}_o(\Omega_o)$ obviously do possess these invariances: $\alpha_\gamma^\Gamma\left(\Pi_o^{-1}(Z)\right) = \Pi_o^{-1}(Z)$ and $\alpha_\tau^s\left(\Pi_o^{-1}(Z)\right) =$

$\Pi_o^{-1}(Z)$ for $Z \in \mathcal{B}(\Omega_o)$. We may ask whether the POV measure M_o from (3.15) displays the invariances (3.21), (3.22) of $\mathbf{L}$, that is, whether for $Z \in \mathcal{B}_o(\Omega_o)$,

$$U_\gamma^\Gamma M_o(Z) \left(U_\gamma^\Gamma\right)^{-1} = M_o(Z) \tag{3.25}$$

$$U_\tau^s M_o(Z) \left(U_\tau^s\right)^{-1} = M_o(Z) \tag{3.26}$$

One can easily verify that the covariance properties (2.39), (2.40) of the phase space POV measure G are preserved if the actions α_τ^s, α_γ^Γ are combined with the application of the unitary transformations; this is to say that the following POV measures are again phase space observables:

$$Z \mapsto G_\tau^s(Z) := U_\tau^s G\left(\alpha_\tau^s(Z)\right) \left(U_\tau^s\right)^{-1} \tag{3.27}$$

$$Z \mapsto G_\gamma^\Gamma(Z) := U_\gamma^\Gamma G\left(\alpha_\gamma^\Gamma(Z)\right) \left(U_\gamma^\Gamma\right)^{-1} \tag{3.28}$$

Thus the operations of scaling and rotating in phase space do not alter the character of being a phase space POV measure; but in general one ends up with a modified observable as, for instance, scaling alters the intrinsic measuring inaccuracies of the marginal position and momentum observables. This is no longer true for the marginal angular momentum. In that case one may use the fact that the sets in $\mathcal{B}_o(\Omega_o)$ are invariant under the actions (3.23), (3.24) to simplify the above expressions for G_τ^s, G_γ^Γ or the ensuing analogous entities $M_{o,\tau}^s$, $M_{o,\gamma}^\Gamma$.

With these considerations we have not yet determined whether there exist angular momentum POV measures that are invariant under the two types of transformations. This question may be tackled to some degree if the underlying phase space POV measure is assumed to be of the form $G = G_{S_0}$ with some positive trace-one operator S_0. Indeed if the operator S_0 is any one of the (harmonic oscillator) eigenstates of the generator of (3.17), or a mixture of them, then it commutes with all U_γ^Γ and consequently,

$$M_{o,\gamma}^\Gamma(Z) = M_o(Z) \tag{3.29}$$

We shall henceforth restrict our considerations to that case. On the other hand there is no scale-invariant state so that a similar result cannot be obtained for the scale transformation.

The above symmetry considerations provide information on the structure of the marginal POV measures of M_o associated with the Cartesian components L_k of (3.16). Let $\Pi_3 : \Omega_o \to \mathbf{R}$ to be the coordinate projection onto the third component of the angular momentum variable, that is, $\Pi_3(\omega) = (\mathbf{q} \times \mathbf{p})_3$. We then get the following real POV measure:

$$X \mapsto M_o^{(3)}(X) := M_o\left(\Pi_3^{-1}(X)\right) = G_{S_0}\left(\{(\mathbf{q},\mathbf{p}) \,|\, (\mathbf{q} \times \mathbf{p})_3 \in X\}\right) \tag{3.30}$$

Note that the sets $\Pi_3^{-1}(X)$ are in $\mathcal{B}_o(\Omega_o)$ as they can be written as $\Lambda^{-1}(Z_X)$ with $Z_X = \{\mathbf{q} \times \mathbf{p} \,|\, (\mathbf{q} \times \mathbf{p})_3 \in X\}$. This ensures that the definition (3.30) is meaningful.

Exploiting the symmetries of this POV measure, we show that all effects $M_o^{(3)}(X)$ are smearings of L_3.

The quantity $(\mathbf{q} \times \mathbf{p})_3$, and therefore the set $\Pi_3^{-1}(X)$, is invariant under rotations about the axis given by the unit vector $\mathbf{e}_3$, as well as under translations and boost in that direction. Consequently $M_o^{(3)}(X)$ commutes with the generating operators L_3, P_3, and Q_3. Due to the irreducibility of the subrepresentation of phase space translations for one degree of freedom (q_3), the effect $M_o^{(3)}(X)$ acts on $L^2(\mathbf{R}^3) \simeq L^2(\mathbf{R}^2, dq_1 dq_2) \otimes L^2(\mathbf{R}, dq_3)$ as an operator of the form $F_X \otimes I_{L^2(\mathbf{R}, dq_3)}$. According to the above analysis of the rotations in phase space, the property (3.29) ensures that $M_o^{(3)}$ commutes not only with L_3 but also with the operator $H_{12} := \frac{1}{2}(Q_1^2 + Q_2^2 + P_1^2 + P_2^2)$. Since the operators L_3, H_{12} form a complete set of commuting self-adjoint operators on $L^2(\mathbf{R}^2, dq_1 dq_2) \equiv L^2(\mathbf{R}^+ \times [0, 2\pi), \rho d\rho d\phi)$, it follows that $M_o^{(3)}(X)$ is a function of them, $M_o^{(3)}(X) = f_X(L_3, H_{12})$. It can be shown that this POV measure is not, in general, scale invariant. This means that the dependence on H_{12} cannot be eliminated. Nevertheless the POV measure (3.30) is commutative and satisfies $\int x dM_o^{(3)} = L_3$; therefore it corresponds to a smeared version of the sharp angular momentum observable. In this way the POV measure (3.15) is found to be a joint observable for arbitrary components $L_{\mathbf{n}} = \mathbf{n} \cdot \mathbf{L}$ of the orbital angular momentum. Moreover it is remarkable that in the covariance approach pursued here angular momentum emerges as a function on phase space, or more precisely, as a function of some phase space observable, quite in analogy to the corresponding classical quantity.

III.3.3 Spin
a) Covariant spin. In order to complete the construction of the POV measure (3.11), we now proceed to define a POV measure M_s acting on $\mathcal{H}_s$ and rotation covariant on S^2. The sphere S^2 can be identified in a canonical way as a quotient space $\mathcal{R}/H$ of $\mathcal{R} = O(3)$ with respect to some subgroup $H = \mathcal{R}_{\mathbf{n}_o}$ of rotations that leave a given point $\mathbf{n}_o$ fixed, $R\mathbf{n}_o = \mathbf{n}_o$.

The action α_g of $\mathcal{G}_o$ on S^2 deriving from (3.9) is $\alpha_g : \mathbf{n} \mapsto R\mathbf{n}$. For any $\mathbf{n} \in S^2$, let $T_{\mathbf{n}} = P[\psi_{\mathbf{n}}]$ be the eigenprojection of $s_{\mathbf{n}}$ associated with its greatest eigenvalue s. The covariance reads then

$$D^s(R) T_{\mathbf{n}} D^s(R)^{-1} = T_{R^{-1}\mathbf{n}} \tag{3.31}$$

Introducing the rotation invariant measure $d\mu = \frac{2s+1}{4\pi} d\sigma$, where $d\sigma = \sin \vartheta \, d\vartheta \, d\varphi$, on S^2 one obtains a rotation covariant normalised POV measure on the spin Hilbert space:

$$Z \mapsto M_s(Z) = \frac{2s+1}{4\pi} \int_Z T_{\mathbf{n}} \, d\sigma(\mathbf{n}), \qquad Z \in \mathcal{B}(S^2) \tag{3.32}$$

The normalisation is due to the fact that the vectors $\psi_{\mathbf{n}}$ form an overcomplete family of Bloch coherent states.

In addition to the phase space observables, this is another concrete example of a continuous system of covariance. It should be noted that while a finite-dimensional Hilbert space admits only discrete sharp observables, it does host continuous unsharp observables.

To elucidate the relation between the unsharp observable (3.32) with the spin operators $s_{\mathbf{n}}$, let us consider sets $Z \in \mathcal{B}(S^2)$ which are invariant under rotations about the $\mathbf{n}$-axis. Making use of polar coordinates, (ϑ, φ), such sets can be represented as $Z = X \times [0, 2\pi)$, and the resulting POV measure is

$$X \mapsto M_s^{(\mathbf{n})}(X) = \tfrac{2s+1}{4\pi} \int_X d\vartheta \, \sin\vartheta \int_0^{2\pi} T_{\mathbf{n}(\vartheta,\varphi)} \, d\varphi \qquad X \in \mathcal{B}(0,\pi) \qquad (3.33)$$

Commuting with the maximal self-adjoint operator $s_{\mathbf{n}}$, the effects $M_s^{(\mathbf{n})}(X)$ are functions of $s_{\mathbf{n}}$ and constitute thus a smeared version of the sharp spin observable associated with the direction $\mathbf{n}$. Again we see that the fully covariant POV measure (3.32) is a joint observable for the rotation covariant family of unsharp spin observables (3.33). The moment operators derived from these POV measures are the ordinary spin operators:

$$\mathbf{s}' := (s+1) \int_{S_2} \mathbf{n} \, T_{\mathbf{n}} \, d\mu(\mathbf{n}) = \mathbf{s} \qquad (3.34)$$

To see this, one needs to recall that the operators s_k' satisfy the commutation relations $[s_k', s_l] = i\varepsilon_{kln} s_n'$ [following from (3.6)]; these relations are equivalent to the fact that $\mathbf{s}'$ transforms as a vector under rotations: $D^s(R)\mathbf{s}'D^s(R)^{-1} = R\mathbf{s}'$, which is readily verified for $\mathbf{s}'$ from (3.34) in view of (3.31). As shown in Section 3.1, this implies $\mathbf{s}' = c\mathbf{s}$. The constant c is found to be 1.

b) Sharp spin. We shall now review the conventional description of spin observables. This will allow us to lay the grounds for the interpretation of spin POV measures as unsharp spin observables to be applied in the analysis of polarisation measurements. We shall be mostly concerned with spin-$\tfrac{1}{2}$ systems represented by the Hilbert space $\mathbf{C}^2$.

The spin component operators can be realised as $s_j = \tfrac{1}{2}\sigma_j$, where σ_j are the Pauli spin operators,

$$\sigma_j = \sigma_j^*, \qquad \sigma_j\sigma_k = i\,\epsilon_{jkl}\,\sigma_l + \delta_{jk}\,I \qquad (3.35)$$

Any operator A acting on $\mathbf{C}^2$ can be written as

$$A = A(a, \mathbf{a}) = \tfrac{1}{2}(aI + \mathbf{a}\cdot\sigma), \qquad (a, \mathbf{a}) \in \mathbf{C}^4 \qquad (3.36)$$

where $\mathbf{a}\cdot\sigma = a_1\sigma_1 + a_2\sigma_2 + a_3\sigma_3$. Self-adjoint operators are characterised by the condition $(a, \mathbf{a}) \in \mathbf{R}^4$, and positivity corresponds to the requirement $\|\mathbf{a}\| \le a$. In particular the state operators are of the form

$$T = T_{\mathbf{n}} = \tfrac{1}{2}(I + \mathbf{n}\cdot\sigma), \qquad \mathbf{n} \in \mathbf{R}^3, \quad \|\mathbf{n}\| \le 1 \qquad (3.37)$$

One-dimensional projections are those $T_\mathbf{n}$ with $\|\mathbf{n}\| = 1$. They represent vector states as well as sharp properties of the spin system. The normalised eigenvectors of $T_\mathbf{n}$ corresponding to the eigenvalues ± 1 will be denoted $\varphi_{\pm\mathbf{n}}$, or $\mid \pm\mathbf{n}\rangle$, or simply $\varphi_\pm$ if clear from the context. The spectral decomposition of any self-adjoint operator $A(a, \mathbf{a})$, represented by $\mathbf{a} = \lambda\mathbf{n}$ with $a, \lambda \in \mathbf{R}$, $\mathbf{n} \in \mathbf{R}^3$, $\|\mathbf{n}\| = 1$, is given as follows:

$$A(a, \lambda\mathbf{n}) = \tfrac{1}{2}(aI + \lambda\mathbf{n}\cdot\sigma) = \tfrac{1}{2}(a + \lambda)T_\mathbf{n} + \tfrac{1}{2}(a - \lambda)T_{-\mathbf{n}} \tag{3.38}$$

In particular the spin observables are $s_\mathbf{n} = A(0, \mathbf{n})$, with the spectral projections $E^{s_\mathbf{n}}(\{\pm\tfrac{1}{2}\}) \equiv E_\pm^{(\mathbf{n})} = T_{\pm\mathbf{n}}$. Their covariance under the rotation group is expressed in (3.31), with D^s being now $D^{1/2}$, for short W:

$$W_R E_\pm^{(\mathbf{n})} W_R^{-1} = E_\pm^{(R^{-1}\mathbf{n})} \tag{3.39}$$

c) Unsharp spin. The construction of unsharp spin observables is based on the covariance condition (3.39). Thus having in mind a typical spin measurement device, with two possible 'screen' regions representing the 'pointer' and with a unit vector $\mathbf{n} \in \mathbf{R}^3$ representing the orientation of the intended spin quantity, we shall define a family of smeared spin-$\tfrac{1}{2}$ observables as POV measures $F^{(\mathbf{n})}$ on $\Omega = \{-1, +1\}$, that satisfy the covariance condition

$$W_R F_\pm^{(\mathbf{n})} W_R^{-1} = F_\pm^{(R^{-1}\mathbf{n})} \tag{3.40}$$

The effects $F_\pm^{(\mathbf{n})}$ have spectral decomposition $F_\pm^{(\mathbf{n})} = \tfrac{1}{2}(a_\pm I + \lambda_\pm\mathbf{m}\cdot\sigma) = \tfrac{1}{2}(a_\pm + \lambda_\pm)T_\mathbf{m} + \tfrac{1}{2}(a_\pm - \lambda_\pm)T_{-\mathbf{m}}$, with $\mathbf{m} \in S^2$. That the eigenvalues are in $[0, 1]$ is ensured by the conditions $|\lambda_\pm| \le a_\pm$ and $a_\pm + |\lambda_\pm| \le 1$. One may assume that the device has been calibrated so as to yield maximal probability for $F_+^{(\mathbf{n})}$ in the state $T_\mathbf{n}$. This implies that $\lambda_\pm \ge 0$ and $\mathbf{m} = \mathbf{n}$. Moreover, the covariance condition entails that the parameters $a_\pm, \lambda_\pm$ are independent of the orientation $\mathbf{n}$. Finally it is natural to assume that rotating a device measuring $F_\pm^{(\mathbf{n})}$ by an angle of π yields a device that operates exactly the same way but now for a reversed spin; thus the spin observable associated with this rotated device should be $F_\mp^{(\mathbf{n})}$:

$$F_\pm^{(-\mathbf{n})} = F_\mp^{(\mathbf{n})} \tag{3.41}$$

Consequently $a_\pm = 1$, $\lambda_+ = -\lambda_- = \lambda$, and

$$F_+^{(\mathbf{n})} = \tfrac{1}{2}(I + \lambda\mathbf{n}\cdot\sigma), \qquad F_-^{(\mathbf{n})} = \tfrac{1}{2}(I - \lambda\mathbf{n}\cdot\sigma) \tag{3.42}$$

This is a POV measure obtained from the PV measure $E^{s_\mathbf{n}}$ by application of a smearing map of the form (II.2.9), $F_k^{(\mathbf{n})} = \sum_{l=+,-} \lambda_{kl} T_{l\mathbf{n}}$, $k = +, -$, with the stochastic matrix

$$(\lambda_{kl}) = \begin{pmatrix} \tfrac{1}{2}(1 + \lambda) & \tfrac{1}{2}(1 - \lambda) \\ \tfrac{1}{2}(1 - \lambda) & \tfrac{1}{2}(1 + \lambda) \end{pmatrix} \tag{3.43}$$

We conclude that any family of spin observables satisfying the covariance condition (3.40) and the natural requirement (3.41) consists of smeared versions of sharp spin observables, the smearing being characterised by the parameter λ. Furthermore the eigenvalues satisfy $0 \leq \frac{1}{2}(1 - \lambda) < \frac{1}{2} < \frac{1}{2}(1 + \lambda) \leq 1$ whenever $0 < \lambda \leq 1$. Hence the POV measures $\pm 1 \mapsto F_{\pm}^{(\mathbf{n})}$ allow for an interpretation as unsharp observables. The larger eigenvalue $\frac{1}{2}(1 + \lambda) = 1 - \varepsilon$ represents the maximal degree of reality of the unsharp spin property $F_{+}^{(\mathbf{n})}$, while $\varepsilon \in [0, \frac{1}{2})$ is an unsharpness parameter. The sharp spin observables are recovered by putting $\lambda = 1$, or $\varepsilon = 0$.

III.3.4 Angle observables. The parameter set $\Omega = [0, 2\pi)$ of azimuthal angles about a fixed axis $\mathbf{e}_3$, say, admits a natural transitive action of the corresponding subgroup of rotations R_ϕ, namely, $\alpha_{R_\phi}(\phi') = \phi' + \phi$, the addition being modulo 2π. Any POV measure A on $\mathcal{B}(0, 2\pi)$ representing an angle, or phase, observable must therefore satisfy the covariance condition

$$e^{-iJ_3\phi} A(X) e^{iJ_3\phi} = A(X + \phi) \qquad X \in \mathcal{B}(0, 2\pi) \qquad (3.44)$$

An observable of this kind accounts for the (azimuthal) angular orientation of some vectorial physical quantity. The covariance properties of that quantity give rise to further specifications of the associated angle observable. We shall construct several examples. For the sake of simplicity we consider the orbital ($J_3 = L_3$) and spin ($J_3 = s_3$) cases separately.

a) Localisation angle. Besides the decomposition into Cartesian components, any vector quantity can be represented equally well with respect to polar coordinates (r, θ, ϕ). If applied to the position observable $\mathbf{Q}$, or $E = E^{\mathbf{Q}}$ (or to any of its smeared versions) one gets a PV (or POV) measure $A^{\mathbf{Q}}$ associated with the azimuthal angle ϕ. Indeed let us take (Borel) sets Z_X, for $X \in \mathcal{B}(0, 2\pi)$, of the form

$$Z_X = \left\{ \mathbf{q} \in \mathbf{R}^3 \,\middle|\, q_3 \in \mathbf{R}, \; \frac{q_1}{\sqrt{q_1^2 + q_2^2}} = \cos\phi, \; \frac{q_2}{\sqrt{q_1^2 + q_2^2}} = \sin\phi, \; \phi \in X \right\} \quad (3.45)$$

For intervals X this corresponds to partitioning the space into cylindrical slices with central axis along $\mathbf{e}_3$. It is evident that the Z_X form a family of sets that is corvariant under rotations about the $\mathbf{e}_3$-axis, $R_\phi(Z_X) = Z_X + \phi$. This suffices to ensure that the PV measure

$$X \mapsto A^{\mathbf{Q}}(X) := E^{\mathbf{Q}}(Z_X) \qquad (3.46)$$

possesses the covariance (3.44). The operators $A^{\mathbf{Q}}(X)$ act as multiplication operators $\left(A^{\mathbf{Q}}(X)\psi\right)(r, \theta, \phi) = \chi_X(\phi)\, \psi(r, \theta, \phi)$ on the Hilbert space

$$L^2\left(\mathbf{R}^+ \times [0, \pi] \times [0, 2\pi), \, r^2 dr \, d\sigma\right) \simeq L^2\left(\mathbf{R}^+, r^2 dr\right) \otimes L^2\left(S^2, d\sigma\right) \qquad (3.47)$$

There is another instructive way of constructing this angular observable. As is well known, the operator L_3 can be represented as a differential operator on the function space $L^2([0,2\pi), d\phi)$,

$$(L_3\psi)(\phi) = -i\frac{\partial}{\partial\phi}\psi(\phi) \tag{3.48}$$

The spectrum of L_3 is the set of integers $\mathbf{Z}$, and the eigenvectors $|m\rangle$ are given by the functions $\phi \mapsto \psi_m(\phi) = e^{im\phi}$, $m \in \mathbf{Z}$, in $L^2(0,2\pi)$. These properties allow one to introduce an additive unitary group of shift operators,

$$U_k : |m\rangle \mapsto |m+k\rangle \tag{3.49}$$

The self-adjoint generator Φ of this group, $U_k = e^{ik\Phi}$, can be directly determined. Let us introduce the improper eigenvectors of U_k,

$$|\phi) := (2\pi)^{-1/2}\sum_{m=-\infty}^{\infty}e^{-im\phi}\,|m\rangle \qquad U_k|\phi) = e^{ik\phi}\,|\phi) \tag{3.50}$$

Then the following operator is indeed self-adjoint:

$$\Phi := \int_0^{2\pi}\phi\,|\phi)(\phi|\,d\phi = \sum_{m\neq n}\frac{1}{i(n-m)}|m\rangle\langle n| + \pi I \tag{3.51}$$

The associated spectral measure is

$$X \mapsto E^{\Phi}(X) = \int_X|\phi)(\phi|\,d\phi = (2\pi)^{-1}\sum_{n,m}\int_X e^{i(n-m)\phi}\,d\phi\,|m\rangle\langle n| \tag{3.52}$$

A direct application of the Weyl relation

$$e^{iL_3\phi}\,e^{ik\Phi} = e^{ik\phi}\,e^{ik\Phi}\,e^{iL_3\phi} \tag{3.53}$$

shows that the PV measure E^{Φ} possesses the covariance (3.44). Angular momentum and angle are thus established as a pair of canonically conjugate self-adjoint operators.

The measure (3.52) can be extended in an obvious way to a PV measure acting on the Hilbert space (3.47), and this is precisely the observable (3.47):

$$I_{L^2(\mathbf{R}^+)} \otimes I_{L^2(0,\pi)} \otimes E^{\Phi}(X) = A^{\mathbf{Q}}(X) \tag{3.54}$$

The set of coordinates (θ, ϕ), which parametrises the sphere S^2, gives rise to a fully rotation covariant PV measure $O^{\mathbf{Q}} : \mathcal{B}(S^2) \to \mathcal{L}(\mathcal{H})$ which corresponds to directional localisation, or the solid angle observable. Indeed define for $Y \in \mathcal{B}(S^2)$

$$Z_Y = \{\mathbf{q} \in \mathbf{R}^3\,|\,\mathbf{q} = (r,\theta,\phi),\ r \in \mathbf{R}^+,\ (\theta,\phi) \in Y\} \tag{3.55}$$

Then the POV measure

$$Y \mapsto O^{\mathbf{Q}}(Y) := E^{\mathbf{Q}}(Z_Y) \tag{3.56}$$

has the claimed property. Again the projections $O^{\mathbf{Q}}(Y)$ act as multiplication operators on $L^2(S^2, d\sigma)$.

Finally it may be noted that the radial variable r gives rise to a fully rotation invariant PV measure on $\mathbf{R}^+$, with the characteristic covariance behaviour under scale transformations (cf. Section 3.2). We have thus recovered the polar coordinate representation of the sharp localisation observable $E^{\mathbf{Q}}$.

b) Orientation of orbital angular momentum. Instead of position $\mathbf{Q}$ one can consider the orientation of angular momentum $\mathbf{L} = \mathbf{Q} \times \mathbf{P}$. In contrast to the former, the components of $\mathbf{L}$ are mutually noncommuting operators so that one cannot expect to find a commutative POV measure associated with the azimuthal angle. Still it is possible to proceed analogously to the above construction. As pointed out in Section 3.2, the rotation covariant POV measure M_o (3.15) is a joint observable for all components $L_{\mathbf{n}}$. Hence we may introduce a polar cordinate representation (r, θ, ϕ) for the angular momentum variable and define for $X \in \mathcal{B}(0, 2\pi)$

$$Z_X = \left\{ (\mathbf{q}, \mathbf{p}, \mathbf{q} \times \mathbf{p}) \,\middle|\, \mathbf{q} \times \mathbf{p} = (r, \theta, \phi), \ r \in \mathbf{R}^+, \ \theta \in [0, \pi], \ \phi \in X \right\} \tag{3.57}$$

Again we have $\alpha_{R_\phi}(Z_X) = Z_{X+\phi}$ so that the following POV measure is covariant in the sense of (3.44):

$$X \mapsto A^{\mathbf{L}}(X) := M_o(Z_X) \tag{3.58}$$

c) Spin phase. The above considerations also suggest looking for an angular-type variable that is conjugate to a given spin component, the latter acting as a generator of shifts of this angle. Such a quantity can indeed be constructed as a POV measure acting on the spin Hilbert space $\mathcal{H}_s$. Let us consider a complete orthogonal system of s_3-eigenvectors $| m \rangle$. For the operators $s_\pm := s_1 \pm i s_2$ we use the polar decomposition:

$$s_+ = C |s_+| \tag{3.59}$$

where $|s_+|^2 = s_- s_+$ and the partial isometry C is simply

$$C = \sum_{m=-s}^{s-1} |m+1\rangle\langle m| \tag{3.60}$$

The initial and final projections of C are thus

$$C^*C = I - |+s\rangle\langle +s|, \qquad CC^* = I - |-s\rangle\langle -s| \tag{3.61}$$

which show that C is a contraction. Hence there is a unique POV measure S (II.2.25) such that

$$C^n = \int_0^{2\pi} e^{in\phi} \, dS(\phi) \tag{3.62}$$

for $n = 0, 1, 2, \ldots$. We shall call S the *spin phase*. The commutation relation $[C, s_3] = C$ implies

$$e^{-i\phi s_3} C e^{i\phi s_3} = e^{-i\phi} C \tag{3.63}$$

Due to the uniqueness of the POV measure S, the spin phase satisfies the covariance condition

$$e^{-i\phi s_3} S(X) e^{i\phi s_3} = S(X + \phi) \tag{3.64}$$

modulo 2π for arbitrary $X \in \mathcal{B}(0, 2\pi)$. This observable describes the azimuthal orientation of the spin vector. Since it comprises the three noncommuting spin operators, S is necessarily a noncommutative POV measure.

Let $\widetilde{\mathcal{H}} := L^2\big([0, 2\pi), \frac{d\phi}{2\pi}\big)$ and consider the following orthogonal system of normalised vectors $\psi_m \in \widetilde{\mathcal{H}}$, $\phi \mapsto \psi_m(\phi) = e^{-im\phi}$. The map

$$W_s : \mathcal{H}_s \to \widetilde{\mathcal{H}}, \; \psi \mapsto \sum_{m=-s}^{s} \langle m \,|\, \psi \rangle \, \psi_m \tag{3.65}$$

is an isometry from $\mathcal{H}_s$ onto the $(2s+1)$-dimensional subspace $\widetilde{\mathcal{H}}_s := \mathrm{span}\big[\psi_m | m = -s, -s+1, \cdots, s\big]$ of $\widetilde{\mathcal{H}}$. In $\widetilde{\mathcal{H}}_s$ the operator C is the Neumark projection of the unitary operator

$$C_o = \sum_{m=-\infty}^{\infty} |\psi_{m+1}\rangle\langle \psi_m| \tag{3.66}$$

on $\widetilde{\mathcal{H}}$. Since $(C_o \psi)(\phi) = e^{-i\phi}\psi(\phi)$, the spectral measure of C_o is the canonical one $X \mapsto E(X)$, with $(E(X)\psi)(\phi) = \chi_X(\phi)\psi(\phi)$. Projecting E down to $\widetilde{\mathcal{H}}_s$ one gets the spin phase S. An explicit expression of S as a POV measure in $\mathcal{H}_s$ is easily obtained:

$$S(X) = \int_X \sum_{m,n=-s}^{s} e^{i(n-m)\phi} |m\rangle\langle n| \, \frac{d\phi}{2\pi} \tag{3.67}$$

Introducing the family of (normalised) states

$$|\phi\rangle = \frac{1}{\sqrt{2s+1}} \sum_{m=-s}^{s} e^{-im\phi} \, |m\rangle \tag{3.68}$$

S can be written in the form

$$S(X) = \frac{2s+1}{2\pi} \int_X |\phi\rangle\langle\phi| \, d\phi \tag{3.69}$$

It is important to note that S itself is no PV measure. The space $\widetilde{\mathcal{H}}_s$ consists of finite linear combination of trigonometric functions. The idempotency condition would demand that $\chi_X \xi \in \widetilde{\mathcal{H}}_s$ whenever $\xi \in \widetilde{\mathcal{H}}_s$. This is impossible since such functions cannot be represented as finite linear combinations of sine and cosine functions. For the same reasons the spin phase cannot be 'localised' since for any proper subset X of $[0, 2\pi)$ (with $S(X) \neq I$) and all $\xi \in \widetilde{\mathcal{H}}_s$ one has

$$\langle \xi \mid S(X)\xi \rangle < 1 \tag{3.70}$$

There is another way to obtain a phase-shift covariant POV measure acting on $\mathcal{H}_s$. Instead of considering the polar angle marginal of the unsharp observable (3.32), one can alternatively integrate out this variable to obtain a POV measure for the azimuthal angle ϕ. Let $Z \in \mathcal{B}(S^2)$ be of the form $Z_X = \{(\theta, \phi) | 0 \leq \theta \leq \pi, \phi \in X\}$, where $X \in \mathcal{B}(0, 2\pi)$. Then

$$X \mapsto \widetilde{S}(X) := M_s(Z_X) = \tfrac{2s+1}{4\pi} \int_X d\phi \int_0^\pi d\theta \, \sin\theta \, |\theta, \phi\rangle\langle\theta, \phi| \tag{3.71}$$

possesses the required covariance (3.64). There is an obvious formal connection between S and $\widetilde{S}$. The Bloch coherent states appearing in (3.71) can be conveniently generated from one fixed state ψ_o by application of suitable rotations, $\mid \theta, \phi\rangle = W_\theta e^{-i\phi s_3} \psi_o$. Noting that the vectors $\mid \phi\rangle$ from (3.68) satisfy $\mid \phi\rangle = e^{-i\phi s_3} \mid 0\rangle$, we may identify $\psi_o = \mid 0\rangle$ and obtain:

$$\widetilde{S}(X) = \tfrac{1}{2} \int_0^\pi d\theta \, \sin\theta \, W_\theta \, (I \otimes S(X)) \, W_\theta^{-1} \tag{3.73}$$

Thus $\widetilde{S}$ can be viewed as a noisy version of S in the sense of the construction b) given in Section II.2.3.

III.4 Energy and time

The covariance point of view provides also a systematic answer to the long-standing problem of understanding energy and time as a canonically conjugate pair of observables. Indeed soon after the advent of quantum mechanics, it was realised that time cannot, in general, be represented as a self-adjoint operator. This is due to the fact that a Hamiltonian operator with a semibounded spectrum does not admit a group of shifts generated by some self-adjoint operator, which would then be the canonically conjugate time observable. Henceforth it has been a widespread common sense that time is no observable but just a parameter. As a consequence the status of the time-energy uncertainty relation has remained an issue of endless debates. There have been proposals to introduce symmetric operators T that satisfy formally the canonical commutation relation with the Hamiltonian H

$$HT - TH = iI \tag{4.1}$$

These relations are then advocated in order to establish the uncertainty relation

$$\mathrm{Var}(H, \varphi)\, \mathrm{Var}(T, \varphi) \;\geq\; \tfrac{1}{4} \tag{4.2}$$

Such operators T turned out to be non-self-adjoint. It was realised only rather late that this view can be justified systematically if time observables are understood as POV measures that are characterised by their covariance under time translations. Before reviewing the definition of time observables, we shall outline the covariance properties of the Hamiltonian, or energy observable, which is commonly introduced as the generator for the representation of the time translation group.

III.4.1 Energy. Classically the energy of a free particle is a quadratic function of its momentum, or velocity $\mathbf{u} \in \mathbf{R}^3$. Thus the corresponding quantum observable should be defined on the space of variables $e := \lambda u^2 + c$ with $u^2 = \mathbf{u} \cdot \mathbf{u} \in \mathbf{R}^+$ and $\lambda > 0$, c being some fixed numbers. (We shall put $\lambda = \tfrac{1}{2}$ and $c = 0$.) However this set does not host an action α_g of $\mathcal{G}$ such that $e = \tfrac{1}{2}u^2 \mapsto \alpha_g(e) = e' = \tfrac{1}{2}u'^2$. In fact taking into account the action of $\mathcal{G}$ on velocity space, $\mathbf{u} \mapsto \mathbf{u}' = R\mathbf{u} + v$, one has $u'^2 = u^2 + v^2 + 2\mathbf{u} \cdot \mathbf{v}$. Hence the structure of the Galilei group forces one to consider the energy-momentum four-vector if energy is to be introduced as a covariant quantity:

$$\Omega \;:=\; \big\{ (m\mathbf{u}, \tfrac{1}{2}mu^2) \,\big|\, \mathbf{u} \in \mathbf{R}^3 \big\} \tag{4.3}$$

The map

$$\alpha_g : \; (m\mathbf{u}, \tfrac{1}{2}mu^2) \;\mapsto\; \big(m[R\mathbf{u} + \mathbf{v}], \tfrac{1}{2}m[u^2 + v^2 + 2\mathbf{u} \cdot \mathbf{v}]\big) \tag{4.4}$$

yields a transitive action of $\mathcal{G}$ on Ω. In order to express the fact that energy is a non-invertible function of momentum, we introduce first the coordinate projection

$$\Pi_o : \; \Omega \to \mathbf{R}^+, \qquad (m\mathbf{u}, \tfrac{1}{2}mu^2) \;\mapsto\; \tfrac{1}{2}mu^2 \tag{4.5}$$

This allows us to define the following reduced σ-algebra of subsets $\mathcal{B}_o(\Omega) \subseteq \mathcal{B}(\Omega)$ which reflects the loss of information accompanying the transition from the momentum observable to energy:

$$\mathcal{B}_o(\Omega) \;:=\; \big\{ Z \subseteq \Omega \,\big|\, Z = \Pi_o^{-1}(X) = Z_X, \; X \in \mathcal{B}(\mathbf{R}^+) \big\} \tag{4.6}$$

Any energy observable should now be represented as a POV measure $Z \mapsto E(Z)$, $Z \in \mathcal{B}_o(\Omega)$ with the covariance

$$U_g\, E(Z)\, U_g^{-1} \;=\; E\big(\alpha_g(Z)\big) \tag{4.7}$$

Given the (sharp) momentum observable $F^{\mathbf{P}}$, it is straightforward to construct a PV measure on $\mathcal{B}_o(\Omega)$ which is covariant in the sense of Eq. (4.7). Indeed we define a map from momentum space $\mathbf{R}^3$ onto Ω,

$$\Lambda : \; \mathbf{R}^3 \to \Omega, \qquad \mathbf{p} \mapsto (\mathbf{p}, \tfrac{1}{2m}\mathbf{p}^2) \tag{4.8}$$

Then the PV measure in question reads as follows:

$$Z \mapsto E(Z) := F^{\mathbf{P}}\big(\Lambda^{-1}(Z)\big), \qquad Z \in \mathcal{B}_o(\Omega) \tag{4.9}$$

The fact that $Z \in \mathcal{B}_o(\Omega)$ can be expressed in view of (4.5) by writing Z as Z_X, where $X \in \mathcal{B}(\mathbf{R}^+)$. Then the POV measure (4.9) coincides with the spectral measure E^H of the generator (2.20) of time translations, $H = \frac{1}{2m}\mathbf{P}^2$:

$$E\big(\Pi_o^{-1}(X)\big) = E^H(X) \tag{4.10}$$

An entirely analogous procedure yields unsharp energy observables from smeared momentum observables F^f, where it is crucial that the confidence function $f(\mathbf{p})$ is invariant under rotations.

III.4.2 Time. The parameter set of the time translation group is itself a homogeneous space of $\mathcal{G}$. This corresponds to the fact that $\mathcal{Z}$ can be identified as a factor group with respect to the isochronous Galilei group and is therefore homomorphic to $\mathcal{G}$. Indeed we have the action

$$\alpha_g : \mathbf{R} \to \mathbf{R}, \quad t \mapsto t + b \tag{4.11}$$

Any time observable B should possess the following covariance:

$$e^{iHt} B(Z) e^{-iHt} = B(Z + t), \qquad Z \in \mathcal{B}(\mathbf{R}) \tag{4.12}$$

Before we turn to the construction of time observables, some historical and interpretational remarks are in order. As mentioned above time operators have been used early in a formal and non-rigorous manner to motivate the time-energy uncertainty relation. Given a time observable B as a POV measure, one can introduce a time operator $T^B := \int_{\mathbf{R}} t\, dB(t)$ that is symmetric but cannot, in most cases, be self-adjoint. If B were a PV measure then T^B would be self-adjoint. In this way the formal obstacles against considering time as an observable have been overcome, and meanwhile many examples are known. This is satisfactory as the experimental question about the time of occurrence is perfectly legitimate and has been asked already in the early days of quantum mechanics with respect to the decay time of unstable systems.

Again there is not just one unique time observable but rather there are many POV measures satisfying (4.12). The physical relevance of this non-uniqueness will be illustrated by means of the subsequent examples. In general time is measured by making reference to some continuously changing properties of a physical system $\mathcal{S}$ that is thereby treated as a 'clock'. In turn any dynamically changing quantity gives rise to a certain type of experimental question, namely, at what time the quantity assumes a given value. It becomes thus apparent that any *event time* observable will depend on both, the energy observable H of $\mathcal{S}$ as well as the type of event under consideration. An event is understood as the occurrence of a specified value of some quantity. In the following we give up the requirement that the energy should be fully Galilei invariant. That is, we admit the possibility of external fields, which corresponds to the approximation that the system $\mathcal{S}$ in question is part of a larger system but in such a way that the 'rest' is not affected by the interaction with $\mathcal{S}$.

An illustration: the idea of screen observables. There is an important family of Galilei covariant observables that has not received much attention until very recently although it is far closer to experimental practice than most of the quantities usually discussed in textbooks. In fact the registration of particles is carried out with extended detectors which should be formally represented by means of *screen observables*. These account for the time and location of the particle's hitting the screen, i.e., the detector's sensitive surface [1.4, 3.10]. Such observables cannot be described as self-adjoint operators and this may be one reason for their late discovery. Reducing the experimental question as to *when* a particle hits the screen, irrespective of where it does so, yields a time observable associated with any such screen. As an idealised screen is represented by means of a plane, it is obvious that there is a family of screen time observables, labeled with the location and orientation parameters of the associated plane. The explicit construction of such screen observables is based on a fairly abstract formalisation of the classical conception of the time of passage of a particle trajectory through some plane. The solutions turn out technically rather involved. We shall therefore go on to discuss some further examples of time observables in order to show that classical intuition may be used in a direct way to yield explicit constructions. The strategy will be to find, for a given Hamiltonian, a suitable dynamical variable and define the ensuing time of occurrence for its values.

In the sequel four Hamiltonians with essentially different spectra will be discussed. In all cases there exist natural choices of dynamical variables giving rise to event time observables with very distinct properties.

Example 1. The time evolution of a free (spinless) Galilean particle (with one degree of freedom) is determined by the Hamiltonian $H = P^2/2m$. Classically the time of passage of some trajectory $Q(t) = \frac{1}{m}Pt + Q_o$ through the plane $Q = 0$, say, would be defined implicitly by the relation $Q(t) = 0$ and explicitly as $t = -\frac{mQ_o}{P}$. Alternatively, one may view $t = \frac{mQ_o}{P}$ as the time of travel for the particle to reach the given plane. A first attempt to implement this definition into quantum mechanics would be to read it as a definition of an operator time [3.11],

$$T = \frac{m}{2}\left[QP^{-1} + P^{-1}Q\right] \tag{4.13}$$

While this operator formally obeys the canonical commutation relation (4.1), it is only symmetric without admitting a self-adjoint extension. Hence T does not meet the requirements of the ordinary definition of an observable. Yet there exists a POV measure B_o that does possess the covariance (4.12) and reproduces T as its moment operator. To see this we note first that in the momentum representation T is found to coincide with $T' := m \operatorname{sgn}(p) |p|^{-1/2} i \frac{\partial}{\partial p} |p|^{-1/2}$. In turn, this operator is known to be a maximal symmetric operator arising from the POV measure $Z \mapsto B_o(Z)$, where

$$\langle \psi \,|\, B_o(Z)\psi \rangle \;=\; \frac{1}{2\pi} \int_Z dt \left| \int dp \, (|p|/m)^{1/2} \, e^{itp^2/2m} \, \tilde{\psi}(p) \right|^2 \qquad (4.14)$$

In fact one readily verifies that T' is the first moment of the measure (4.14), its maximal domain being $\{\psi | \int t^2 \, \langle \psi \,|\, B_o(dt)\psi \rangle < \infty\}$ [1.12]. Such covariant POV measures can be constructed systematically by application of *covariant* Neumark extensions of the Hilbert space, thereby lifting the semiboundedness of H. The resulting time spectral measure is then restricted to the original Hilbert space, yielding the sought time observables [3.10].

Example 2. The definition of a time observable becomes much easier in the case of a freely falling particle in a homogeneous force field: $H = P^2/2m + mgQ$. The equation of motion for the momentum variable becomes $P(t) = mgIt + P_o$, so $T := \frac{1}{mg} P$ will do as a self-adjoint operator canonically conjugate to H. Moreover H and T satisfy the Weyl commutation relation

$$e^{iHb} \, e^{iTh} \;=\; e^{ibh} \, e^{iTh} \, e^{iHb} \qquad (4.18)$$

and the spectral measure of T possesses the covariance property (4.12). The present Hamiltonian is unbounded, its spectrum being absolutely continuous and covering the real line. This explains why T can act as a shift generator for H. Further, T being a function of P, it is again possible to derive time observables from phase space POV measures; the resulting time POV measures are functions of the marginal unsharp momentum observables.

Example 3. Another type of example is that of a periodic time observable for the harmonic oscillator, $H = \frac{1}{2} [P^2 + Q^2]$. The spectrum of H is discrete and consists of nonnegative, equidistant values so that one can at best expect a shift semigroup and certainly no self-adjoint operator T satisfying the Weyl relation (4.18). Yet classical reasoning offers a dynamical quantity that grows linearly in time modulo the oscillator's period: the phase, defined as the angular coordinate in phase space. Moreover this quantity is covariant under the group of time shifts modulo the period. This reasoning yields a very natural and unifying approach to oscillator phase observables, which we shall develop in the next section.

Example 4. Finally even a finite quantum system with a bounded, discrete Hamiltonian may permit a (periodic) time observable. Let $H = \beta s_3$, where s_3 is the third component of the spin of a spin-s system. The vectors

$$|\vartheta\rangle \;=\; \sum_{m=-s}^{s} e^{im\beta\vartheta} \, |m\rangle, \qquad \vartheta \in [0, 2\pi) \qquad (4.19)$$

form a time shift-covariant family as

$$e^{iH\tau} \mid \vartheta \rangle = \mid \vartheta + \tau \rangle \qquad (4.20)$$

(addition being modulo 2π). It follows that the effects

$$B(Z) = (2\pi)^{-1} \int_Z d\vartheta \mid \vartheta \rangle \langle \vartheta \mid = (2\pi)^{-1} \int_Z \sum_{m,n} e^{i(m-n)\beta\vartheta} \mid m \rangle \langle n \mid \qquad (4.21)$$

with $Z \in \mathcal{B}(0, 2\pi))$, constitute a time shift-covariant, normalised POV measure. Thus the dynamical quantity 'spin phase' serves as a periodic quantum clock [3.12] if the Hamiltonian is proportional to the corresponding spin component.

We have thus seen that defining time observables as time shift-covariant POV measures opens up the possibility of constructing such observables very much in the spirit of the experimental question concerning the time of occurrence (of some event). Depending on the particular Hamiltonian there seem to exist choices of dynamical variables whose change in time allows for particularly simple and intuitively appealing time observables. It appears therefore that any physical system has some dynamical quantities which display its time evolution in a characteristic way. In principle it should be possible to associate a time of occurrence POV measure with any conceivable nonstationary quantity, though it does not seem easy to devise a general procedure of construction [3.13]. Future investigations into this topic should lead to experimental proposals for measuring event times.

III.5 Photon observables
The concept of observable as a POV measure has found ample applications in other important areas of quantum physics, such as field theory and Poincaré relativistic quantum mechanics. We shall give an illustration of these achievements with regard to two rather distinct ways of describing photons, either as mass-zero, spin-one irreducible representations of the Poincaré group, or as occupations of electromagnetic field modes. The latter method has been systematically developed in the detection theory for quantum fields. Early steps of these investigations are documented in the monographs of Davies [1.9] and Helstrom [1.11]; more recent research has led to quantum theories of stochastic processes and open systems [3.14–16]. As these results are beyond the scope of the present text, we shall select some examples of photon observables that are currently under investgation in quantum optics. These will be studied in the following two subsections, while the last subsection is devoted to the problem of photon localisation.

III.5.1 Photon number and phase. We use freely the elementary tools of quantum electrodynamics, without aiming at a systematic picture. In particular a free field mode is assumed to be represented by its annihilation and creation operators satisfying the boson commutation relation. This allows one to make use of the formal equivalence to the quantum mechanical harmonic oscillator.

The classical mechanical theory of the harmonic oscillator can be formulated either in terms of position and momentum or by means of amplitude and phase. On the other hand in quantum mechanics a harmonic oscillator (of unit mass) is conveniently described by the operators $a = \frac{1}{\sqrt{2\omega}}(\omega Q + iP)$ and $a^* = \frac{1}{\sqrt{2\omega}}(\omega Q - iP)$. The energy operator assumes the form

$$H = \tfrac{1}{2}\left(P^2 + \omega^2 Q^2\right) = \omega\left(a^* a + \tfrac{1}{2}\right) \tag{5.1}$$

The operator $N = a^* a$ has the nondegenerate eigenvalues $n = 0, 1, 2, \ldots$ corresponding to the equation $N|n\rangle = n|n\rangle$, with the eigenstates $|n\rangle = \frac{1}{\sqrt{n!}} a^{*n}|0\rangle$ and the ground state characterised by $a|0\rangle = 0$.

The quantum theoretic description of a single-mode electromagnetic field leads to the introduction of the photon annihilation and creation operators a, a^* and the ensuing photon number operator $N = a^* a$, which are formally equivalent to the above oscillator operators. In particular a and a^* can be decomposed into real and imaginary parts

$$
\begin{aligned}
a &= \frac{1}{\sqrt{2}}\left(a^q + i a^p\right), \quad a^* = \frac{1}{\sqrt{2}}\left(a^q - i a^p\right) \\
a^q &= \frac{1}{\sqrt{2}}\left(a^* + a\right), \quad a^p = \frac{i}{\sqrt{2}}\left(a^* - a\right)
\end{aligned}
\tag{5.2}
$$

Operators a^q and a^p are called quadrature components of the field mode. They are a Schrödinger couple and therefore unitarily equivalent to position and momentum operators. In particular, the Bose commutation relation

$$[a, a^*] = I \tag{5.3}$$

is equivalent to the canonical commutation relation

$$[a^q, a^p] = iI \tag{5.4}$$

A long-standing problem of quantum optics has been the question whether there exists a phase observable that is canonically conjugate to the number observable for a single-mode field [3.17]. There is now a satisfactory solution in terms of POV measures, which also meets the experimental demands for phase measurements. This solution starts with giving a precise meaning to the quantum counterpart of the classical phase-amplitude pair. Let us write the position and momentum variables as complex numbers $z = (\omega q + ip)/\sqrt{2\omega} = e^{i\varphi}|z|$. Then $|z| = \sqrt{E/\omega}$, where E is the energy of the classical oscillator. Comparing z with the operator a suggests searching for the operator analogue of the polar decomposition of the complex number z in the form $a = e^{i\Phi}|a|$, where $|a| = \sqrt{a^* a} = \sqrt{N}$.

Let $a = V|a| = V\sqrt{N}$ be the polar decomposition of a. From the properties of a and a^* one can infer that $V|n\rangle = |n-1\rangle$, $V|0\rangle = 0$ and $V^*|n\rangle = |n+1\rangle$. Therefore $VV^* = I$ and $V^*V = I - |0\rangle\langle 0|$ which shows that V is a partial isometry but not unitary. Hence there is no self-adjoint operator Φ such that $a = e^{i\Phi}|a|$. Apart from this well known argument the lacking unitarity of V can also be understood from the fact that N is non-negative. Indeed the commutation relation (5.3) yields

$$VNV^* = N + I \tag{5.5}$$

Thus if V were unitary, then N and $N+I$ would be unitarily equivalent and should therefore have the same spectra, which is wrong. This observation has led to the consideration of a 'duplication' of the Hilbert space by extending N to an operator N' whose spectrum consists of all integers [3.18]. In this way a Weyl pair (N', Φ') can be defined in formal analogy to the angular momentum and angle pair [Section 3.4]. In view of Neumark's theorem it is obvious that the projection of the spectral measure of Φ' onto the subspace corresponding to non-negative eigenvalues of N is a POV measure and no longer a PV measure.

In order to construct a phase observable we make use of the fact that the operator V is a contraction so that there is a unique POV measure M satisfying

$$V^n = \int_0^{2\pi} e^{in\phi}\, dM(\phi) \tag{5.6}$$

for $n = 0, 1, 2, \ldots$ (Section II.2.5). Since V has the structure

$$V = \sum_{n=1}^{\infty} |n-1\rangle\langle n| \tag{5.7}$$

one obtains

$$e^{i\phi N} V e^{-i\phi N} = e^{-i\phi} V \tag{5.8}$$

The uniqueness of M then entails

$$e^{i\phi N} M(X) e^{-i\phi N} = M(X + \phi) \tag{5.9}$$

for $X \in \mathcal{B}(0, 2\pi)$, $X + \phi = \{\phi' + \phi \,(\mathrm{mod}\, 2\pi) \mid \phi' \in X\}$, which is to say that M is covariant under the shifts generated by N. Following the relativistic approach, we adopt the covariance condition (5.9) as the defining property of a *phase observable*. This condition does not single out the POV measure M as the only phase observable. On the contrary, there are infinitely many phase shift covariant POV measures; and it will become apparent that they should be associated with different possible measurement schemes offered in the framework of measurement theory. Since N is proportional to the energy observable it is natural to interpret the phase as a (periodic) time observable in the sense of Section 4.

The mapping V has a natural dilation to a unitary operator by extending the sum (5.7) to run over all integers, with a corresponding extension of the Hilbert space and the system of basis vectors $|n\rangle$. This dilation leads to an explicit form of M as the Neumark projection of the corresponding spectral measure. Consider the Hilbert space $L^2(0, 2\pi)$, with the basis $\xi_k(\phi) = \frac{1}{\sqrt{2\pi}} e^{-ik\phi}$, $k \in \mathbf{Z}$. The space $\mathcal{H}$ can be mapped onto a subspace $\mathcal{H}^2$ of $L^2(0, 2\pi)$ by the isometry

$$W : \mathcal{H} \to \mathcal{H}^2, \quad |n\rangle \mapsto W(|n\rangle) = \xi_n$$

$$(W\psi)(\phi) \equiv \psi(\phi) := \sum_{n=0}^{\infty} \frac{1}{\sqrt{2\pi}} e^{-in\phi} \langle n | \psi \rangle \tag{5.10}$$

Clearly $W = \sum_{n=0}^{\infty} |\xi_n\rangle\langle n|$ so that $WVW^* = \sum_{n=1}^{\infty} |\xi_{n-1}\rangle\langle \xi_n|$. Let V_o be the unitary extension of WVW^*, $V_o = \sum_{k=-\infty}^{\infty} |\xi_{k-1}\rangle\langle \xi_k|$. Since $(V_o\psi)(\phi) = e^{i\phi}\psi(\phi)$, the spectral measure E of V_o is the canonical one, with $E(X)\psi = \chi_X \cdot \psi$. For any $\psi, \eta \in \mathcal{H}$ and all $X \in \mathcal{B}(0, 2\pi)$ one has

$$\langle \eta | E(X)\psi \rangle = \int_X \overline{(W\eta)(\phi)}(W\psi)(\phi)\, d\phi \tag{5.11}$$

$$= \sum_{n,m=0}^{\infty} \frac{1}{2\pi} \int_X e^{i(n-m)\phi}\, d\phi \, \langle \eta | n \rangle \langle m | \psi \rangle =: \langle \eta | M(X)\psi \rangle$$

Therefore

$$M(X) = \sum_{n,m=0}^{\infty} \frac{1}{2\pi} \int_X e^{i(n-m)\phi}\, d\phi \, |n\rangle\langle m| \tag{5.12}$$

The elements of $\mathcal{H}^2$ arise as the boundary functions of analytic functions on the unit disc, where $\psi(\phi) = \lim_{r \to 1} \psi(re^{i\phi})$ exists in the L^2-sense, that is,

$$\lim_{r \to 1} \int_0^{2\pi} |\psi(\phi) - \psi(re^{i\phi})|^2\, d\phi = 0 \tag{5.13}$$

$\mathcal{H}^2$ is called the Hardy class on the unit disk. The analytic functions $\psi(re^{i\phi})$ do not vanish on sets with positive Lebesgue measure. The same is true for the limit functions $\psi(\phi)$. Thus $\chi_X(\phi)\,\psi(\phi)$ does not belong to $\mathcal{H}^2$ whenever $X \in \mathcal{B}(0, 2\pi)$ is such that its complement set has positive Lebesgue measure. It follows that M is no PV measure. This result is also directly confirmed by the observation that M, as given in (5.12), is a noncommutative measure. As in the case of the spin phase, M cannot be localised, i.e., for any vector state ψ and any set X with $M(X) \neq I$ one has

$$\langle \psi | M(X)\psi \rangle < 1 \tag{5.14}$$

The noncommutativity of M can be made explicit by writing it symbolically as

$$M(X) = \frac{1}{2\pi} \int_X |\phi\rangle\langle\phi|\, d\phi \tag{5.15}$$

The entities

$$|\phi) := \sum_{n=0}^{\infty} e^{in\phi} \, | \, n\rangle \tag{5.16}$$

form a shift covariant family of (improper) eigenvectors of V, i.e., $V|\phi) = e^{i\phi}|\phi)$, and these do not form an orthogonal system:

$$(\phi'|\phi) = \sum_{n=0}^{\infty} e^{in(\phi-\phi')} = \pi\delta(\phi'-\phi) - \tfrac{i}{2}\cot\left(\tfrac{\phi'-\phi}{2}\right) + \tfrac{1}{2} \tag{5.17}$$

The orthogonality is restored as soon as the sum is extended to contain also all negative values of n, thus yielding $|\phi)$ as the eigenvectors of the extended phase operator Φ'.

The term $(\phi \mid n)$ can be interpreted as the amplitude of a number eigenstate in the phase representation (5.10), showing the complete indeterminacy of the phase. This illustrates the complementarity of the number-phase pair. One may also write down the phase probabilities in a number state to verify that these correspond to the uniform distribution:

$$\langle n \mid M(X)|n\rangle = \frac{1}{2\pi}\int_X d\phi \tag{5.18}$$

Since the phase spectrum was chosen to be the interval $[0, 2\pi)$, it is clear that the first moment of this distribution must have the value π.

We shall show now that the phase observable (5.12) provides a unification of the various approaches to the phase concept discussed in the literature. First of all, since the space of values of M is a bounded subset of the real line all its moment operators

$$M^{(k)} := \int_0^{2\pi} \phi^k \, M(d\phi) \tag{5.19}$$

are bounded self-adjoint operators. In particular its first and the second moment operators are

$$M^{(1)} = -i \sum_{m\neq n=0}^{\infty} (n-m)^{-1} |n\rangle\langle m| + \pi I \tag{5.20}$$

$$M^{(2)} = \frac{4}{3}\pi^2 I + \sum_{m\neq n=0}^{\infty} \left(-i2\pi(n-m)^{-1} + 2(n-m)^{-2}\right)|n\rangle\langle m| \tag{5.21}$$

Clearly $(M^{(1)})^2 \neq M^{(2)}$, which is yet another way of saying that M is no PV measure.

It is interesting to observe that, as a reflection of the covariance property, the first moment operator $M^{(1)}$ satisfies the Heisenberg commutation relation with the number operator,

$$[M^{(1)}, N]\psi = i\psi, \quad \psi \in \mathcal{D} \tag{5.22}$$

in a dense domain $\mathcal{D} = \{\psi \in \mathcal{D}(N) \,|\, \sum_{n=0}^{\infty} \langle n \,|\, \psi \rangle = 0\}$. For this reason the operator $\Phi_o := M^{(1)}$ has been considered as a candidate of a phase operator (see, e.g., [3.19]). This 'phase operator' has, however, only a limited use. First of all, since M is not a PV measure, the powers $(M^{(1)})^k$ do not give the moment operators $M^{(k)}$, except for $k = 0, 1$. Second, it carries an essential nonuniqueness. In fact one may define a whole family of phase operators Φ_ϕ, where the label ϕ indicates the choice of the origin of the phase parameter scale within $[0, 2\pi)$:

$$\Phi_\phi := e^{i\phi N} \Phi_o \, e^{-i\phi N} = \Phi_o - \phi I + 2\pi M\big([0, \phi]\big) \tag{5.23}$$

These operators, which have the same spectra as Φ_o, form a shift covariant non-commuting family: $e^{i\phi' N} \Phi_\phi e^{-i\phi' N} = \Phi_{\phi+\phi'}$

For a bounded measurable real-valued function f on $[0, 2\pi)$, the integral

$$B(f) := \int_0^{2\pi} f(\phi) \, dM(\phi) \tag{5.24}$$

defines a bounded self-adjoint operator. Obviously the operators V and Φ_ϕ can be recovered from the POV measure M in this way. As another example the sine and cosine functions give

$$B(\sin) = \frac{i}{2}(V^* - V) =: S, \qquad B(\cos) = \frac{1}{2}(V^* + V) =: C \tag{5.25}$$

Also these operators have been studied as candidates for representing the phase [3.20], but this leaves one in the unsatisfactory situation of circumscribing the phase by means of two noncommuting operators $\big([C, S] = \frac{i}{2}[V, V^*] = \frac{i}{2}|0\rangle\langle 0|\big)$, rather than representing it as one single quantum observable.

There is another approach towards obtaining a self-adjoint phase operator that starts with reducing the Hilbert space to an s-dimensional subspace $\mathcal{H}_s$, spanned by the number states $|\,0\rangle, |\,1\rangle, \ldots, |\,s-1\rangle$ [3.21]. In these spaces one can define shift covariant families of states $|\,\phi\rangle_s$ for $\phi \in [0, 2\pi)$ as follows:

$$|\,\phi\rangle_s := \frac{1}{\sqrt{s}} \sum_{n=0}^{s-1} e^{in\phi} \,|\,n\rangle \tag{5.26}$$

Then $e^{i\phi' N} \,|\,\phi\rangle_s = |\,\phi + \phi'\rangle_s$. Selecting a sequence of phase values $\phi_k = 2\pi\frac{k}{s}$, $k = 0, 1, \ldots, s-1$ yields a complete orthogonal system of normalised vectors $|\,\phi_k\rangle_s$ in $\mathcal{H}_s$. Therefore one may introduce a discretised self-adjoint phase operator in $\mathcal{H}_s$,

$$\Phi_s := \sum_{k=0}^{s-1} \phi_k \,|\,\phi_k\rangle_{ss}\langle\phi_k\,| \tag{5.27}$$

This approach seems at first sight to have some drawbacks. The operator Φ_s is covariant only under a finite group of phase shifts and is thus not conjugate to the number operator. Still, as with the above operator Φ_o, there exists a shift covariant family of discrete phase operators in $\mathcal{H}_s$. It is obvious that the product $a_s := e^{i\Phi_s}\sqrt{N}$ cannot coincide with the annihilation operator a and certainly a_s and a_s^* do not satisfy the Bose commutation relation (5.3). Nevertheless one may show that this approach affords a description of phase properties that can be regarded as satisfactory for all practical purposes. Moreover in the limit $s \to \infty$, the usual formal relationships between the various operators are recovered with increasing accuracy. In particular the commutator $[a_s, a_s^*] = \sum_{n=0}^{s} |n\rangle\langle n| - s|s\rangle\langle s|$ approaches $[a, a^*] = I$ weakly. Furthermore the spectrum of Φ_s becomes more and more 'dense' in $[0, 2\pi]$ and this operator approaches Φ_o weakly [3.22]: for any $\eta, \xi \in \mathcal{H}$,

$$
\begin{aligned}
\langle \xi | \Phi_s \eta \rangle &= \sum_{k=0}^{s-1} \phi_k \langle \xi | \phi_k \rangle_s \langle \phi_k | \eta \rangle \\
&= \frac{1}{s} \sum_{k=0}^{s-1} \phi_k \sum_{n,m \le s-1} e^{i(n-m)\phi_k} \langle \xi | n \rangle \langle m | \eta \rangle \\
&\to \frac{1}{2\pi} \int_0^{2\pi} d\phi\, \phi \sum_{n,m=0}^{\infty} e^{i(n-m)\phi} \langle \xi | n \rangle \langle m | \eta \rangle = \langle \xi | \Phi_o \eta \rangle
\end{aligned}
\tag{5.28}
$$

Here one makes use of the substitution $\frac{1}{s}\sum_{k=0}^{s-1} \to \frac{1}{2\pi}\int_0^{2\pi} d\phi$ as $s \to \infty$. The last equality is obtained by carrying out the ϕ-integration.

In this section we have defined a phase observable as a POV measure satisfying the shift covariance condition (5.9). This phase observable M is the POV measure associated with the polar decomposition of $a = V\sqrt{N}$. Further motivation for the characterisation of the phase via covariance derives from the phase space description of a single-mode photon field which yields an intuitively appealing connection with the phase of a classical field. In the following subsection we introduce a new class of phase observables, which emerge as 'noisy' versions of M.

III.5.2 Joint observables for the quadrature components. The analogy between the harmonic oscillator and the single-mode electromagnetic field allows one to transcribe the phase space picture formulated in Section 2.4. First let us introduce phase space parameters q, p and define $z = (q + ip)/\sqrt{2}$. With this notation the phase space translations can be represented in the following form:

$$
\begin{aligned}
D_z &:= \exp(za^* - \bar{z}a) = \exp(-iqP + ipQ)) = W_{qp} \\
D_z\, a\, D_z^{-1} &= a - zI
\end{aligned}
\tag{5.29}
$$

The operator a has an overcomplete system of eigenvectors associated with the nondegenerate eigenvalues z. Indeed let ψ be a unit vector such that $a\psi = z\psi$.

This can be written as $D_z a D_z^{-1} \psi = O$ so that $D_z^{-1} \psi$ is proportional to the vacuum state $| \, 0 \rangle$. In this way one is led to the family of *coherent states* $\{| \, z \rangle \, | \, z \in \mathbf{C}\}$ satisfying

$$| \, z \rangle \,=\, D_z \, | \, 0 \rangle, \quad a \, | \, z \rangle \,=\, z \, | \, z \rangle \tag{5.30}$$

The overcompleteness relation is now a direct consequence of the normalisation condition (2.45):

$$\frac{1}{\pi} \int_{\mathbf{C}} d^2 z \, | z \rangle \langle z | \,=\, I \tag{5.31}$$

Here we have used the identities $d^2 z = \frac{1}{2} \, dq \, dp$ and $| z \rangle \langle z | = W_{qp} \, | 0 \rangle \langle 0 | \, W_{qp}^{-1}$. In this way we obtain a POV measure representing a joint observable for the quadrature components of a,

$$A(Z) \,=\, \frac{1}{\pi} \int_Z d^2 z \, | z \rangle \langle z |, \quad Z \in \mathcal{B}(\mathbf{C}) \tag{5.32}$$

Instead of using the vacuum state $| 0 \rangle \langle 0 |$ one may take any other positive trace one operator T_o and integrate over $T_z = D_z \, T_o \, D_z^{-1}$ to obtain further joint observables for the quadrature components. In particular the choice $| n \rangle \langle n |$ gives the joint observable

$$A^{|n\rangle}(Z) \,=\, \frac{1}{\pi} \int_Z d^2 z \, D_z | n \rangle \langle n | D_z^*, \quad Z \in \mathcal{B}(\mathbf{C}) \tag{5.33}$$

We list some useful properties of coherent states. Their name derives from the fact that the eigenvalue equation (5.30) is a necessary and sufficient condition for the factorisation property

$$\langle \psi \, | \, a^* a \psi \rangle \,=\, \langle \psi \, | \, a^* \psi \rangle \, \langle \psi \, | \, a \psi \rangle \tag{5.34}$$

It is obvious that coherent states give rise to factorisations to all orders of the expectation values $\langle \psi \, | \, a^{*n} a^n \psi \rangle$ and therefore to field coherence [3.23] in all orders. As an application of the Baker-Campbell-Hausdorff formula

$$e^{A+B} \,=\, e^{-\frac{1}{2}[A,B]} \, e^A \, e^B \quad \text{whenever} \quad [A,[A,B]] = [B,[A,B]] = 0 \tag{5.35}$$

the shift operator can be decomposed into a product form

$$D_z \,=\, e^{-\frac{1}{2}|z|^2} \, \exp(za^*) \, \exp(-\bar{z}a) \tag{5.36}$$

This allows one to expand the coherent states as series of number states:

$$| \, z \rangle \,=\, D_z \, | \, 0 \rangle \,=\, e^{-\frac{1}{2}|z|^2} \sum_{n=0}^{\infty} \frac{1}{n!} z^n a^{*n} \, | \, 0 \rangle \,=\, e^{-\frac{1}{2}|z|^2} \sum_{n=0}^{\infty} \frac{1}{\sqrt{n!}} z^n \, | \, n \rangle \tag{5.37}$$

Coherent states are thus seen to be among those yielding Poisson number statistics, with the mean value $\langle z \, | \, N | z \rangle = |z|^2$.

It is interesting to study two types of marginal observables of the POV measure (5.32). These can be defined with reference to different coordinates of the complex plane. First, one may choose Cartesian coordinates given by the real and imaginary parts (q, p) of the variable z. This gives rise to smeared versions of the quadrature observables, in analogy to unsharp position and momentum observables. Second, one obtains a new pair of conjugate observables if the marginal integrations are carried out with respect to coordinates $(r^2, \phi) \in [0, \infty) \times [0, 2\pi)$, defined via $z = re^{i\phi}$, $r \geq 0$. In this case one is dealing with sets of the form $Z_R = R \times [0, 2\pi)$ or $Z_F = [0, \infty) \times F$. Taking into account that $\frac{1}{\pi} d^2 z = \frac{1}{2\pi} d\phi \, dr^2$ and applying (5.37), one computes

$$
\begin{aligned}
A_N(R) \; &:= \; A(Z_R) \; = \; \int_R dr^2 \, \frac{1}{2\pi} \int_0^{2\pi} d\phi \, D_z \, |0\rangle\langle 0| \, D_z^* \\
&= \; \int_R dr^2 \, e^{-r^2} \sum_{m,n=0}^{\infty} \frac{1}{2\pi} \int_0^{2\pi} d\phi \, e^{i(n-m)\phi} \, \frac{r^{n+m}}{\sqrt{n!m!}} \, |n\rangle\langle m|
\end{aligned}
\tag{5.38}
$$

This yields

$$
A_N(R) \; = \; \sum_{n=0}^{\infty} \int_R p_n(r^2) dr^2 \, |n\rangle\langle n|, \qquad p_n(r^2) \; = \; \frac{1}{n!} \, e^{-r^2} r^{2n}
\tag{5.39}
$$

Here $p_n(r^2)$ is a discrete-to-continuous transition probability density in the sense of part a) of Section II.2.3. A similar result can be obtained for the observable (5.33). Thus the POV measures $A_N^{|n\rangle}$ are continuously smeared number observables. In order to interpret their measurements as unsharp photon number measurements, one would proceed by introducing a calibration by associating with each number m a set $R_m = \{r^2 | m - \frac{1}{2} \leq r^2 < m + \frac{1}{2}\}$ for $m = 1, 2, \ldots$, and $R_0 = [0, \frac{1}{2})$. Then

$$
A_N(R_m) \; = \; \sum_{n=0}^{\infty} \lambda_{mn} \, |n\rangle\langle n|, \qquad \lambda_{mn} \; = \; \int_{R_m} p_n(r^2) \, dr^2
\tag{5.40}
$$

which demonstrates the interpretation of λ_{mn} as a confidence measure for inferring a number n given a registered set R_m. For our subsequent use we note already here that the first moment of the unsharp number observable A_N is $N + I$. In fact it can be shown [3.24] that the first moments of the observables $A_N^{|n\rangle}$ are

$$
\int_0^{\infty} r^2 \, dA_N^{|n\rangle}(r^2) \; = \; N + (n+1)I
\tag{5.41}
$$

This means that on the statistical level of first moments, the unsharp number observable $A_N^{|n\rangle}$ equals the number observable apart from the positive constant $n + 1$ indicating noise in the form of additional photons.

We investigate next the second marginal observable of the joint observable A:

$$A_{ph}(F) := A(Z_F) = \int_0^\infty dr^2 (2\pi)^{-1} \int_F d\phi \, |z\rangle\langle z|$$

$$= \sum_{n,m=0}^\infty \int_0^\infty dr^2 \, e^{-r^2} \frac{r^{n+m}}{\sqrt{n!m!}} \frac{1}{2\pi} \int_F d\phi \, e^{i(n-m)\phi} \, |n\rangle\langle m| \tag{5.42}$$

The value space of this POV measure is that of a phase observable. There is a close formal relationship between A_{ph} and the phase observable M, Eq. (5.12). Indeed let us introduce the operators

$$T_r = \sum_{n=0}^\infty \frac{1}{\sqrt{n!}} e^{-\frac{1}{2}r^2} r^n \, |n\rangle\langle n| \tag{5.43}$$

which satisfy the normalisation condition

$$\int_0^\infty dr^2 \, T_r^* T_r = I \tag{5.44}$$

This allows us to define a (completely positive, unital and normal) map on $\mathcal{L}(\mathcal{H})$, $B \mapsto \int_0^\infty dr^2 \, T_r \, B \, T_r^* =: V^*(B)$. It then follows readily that

$$A_{ph}(F) = \int_0^\infty dr^2 \, T_r \, M(F) \, T_r^* = V^*(M(F)) \tag{5.45}$$

The observable A_{ph} therefore represents a 'noisy' phase measurement in the sense of part b) of Section II.2.3. This interpretation is further supported by the fact that the T_r commute with the number operator N so that A_{ph} is seen to satisfy the covariance relation (5.9). In this way we have achieved a general method of producing new phase observables starting with M: one can choose any family of bounded operators $\{T_\lambda | \lambda \in \Lambda\}$ commuting with N and such that $\sum_\lambda T_\lambda T_\lambda^* = I$. Then $\widetilde{M}(F) = \sum_\lambda T_\lambda M(F) T_\lambda^*$ is a phase observable in the sense of shift covariance. Finally since $A_{ph}^{(1)} = V^*(M^{(1)})$, this 'noisy phase operator' also forms a Heisenberg pair with the number: $[A_{ph}^{(1)}, N] = \int dr^2 \, T_r^* \, [M^{(1)}, N] \, T_r = iI$.

Another type of generalisation is obtained if the (vacuum) state operator $S_o = |0\rangle\langle 0|$ underlying the POV measure (5.32) is replaced with any number state $S_o^{(n)} = |n\rangle\langle n|$ or a mixture of them [3.25]. The phase shift invariance of these states ensures that of the ensuing phase POV measures $A_{ph}^{|n\rangle}$. It is an important open question which of these phase POV measures can be realised experimentally. We return to this issue in Sections VII.3.4 and VII.3.7.

The 'phase' space picture of a single-mode photon field provides a justification of the term 'phase observable' in an intuitively appealing way. This is evident in view of the well known observation that coherent states $\mid z \rangle$ with a large (squared) amplitude modulus $|z|^2 = \langle z \mid N \mid z \rangle$ yield the best available approximation to classical periodically oscillating fields, with stationary and relatively small variances of the electric and magnetic field operators. Such 'macroscopic' coherent states $\mid z(t) \rangle$ can be visualised in the phase space picture by means of circular error regions of unit area, centered at the point $z(t)$ which rotates at the mode frequency around the coordinate origin. The phase of such a field is indeterminate by an order $|z|^{-1} = \langle N \rangle^{-1/2}$, corresponding to the aperture of the error region with respect to the origin. In this way a coherent state furnishes a quantum clock whose phase serves as a periodic time variable.

III.5.3 Photon localisation. The term 'photon' does not correspond to a unique notion in the literature. There are several distinct sets of ideas of what a photon might be, and these are not easily reconciled with each other. First the historical use of the word photon in atomic physics and chemistry originated from Planck's and Einstein's 'quanta of energy' as they appear in the treatments of blackbody radiation or the photoelectric effect. It refers to the units of energy emitted or absorbed by atomic or molecular systems which are observed in spectroscopy. A more technical formulation describes a photon as a countable quantum of an electromagnetic field mode. Denoting the field mode operators as a, a^*, then the eigenstates $\mid n \rangle$ of the number operator $N = a^* a$ are interpreted as corresponding to a field state consisting of n photons occupying that mode. If the mode in question is that of a plane wave with a definite value of polarisation, then photons are viewed as objects with a definite momentum (and polarisation); due to the Fourier relationship between momentum and configuration variables, such objects cannot be localised in space. On the other hand, many experiments operate with pulses of light, also often called photons and described by means of fairly well localised wave packets.

It is this last notion of a photon as a localisable system which shall be analysed in the following. Without going into details, we indicate how a one-photon state corresponding to a wave packet is constructed within the Fock space. To this end one must form superpositions of plane-wave modes, with associated field operators $a_\lambda(\mathbf{p})$ and (left and right circular) polarisation vectors $\mathbf{e}_\lambda(\mathbf{p})$ $(\lambda = +, -)$. Here we assume the conditions of the Coulomb gauge which entails that $\mathbf{p} \cdot \mathbf{e}_\lambda(\mathbf{p}) = 0$. For any transversal vector function $\mathbf{A} : \mathbf{p} \mapsto \mathbf{A}(\mathbf{p})$, $\mathbf{A} \in \widetilde{\mathcal{H}} := L^2(\mathbf{R}^3, \mathbf{C}^3)$, one can introduce a field operator

$$\Psi_\mathbf{A}^* := \int d\mathbf{p}\, \mathbf{A}(\mathbf{p}) \cdot \left[\mathbf{e}_+(\mathbf{p})\, a_+^*(\mathbf{p}) + \mathbf{e}_-^*(\mathbf{p})\, a_-(\mathbf{p})^* \right] \qquad (5.46)$$

This yields a subspace of one-photon field states in the Fock space associated with the functions $\mathbf{A}$:

$$\mid 1, \mathbf{A} \rangle := \Psi_\mathbf{A}^* \mid 0 \rangle \qquad (5.47)$$

It can be shown that these states are eigenvectors of the number operator $N :=$ $\int \mathbf{dp} \sum_\lambda a_\lambda^*(\mathbf{p}) a_\lambda(\mathbf{p})$ corresponding to the eigenvalue 1. The square of the norm of $\mid 1, \mathbf{A} \rangle$ is readily found to be of the form

$$\langle 1, \mathbf{A} \mid 1, \mathbf{A} \rangle = \int \mathbf{dp} \, \overline{\mathbf{A}(\mathbf{p})} \cdot \mathbf{A}(\mathbf{p}) \tag{5.48}$$

so that Eq. (5.42) indeed establishes an isometry between the space of one-photon states and the subspace of transversal vector functions in $\widetilde{\mathcal{H}}$. This subspace $\mathcal{H}_{0,1}$ is known to host a mass-zero, spin-one projective representation of the Poincaré group, thus describing the photon as an elementary system.

Transferring the space-time evolution of the field operators to the states (5.47) (establishing the Schrödinger picture), one finds that the Fourier-transformed 'wave functions' $\widehat{\mathbf{A}}(\mathbf{x}, t)$ satisfy the vacuum wave equation together with the Coulomb gauge condition. In this way the one-photon sector of the Fock space can be identified with the space of vector potential functions of classical electrodynamics.

The problem of the localisability of a photon was discovered when Wigner and Newton [3.26] tried to define Poincaré-relativistic position operators and found that such entities did not exist in the case of massless objects. Later Wightman [3.4] extended the search for a notion of localisation based on a PV measure, and he again obtained a negative result: there exists no Euclidean system of imprimitivity that could be implemented into a mass-0, spin-1 representation. This means that photons are not objects that could be sharply localised. Nevertheless it is possible to construct Euclidean systems of covariance for the photon describing unsharp localisation. Here we shall only give an idea of the difficulties involved and sketch their possible solutions. Details can be found in [3.27].

The naive approach towards defining an Euclidean covariant position operator in $\mathcal{H}_{0,1}$ would be to consider the operators $\widetilde{Q}_k = i \frac{\partial}{\partial p_k}$. While these are self-adjoint operators in $\widetilde{\mathcal{H}}$, their application to $\mathbf{A} \in \mathcal{H}_{01}$ does not lead to vectors in that subspace as $\mathbf{p} \cdot \widetilde{Q}_k \mathbf{A}(\mathbf{p}) = i \frac{\partial}{\partial p_k} (\mathbf{p} \cdot \mathbf{A}(\mathbf{p})) - i A_k(\mathbf{p})$ is a nonzero function whenever $\mathbf{p} \cdot \mathbf{A}(\mathbf{p}) = 0$. Nevertheless Neumark's theorem suggests to define $Q_k := \Pi \widetilde{Q}_k \big|_{\mathcal{H}_{01}}$ which indeed turns out to be a triple of self-adjoint operators on $\mathcal{H}_{01}$ satisfying the Euclidean covariance; however, these operators are mutually noncommuting and cannot therefore be derived from a common spectral measure. This is the price to be paid for the covariance. This fact can be understood if reformulated in terms of a suitable covariant localisation POV measure, to be established next.

The Hilbert space $\widetilde{\mathcal{H}}$ hosts a representation of the Euclidean group,

$$(\widetilde{W}_{\mathbf{a}, R} \mathbf{A})(\mathbf{p}) = e^{-i\mathbf{a} \cdot \mathbf{p}} R \mathbf{A}(R^{-1} \mathbf{p}) \tag{5.49}$$

with respect to which the spectral measure $\widetilde{E}$ of $\widetilde{\mathbf{Q}}$ is covariant. Furthermore, the transversality condition is Euclidean invariant so that the above representation reduces the subspace $\mathcal{H}_{01}$, restriction $W_{\mathbf{a},R}$ of $\widetilde{W}_{\mathbf{a},R}$ being thus a representation on this space. This suffices to verify that the Neumark projection E of $\widetilde{E}$,

$$X \mapsto E(X) := \Pi\widetilde{E}(X)\big|_{\mathcal{H}_{01}} \tag{5.50}$$

constitutes a Euclidean covariant localisation observable. Moreover the operators Q_k arise as the moments of this POV measure and are therefore seen to represent coexistent observables. To appreciate the physical interpretation of this localisation concept, one may proceed further to verify that the operators Q_k are canonically conjugate to the momentum operators P_k (defined as multiplication operators) and satisfy the Heisenberg uncertainty relation. Furthermore, with respect to the Hamiltonian $H = |\mathbf{P}|$, one can define the velocity operator $\mathbf{V} = \mathbf{P}/H$ to show that in the Heisenberg picture $\mathbf{Q}(t) = \mathbf{Q} + \mathbf{V}t$ and finally $|\langle\mathbf{V}\rangle_{\mathbf{A}}| < 1$. Similarly the spreads of Q_k are also found to grow at a velocity less than the velocity of light. Thus the photon 'wave packets' turn out to propagate within the limits demanded by relativistic causality. The fact that they still have infinite 'tails' may not be too disturbing as a naive interpretation of the functions $\widehat{\mathbf{A}}(\mathbf{x})$ as probabilty amplitudes is incorrect. Intuitively the principal unsharpness of relativistic localisation concepts corresponds to the fact that sharp position measurements would involve arbitrarily high energy and therefore the possibility of particle creation. The only way to stick to a one-particle description is by admitting unsharpness.

Finally it may be noted that the above construction does by no means constitute a unique solution of the localisation problem for photons. There are infinitely many different ways of embedding the Hilbert space $\mathcal{H}_{01}$ into larger spaces $\widetilde{\mathcal{H}}$ hosting a Euclidean system of covariance describing a sharp localisation observable. The ensuing Neumark projections will in general lead to localisation observables in $\mathcal{H}_{01}$ different from the above one. A coherent way of formulating localisation observables is based on a manifestly Poincaré covariant phase space representation of relativistic quantum particles [3.28]. In order to decide which of those localisation concepts are physically realisable, one would need to investigate the connections with photodetection theory which shall not be pursued here.

IV. Measurements

The quantum theory of measurement considers measurements as physical processes subject to the laws of quantum physics. The minimal interpretation of quantum mechanics dictates the basic requirement for a process to be a measurement: the probability reproducibility condition. The general measurement theory, as outlined in Section II.3, provides information about the abstract structure of measuring processes and shows, in particular, that every observable of a quantum system can be measured in principle. However, these results form only the starting point for a theory of real measurements, which has to be concerned with concrete observables and the actually available interactions. This chapter goes beyond the abstract theory in three different directions. First it investigates the measurability of continuous observables; then it addresses the question of measuring pairs of observables, and finally the role of the universal conservation laws as constraints on measurement interactions is elucidated. In each case there arises in the first instance a no-go result which is then turned into a new positive feature by reformulating the issue in terms of POV measures.

IV.1 Continuous observables

Two of the most common assumptions on measurements are their *discreteness* and *repeatability*: a pointer observable can assume at most countably many distinct values, and upon repetition the same outcome occurs. These two possible features of measurements, which have both practical and foundational dimensions, are closely related to each other (Section II.3.5). Any observable admitting a repeatable measurement is necessarily discrete. This fact has an important implication on the measurability of *continuous* observables.

M1. *No continuous observable admits a repeatable measurement.*

This result poses a dilemma: many basic observables of a physical system are continuous; but only discrete observables can be measured (in a repeatable way). We shall study in greater detail the conflict between repeatablilty and continuity, its consequences for an operational definition of the localisation observable, and the possible resolutions of this problem.

Let us recall the main arguments indicating that the discreteness assumption is mandatory. First, there is the *pragmatic argument* that any physical experiment is designed to yield definite outcomes out of a collection of alternatives. These outcomes must be described by essentially finite means, either by digital recordings, or by estimating a pointer position in terms of a rational number on an apparently continuous scale. Secondly, the *statistical evaluation* of experiments is based on counting frequencies of countably many mutually exclusive events. Hence a partition of the pointer value space must be fixed in order to define the outcomes to be recorded. Finally the fact that the pointer observable ultimately assumes a *definite value* must be interpreted in terms of a repeatable measurement of the pointer, so that the pointer observable itself, or at least one of its actually used coarse-grained versions, must be discrete.

The repeatability assumption was often argued to be irrelevant since in many realistic experiments the measured system is simply destroyed so that there is no way to repeat the same measurement. Though this situation may occur frequently, it does not always do so. Recent advances in ultrahigh technology have provided the means to perform experiments with individual objects without immediately destroying them. Amazing examples of these new possibilities are the neutron interferometry [4.1], ion traps [2.12], electronic holography [4.2], and the one-atom micro-maser [4.3]. The repeatability assumption is thereby made amenable to experimental testing. On the other hand it was always taken for granted that physical systems can be prepared in well defined states. In fact without such preparatory measurement procedures there were no reproducibility ensuring reliable measurement statistics and hence no physical experience.

The notion of repeatability has also been criticised due to its alleged reference to the 'collapse postulate' (see, e.g., [4.4]). The repeatability condition (II.3.33) presumes that a measurement induces an acausal state transition, $T \mapsto T_X$, into a final state conditional on a pointer reading $f^{-1}(X)$. But there seems to be no way to explain, within quantum mechanics, the factual occurrence of such a reading. Hence the repeatability assumption, and thus its consequence, the discreteness of the measured observables, appears questionable. There are some apparently weaker formulations of the repeatability requirement which seem to avoid this critique. As these alternatives turn out formally equivalent [4.5] to the original repeatability condition in the case of sharp observables, the conclusion (M1) remains inescapable.

M2. *For any measurement $\mathcal{M}$ of a sharp observable E the following conditions are equivalent:*

(1) *$\mathcal{M}$ is value reproducible:*
    for any X and T, if $p_T^E(X) = 1$, then $p_{T_\Omega}^E(X) = 1$;
(2) *$\mathcal{M}$ is of the first kind:*
    for any X and T, $p_T^E(X) = p_{T_\Omega}^E(X)$;
(3) *$\mathcal{M}$ is repeatable:*
    for any X and T, if $p_T^E(X) \neq 0$, then $p_{T_X}^E(X) = 1$.
Any one of these conditions implies that E is a discrete observable.

Repeatable measurements are *preparatory* in the sense that they bring the system into a state in which it has with certainty the registered property. Indeed for any repeatable measurement of an observable E, one has $E(X)T_X = T_X$, or

$$T_X = E(X)T_X E(X) = \frac{1}{p_T^E(X)} E(X) T_\Omega E(X) \qquad (1.1)$$

for all X and T. It is another question whether this state has even the Lüders form,

$$T_X = \frac{1}{p_T^E(X)} E(X)^{1/2} T E(X)^{1/2} \qquad (1.2)$$

While T_X is the final state of the object system conditional on a pointer reading $f^{-1}(X)$, the state $E(X) T_\Omega E(X)/p_T^E(X)$ is the final state of $\mathcal{S}$ under the condition that the measured observable E has a value X [4.6]. In the case of a Lüders measurement this latter conditional state is obtained directly from the initial state of the system. This is, however, a strong assumption since the validity of (1.2) for a given set X and for all states T implies that this set is essentially a one-point set, that is, $E(X) = E(\{\omega\})$ for some 'eigenvalue' $\omega \in \Omega$ [2.14]. As continuous observables have no eigenvalues, there is the no-go result:

M3. *A continuous observable admits no measurement $\mathcal{M}$ such that the associated state transformer contains a Lüders operation.*

This verdict causes considerable difficulties in understanding the operational definition of continuous observables. To illustrate this problem consider a Cartesian component Q of the localisation observable of a spin-0 object. Q is continuous and does not admit any value reproducible, or first kind, or repeatable measurements. In addition, (M3) rules out the most obvious attempt for its operational definition.

To see this, note that the spectral measure $X \mapsto E^Q(X)$ is completely determined by the mapping $I \mapsto E^Q(I)$ on the closed intervals [2.6]. Thus for the definition of Q it suffices to specify measurements of the localisation properties $E^Q(I)$. To this end one could consider the Lüders operation $T \mapsto \mathcal{I}_I(T) := E^Q(I)TE^Q(I)$ associated with the simple observable $\chi_I(Q)$. A prototypical arrangement modelled by this state transformation is a diaphragm with a slit I. The natural question then is whether the mapping $I \mapsto \mathcal{I}_I$ extends to a state transformer of Q, that is, whether by varying the slit location and width in the diaphragm one can define Q. The answer to this question is negative. If there were a state transformer $\mathcal{I}^Q$ such that $\mathcal{I}_I^Q = \mathcal{I}_I$ for all closed intervals I, then due to the additivity one would have

$$E^Q(I) T E^Q(I) = E^Q(I_1) T E^Q(I_1) + E^Q(I_2) T E^Q(I_2) \qquad (1.3)$$

for all states T and for any partition of I into disjoint subintervals I_1 and I_2. But this is false for vector states φ with $p_\varphi^Q(I_1) \neq 0 \neq p_\varphi^Q(I_2)$. The invalidity of (1.3) is well confirmed by the occurrence of interference effects. Result (M3) already implies that there is no Q-compatible state transformer $\mathcal{I}^Q$ for which $\mathcal{I}_I^Q = \mathcal{I}_I$ would hold even for a single interval I.

The fact that position Q admits neither ideal nor repeatable measurements causes problems in understanding localisation as a possible property of a particle. Hence it appears that the definition of a particle by means of a localisation system of covariance lacks an operational foundation in quantum mechanics.

The usual way out of the continuity-repeatability conflict consists in taking recourse to discretising the continuous observables. This method was already suggested by von Neumann [2.3], and it meets the pragmatic and statistical needs of quantum mechanics. Considering again the observable Q, any partition (I_i) of $\mathbf{R}$ (a reading scale) induces a coarse-grained, discrete position observable $i \mapsto E^Q(I_i)$ which clearly admits a Lüders measurement. Moreover the whole position measurement statistics, $p_T^Q(X), X \in \mathcal{B}(\mathbf{R})$, can for every state T be recovered from the statistics of these coarse-grained observables, when varying over all possible reading scales. The continuity of Q is now reflected in the fact that measuring this observable requires measuring more than countably many of its discretised versions. Another serious drawback of this procedure is that it destroys the natural covariance of the observable in question. For the localisation observable it means nothing less than abolishing its Euclidean covariance. Other approaches have therefore been investigated, among them those aiming at a relaxation of repeatability such as to make it compatible with the covariance requirements. Those attempts still must take into account the pragmatic and statistical needs that ultimately require a discretisation.

Two natural ways of weakening repeatability into 'approximate repeatability' are given with the notions of δ-repeatable and ε-preparatory measurements which merge into the concept of (ε, δ)-repeatability. We formulate these notions for position measurements. For any $\delta > 0$, define the δ-neighbourhood of a set X as $X_\delta := \{x \in \mathbf{R} : |x - x'| \le \delta \text{ for some } x' \in X\}$. Let ε be a number such that $0 \le \varepsilon < \frac{1}{2}$. A measurement of the position observable Q is

δ-*repeatable* if for any X and T,

$$p_T^Q(X) \ne 0 \implies p_{T_X}^Q(X_\delta) = 1 \tag{1.4}$$

ε-*preparatory* if for any X and T,

$$p_T^Q(X) \ne 0 \implies p_{T_X}^Q(X) \ge 1 - \varepsilon \tag{1.5}$$

(ε, δ)-*repeatable* if for any X and T,

$$p_T^Q(X) \ne 0 \implies p_{T_X}^Q(X_\delta) \ge 1 - \varepsilon \tag{1.6}$$

Consider the state transformer of Equation (II.3.7). Fix a $\delta > 0$, and choose a partition (I_i) of $\mathbf{R}$ into intervals of lengths $|I_i| < \delta$. For every i, let T_i be a state that is localised in I_i, $\text{tr}\big[T_i E^Q(I_i)\big] = 1$. Then for any X, $\text{tr}\big[T_i E^Q(X_\delta)\big] = 1$

whenever $I_i \cap X \neq \emptyset$. This observation allows one to confirm that the Q-compatible (completely positive) state transformer

$$\mathcal{I}_X^Q(T) := \sum \mathrm{tr}\big[TE^Q(X \cap I_i)\big] T_i = \sum p_T^Q(X \cap I_i) T_i \qquad (1.7)$$

is δ-repeatable. This state transformer is repeatable with respect to the discrete observable $I_i \mapsto E^Q(I_i)$. Although $E^Q(I_i)T_i = T_i$, one cannot have, in general, $T_i = E^Q(I_i)TE^Q(I_i)/p_T^Q(I_i)$. Apart from its somewhat artificial nature, a deficiency of this state transformer may be seen in its lacking translation invariance; that is, it does not satisfy the covariance condition

$$U_a \left[\mathcal{I}_X^Q (U_a^{-1} T U_a) \right] U_a^{-1} = \mathcal{I}_{X+a}^Q(T) \qquad (1.8)$$

with respect to space translations. In order to obtain a translation invariant (completely positive) Q-compatible state transformer, take a state T^0 and denote $T_q^0 := U_q T^0 U_q^{-1}$. It is easy to confirm that the following state transformer has the required properties:

$$\mathcal{I}_X^Q(T) := \int_X \mathrm{tr}\big[TE^Q(dq)\big] T_q^0 = \int_X dp_T^Q(q) T_q^0 \qquad (1.9)$$

That any such state transformer is of this form is shown in [3.9]. Fix $\delta > 0$ and choose T^0 to be localised within $[-\delta, \delta]$. For any set X one has $\mathrm{tr}\big[T_q^0 E^Q(X_\delta)\big] = 1$ whenever $q \in X$, and this is easily seen to ensure the δ-repeatability.

Like (1.7), the state transformer (1.9) may also appear artificial. The final state of the object depends only on its position distribution before the measurement and on the fixed state T^0. For any initial state T the postmeasurement state is a mixture of one and the same set of component states T_q^0. One may therefore question the feasibility of such a measurement.

It is instructive to consider once more the standard position measurement model of Section II.3.4. The object's position Q is correlated with the position Q_A of the apparatus via the interaction

$$U = e^{-i\lambda Q \otimes P_A} \qquad (1.10)$$

where P_A is the apparatus momentum. The measured observable is the unsharp position E^e, with the confidence function e depending on the initial state ϕ of the apparatus and the coupling constant λ. The state transformer induced by this measurement is

$$\mathcal{I}_X^Q(T) = \int_X K_q T K_q^* \, dq, \qquad K_q = \sqrt{\lambda}\,\phi\big(-\lambda(Q - q)\big) \qquad (1.11)$$

and it was found to be of the first kind: the measurement preserves the outcome statistics of E^e. This state transformer is also manifestly translation covariant. But it is neither repeatable, nor δ-repeatable, nor ε-preparatory, in general. However, choosing ϕ such that e is localised within some interval I, then δ-repeatability holds for any δ for which $I \subset (-\delta, \delta)$.

There are basically two ways to discretise this measurement model. First one may consider, instead of Q, a coarse-grained discretised position, like $f(Q) = \sum f(i) E^Q(I_i)$ associated with a partition (I_i) of $\mathbf{R}$. Adjusting the coupling accordingly, $U = \exp\left(-i\lambda f(Q) \otimes P_A\right)$, one can choose the initial state of the apparatus such that the measurement is a repeatable measurement of $f(Q)$, cf. Section II.3.4.

An operationally more satisfactory discretisation is the one in which the pointer values are registered with respect to a discrete reading scale. Introducing a partition (I_i), $I_i = [i, i+1)$, the measured observable is the discrete unsharp position $i \mapsto E_i = E^e(I_i)$. In general none of the effects E_i has eigenvalue 1. And even if they have, there is still no way to choose ϕ such that the measurement were repeatable. As an example, if ϕ is concentrated on the interval $[0, \lambda)$, say $\phi = \frac{1}{\sqrt{\lambda}} \chi_{[0,\lambda)}$, then

$$E_i \;=\; (Q - i + 1)\, E^Q\left(I_{i-1}\right) + (i + 1 - Q)\, E^Q\left(I_i\right) \tag{1.12}$$

and

$$K_q \;=\; E^Q\left((q - 1, q]\right) \tag{1.13}$$

so that

$$\mathcal{I}_{I_i}^Q(T) \;=\; \int_i^{i+1} E^Q\left((q - 1, q]\right) T\, E^Q\left((q - 1, q]\right) dq \tag{1.14}$$

Though the measurement is not repeatable, it is, for instance, 1-repeatable:

$$\mathrm{tr}\left[E\left(I_{i,1}\right) \mathcal{I}_{I_i}^Q(T)\right] \;=\; \mathrm{tr}\left[\mathcal{I}^Q(I_i)(T)\right] \tag{1.15}$$

This follows from the observation that $I_{i,1} = I_{i-1} \cup I_i \cup I_{i+1}$ and $E_i E_{i,1} = E_i$, so that the left hand side of (1.15) becomes $\mathrm{tr}\left[T E\left(I_{i,1}\right) E\left(I_i\right)\right] = \mathrm{tr}\left[T E_i\right]$.

Finally we investigate the repeatability question for a phase space observable G (III.2.42). For simplicity assume that the generating operator is a projection $P[\xi]$, and denote $T_{qp}^\xi = P[W_{qp}\xi]$. Then the following is a (completely positive) covariant, G-compatible state transformer:

$$\mathcal{I}_Z^G(T) \;=\; \frac{1}{2\pi} \int_Z T_{qp}^\xi\, T\, T_{qp}^\xi\, dq dp \;=\; \frac{1}{2\pi} \int_Z \mathrm{tr}\left[T_{qp}^\xi\, T\right] T_{qp}^\xi\, dq dp \tag{1.16}$$

Application of the Paley-Wiener theorem shows that this phase space state transformer cannot be δ-repeatable. Neither can it be ε-preparatory. However, for given ε one can always choose a δ such that this state transformer is (ε, δ)-repeatable [4.7]. In Sections VI.2 and VII.3.7 the state transformer (1.16) will be recovered from realistic models of joint position-momentum measurements.

IV.2 Pairs of observables

Some observables of a physical system, such as the energy of a hydrogen atom and one of its angular momentum components, can be measured together, while others cannot. Examples of such noncoexistent observables are any two angular momentum components, or pairs of canonically conjugate observables. The non-coexistence of pairs of observables and its possible relaxation are central issues of quantum mechanics and its interpretation. Recent experimental advances have provided the means to perform joint measurements of complementary properties of individual objects, like neutrons or photons. It is therefore rewarding to see how these new possibilities are anticipated in the notion of coexistence, and how they can be explained with unsharp joint observables.

IV.2.1 Coexistent observables. The notion of coexistence was introduced as a probabilistic expression of the joint measurability of two or more observables (cf. Section II.2.2). We shall now elaborate on the measurement-theoretical content of this concept.

In order to measure jointly two observables E_1 and E_2 of a system $\mathcal{S}$, there must be one single measurement $\mathcal{M}$ which allows one to collect the measurement outcome statistics of both observables. In the language of the quantum theory of measurement this means that there is an apparatus $\mathcal{A}$, initially in a state $T_\mathcal{A}$, a pointer observable $P_\mathcal{A}$, and a measurement coupling V such that for any initial state T of $\mathcal{S}$,

$$
\begin{aligned}
p_T^{E_1}(X) &= p_{\mathcal{R}_\mathcal{A}(V(T \otimes T_\mathcal{A}))}^{P_\mathcal{A}}\big(f_1^{-1}(X)\big) && \text{for all } X \in \mathcal{F}_1 \\
p_T^{E_2}(Y) &= p_{\mathcal{R}_\mathcal{A}(V(T \otimes T_\mathcal{A}))}^{P_\mathcal{A}}\big(f_2^{-1}(Y)\big) && \text{for all } Y \in \mathcal{F}_2
\end{aligned}
\tag{2.1}
$$

where $f_1 : \Omega_\mathcal{A} \to \Omega_1$ and $f_2 : \Omega_\mathcal{A} \to \Omega_2$ are suitable pointer functions relating the pointer readings to the values of E_1 and E_2, respectively. If E is the observable defined by $\mathcal{M}$ then one has

$$
E_1 = E \circ f_1^{-1}, \qquad E_2 = E \circ f_2^{-1}
\tag{2.2}
$$

This shows that E_1 and E_2 are coexistent, and they are in fact coarse-grained versions of E. Conversely if two observables E_1 and E_2 are coexistent, with a joint observable E, then any measurement of E is a joint measurement of E_1 and E_2 in the sense just described.

There is another intuitive idea of 'joint measurability' which refers to the possibility of performing order independent sequential measurements. It was shown in Section II.3.2 that whenever for some E_1- and E_2-measurements the composite state transformers $\mathcal{I}_{12}$ and $\mathcal{I}_{21}$ are equivalent, then the observables are coexistent. On the other hand it appears that the existence of order-independent sequential measurements is not guaranteed by coexistence. It may be noted that the two notions of joint measurability coincide when E_1 and E_2 are discrete sharp observables.

In that case the respective Lüders state transformers [given in (II.3.10)] are indeed commutative: $\mathcal{I}_L^1 \circ \mathcal{I}_L^2 = \mathcal{I}_L^2 \circ \mathcal{I}_L^1$.

Finally we note that both formalisations of joint measurability can be relaxed so as to apply only with respect to a certain subset of possible initial preparations, or to some coarse-grained versions of the observables in question.

IV.2.2 Examples. Let us consider two coexistent sharp observables, represented by the commuting self-adjoint operators $A = \sum_i a_i P_i$ and $B = \sum_j b_j R_j$. A joint observable for A and B is given by the PV measure $(i,j) \mapsto P_i R_j$. The associated Lüders measurement, $\mathcal{I}_{L,Z}(T) = \sum_{(i,j) \in Z} P_i R_j T P_i R_j$, is a joint measurement of A and B in the sense of Eq. (2.1), and it satisfies $\mathcal{I}_L^A \circ \mathcal{I}_L^B = \mathcal{I}_L^B \circ \mathcal{I}_L^A = \mathcal{I}_L$.

That the order independence of sequential measurements of a pair of observables A and B is rather exceptional can be illustrated with reference to state transformers of the form (II.3.7). The resulting sequential state transformers are not order independent and do not define a joint observable. On the contrary they define smeared versions of A and B. For instance, choosing a measurement of A by fixing $X_i = \{a_i\}$ in (II.3.7) and an arbitrary B-measurement, one finds the observable associated with $\mathcal{I}^B \circ \mathcal{I}^A$ to be $X \times Y \mapsto (\mathcal{I}_X^A)^* \left(E^B(Y) \right) = \sum_{a_i \in X} p_{T_i}^B(Y) P_i$. Clearly this differs from the joint observable associated with the joint Lüders measurement.

The canonically conjugate position Q and momentum P of an elementary system are the most prominent pair of noncommuting observables. Their strong non-coexistence derives from the fact that Q and P are complementary and has been often related to the uncertainty relation. In order to clarify the connections between these features of position and momentum we need to study the coexistence properties of these observables.

The strong noncommutativity of position and momentum is expressed in the fundamental exchange relation

$$QP - PQ = iI \tag{2.3}$$

This relation has induced a manifold of investigations, starting with Werner Heisenberg's historic papers of 1925 and 1927 and culminating in a recent monograph by Sakai [4.8]. Furthermore position and momentum are totally noncommutative in the sense that their commutativity domain is the null space,

$$\mathrm{com}(Q,P) = \{0\} \tag{2.4}$$

This is equivalent to the fact that Q and P have a joint probability in no state, that is, the mapping

$$X \times Y \mapsto \mathrm{tr}\left[T E^Q(X) \wedge F^P(Y) \right] \tag{2.5}$$

does not extend to a probability measure on $\mathbf{R}^2$ for any state T. Hence these observables can be measured together in no state, and all of their sequential measurements are order dependent. In spite of their extreme noncoexistence, Q and P do have coexistent coarse-grainings, and some of the smeared versions are even informationally equivalent to the sharp quantities.

Let f and g be any two bounded functions on the real line $\mathbf{R}$. The bounded self-adjoint operators $f(Q)$ and $g(P)$ are coarse-grained versions of the sharp position and momentum observables. Such observables may be coexistent. Indeed

$$f(Q)g(P) - g(P)f(Q) = O \tag{2.6}$$

whenever f and g are both periodic functions with minimal periods α, β satisfying $2\pi/\alpha\beta \in \mathbf{Z} \setminus \{0\}$ [4.9]. Hence such $f(Q)$ and $g(P)$ can be measured together. The physical relevance of this result is well-known in crystallography. The electrons in a crystal can be confined arbitrarily close to the atom sites and yet their momenta may be well concentrated at the reciprocal lattice points. The position and momentum of an electron are thereby determined simultaneously modulo the crystal periodicity.

In the previous section we investigated various state transformers associated with position measurements with the aim to achieve some weakened repeatability. These state transformers can be applied to provide examples of sequential position-momentum measurements. Consider first a Q-compatible state transformer of the form (1.7). When combined sequentially with any momentum measurement, it leads to the following sequential joint observable:

$$X \times Y \mapsto (\mathcal{I}_X^Q)^*(F^P(Y)) = \sum p_{T_i}^P(Y)\,E^Q(X \cap I_i) \tag{2.7}$$

If one performs first a momentum measurement of similar type, and then any position measurement, the resulting sequential joint observable is

$$Y \times X \mapsto (\mathcal{I}_Y^P)^*(E^Q(X)) = \sum p_{T_j}^Q(X)\,F^P(Y \cap I_j) \tag{2.8}$$

The marginal observables are $X \mapsto E^Q(X)$ and $Y \mapsto \sum p_{T_i}^P(Y)E^Q(I_i)$ in the first case, and $Y \mapsto F^P(Y)$ and $X \mapsto \sum p_{T_j}^Q(X)F^P(I_j)$ in the second case. Evidently these two sequential measurements are not equivalent so that the sequential joint observables are no joint observables in the sense of coexistence. Position and momentum observables are not coexistent as long as one of them is a sharp quantity.

A more interesting and more realistic case arises from the measurement (1.11) of the unsharp position E^e. The sequential joint observable obtained when this measurement is followed by any (sharp) momentum measurement is

$$X \times Y \mapsto (\mathcal{I}_X^Q)^*(F^P(Y)) = \int_X K_q^*\,F^P(Y)\,K_q\,dq \tag{2.9}$$

The marginals of this sequential joint observable are

$$X \mapsto \int_X K_q^* K_q = E^e(X) \tag{2.10}$$

with $e(q) = \lambda |\phi(-\lambda q)|^2$, and

$$Y \mapsto \int_{\mathbf{R}} K_q^* F^P(Y) K_q \, dq = F^f(Y) \tag{2.11}$$

with $f(p) = \frac{1}{\lambda} |\hat{\phi}(-\frac{p}{\lambda})|^2$. This result exhibits some important features. First of all it shows that unsharp position E^e and unsharp momentum F^f, with their Fourier related confidence functions e and f, are coexistent. Any measurement of the observable (2.9) is a joint measurement of these observables, meaning that it yields the same statistics as the above sequential unsharp position-sharp momentum, or unsharp momentum-sharp position measurement. It can be shown that the observable (2.9) is actually a phase space observable G of the form of Eq. (III.2.42), with $P[\phi]$ as the generating density operator.

IV.2.3 Complementary observables. Like the concept of coexistence, the notion of complementary observables has both probabilistic and measurement theoretical aspects. The 'mutual exclusiveness' associated with the term complementarity refers to the possibilities of predicting measurement outcomes, as well as to the value determinations. Both of these aspects were discussed already by Niels Bohr and Wolfgang Pauli. We shall review below two formalisations, referred to as (measurement-theoretical) complementarity and probabilistic complementarity. The former implies noncoexistence. In the case of sharp observables the two formulations turn out equivalent. However, for unsharp observables the measurement-theoretical notion of complementarity is stronger than the probabilistic one. This fact opens up the possibility that probabilistically complementary observables can be coexistent, and it offers an explanation for the simultaneous measurability of complementary observables, like position and momentum, which has recently become a subject of experimental investigations (Section VII.4).

The predictions of measurement outcomes for two observables are mutually exclusive if probability one for some outcome of one observable entails that none of the outcomes of the other one can be predicted with certainty. Observables E_1 and E_2 are called *probabilistically complementary* if they share the following property:

$$\begin{aligned}
p_T^{E_1}(X) = 1 &\implies 0 < p_T^{E_2}(Y) < 1 \\
p_T^{E_2}(Y) = 1 &\implies 0 < p_T^{E_1}(X) < 1
\end{aligned} \tag{2.12}$$

for any state T and all bounded sets $X \in \mathcal{F}_1$, $Y \in \mathcal{F}_2$ with $E_1(X) \neq I \neq E_2(Y)$.

Assume that for observables E_1 and E_2 the probabilities $p_T^{E_1}(X)$ and $p_T^{E_2}(Y)$ both equal one for some state T and some sets X and Y. This entails $E_1(X)\varphi = \varphi$ and $E_2(Y)\varphi = \varphi$ for some unit vector φ, that is, $P[\varphi] \leq E_1(X)$ and $P[\varphi] \leq E_2(Y)$. In that case $E_1(X)$ and $E_2(Y)$ have a nonzero lower bound. If these effects are projection operators, then $E_1(X)\varphi = \varphi$ and $E_2(Y)\varphi = \varphi$ holds exactly when $\varphi \in E_1(X)(\mathcal{H}) \cap E_2(Y)(\mathcal{H})$. Treating similarly the two other cases excluded by (2.12), one observes, first of all, that the probabilistic complementarity of a given pair of sharp observables is equivalent to the disjointness of their spectral projections:

$$
\begin{aligned}
E_1(X) \wedge E_2(Y) &= O \\
E_1(X) \wedge E_2(\Omega_2 \setminus Y) &= O \\
E_1(\Omega_1 \setminus X) \wedge E_2(Y) &= O
\end{aligned}
\tag{2.13}
$$

for all bounded sets X, Y with $O \neq E_1(X) \neq I$, $O \neq E_2(Y) \neq I$. The above considerations also show that for unsharp observables condition (2.13) always implies (2.12), but need not be implied by that. We take (2.13) as the formal definition of the *complementarity* of observables E_1 and E_2. Thus any two complementary observables are also probabilistically complementary, but not necessarily vice versa.

The prototypical pair of complementary observables are the canonically conjugate sharp position and momentum observables Q and P. On the other hand, among the unsharp position and momentum pairs, which are all probabilistically complementary, there are coexistent pairs, thus breaking the complementarity. We return to this example subsequently.

We mention in passing that in the literature one finds yet another version of complementarity which we shall term *value complementarity* [4.10]. This refers to the case when certain predictability of some value of E_1 implies that all values of E_2 are equally likely. This concept is not rigorously applicable in the case of continuous observables as these have no proper eigenstates. Nevertheless it is applied also to position and momentum in the intuitive sense that, e.g., a sharp momentum (plane wave) state goes along with a uniform position 'distribution'. Examples of value complementary observables are orthogonal spin components s_x, s_y, or the canonically conjugate spin and spin phase (cf. Section III.3.4), or the number and phase observable (cf. Section III.5.1). In the last two cases it is clear that a sharp spin or number value entails the uniform distribution for the (bounded continuous) phase observable. Conversely, however, the phase states are nonorthogonal in both cases and improper in the latter so that there is no sharp phase value in any state. Like (measurement theoretical) complementarity, value complementarity implies probabilistic complementarity. But value complementarity does not compare to complementarity. Indeed any pair of spin components is complementary but not value complementary unless the pair is orthogonal; and the spin and spin phase pair E^{s_z}, S is value complementary but not complementary: for any eigenstate φ_m of s_z there is a positive number $\lambda < 1$ such that $\lambda P[\varphi_m]$ is a lower bound to $P[\varphi_m]$

as well as to any (nonzero) $S(X)$. This follows from the fact that $\lambda P[\varphi]$ is a lower bound of an effect a for some positive λ exactly when φ is in the range of $\sqrt{a}$; in the present case of $a = S(X)$ this range is the whole Hilbert space since for any state $0 < \langle \varphi \,|\, S(X)\varphi \rangle < 1$.

Consider a pair of observables E_1 and E_2 which are probabilistically complementary but coexistent. Let E be a joint observable, so that for any $X \in \mathcal{F}_1$, $E_1(X) = E(Z_X)$ for some $Z_X \in \mathcal{F}$, and for any $Y \in \mathcal{F}_2$, $E_2(Y) = E(Z_Y)$ for some $Z_Y \in \mathcal{F}$. Let $\mathcal{I}$ be any E-compatible state transformer, representing thus a joint measurement of E_1 and E_2. Assume that this measurement is repeatable. Let X and Y be any two bounded sets for which $O \neq E_1(X) \neq I$ and $O \neq E_2(Y) \neq I$, and let T be a state for which $p_T^{E_2}(Y) \neq 0$. If $T_Y := \mathcal{I}_{Z_Y}(T)/p_T^{E_2}(Y)$, we have $p_{T_Y}^{E_2}(Y) = 1$, and

$$p_{T_Y}^{E_1}(X) \;=\; \mathrm{tr}\big[\mathcal{I}_{Z_X}(T_Y)\big] \;=\; p_T^{E_2}(Y)^{-1}\,\mathrm{tr}\big[\mathcal{I}_{Z_X \cap Z_Y}(T)\big] \tag{2.14}$$

If $\mathrm{tr}\big[\mathcal{I}_{Z_X \cap Z_Y}(T)\big] = 0$, then $p_{T_Y}^{E_2}(Y) = 1$ and $p_{T_Y}^{E_1}(X) = 0$, which is excluded by the second line of Eq. (2.12). On the other hand if $\mathrm{tr}\big[\mathcal{I}_{Z_X \cap Z_Y}(T)\big] \neq 0$, then in the state $T' := \mathcal{I}_{Z_X \cap Z_Y}(T)/p_T^E(Z_X \cap Z_Y)$ one has $p_{T'}^{E_1}(X) = 1$ and $p_{T'}^{E_2}(Y) = 1$, which is again excluded by (2.12). It follows that $\mathcal{I}$ cannot be repeatable, and we have the following result.

M4. *Probabilistically complementary observables do not admit any repeatable joint measurements.*

Consider now a pair of complementary observables E_1 and E_2. Conditions (2.13) then imply that these observables cannot be coexistent. Indeed assuming they were coexistent, let E be a joint observable. By the additivity of measures, and with the above notations, $E(Z_X \cup Z_Y) + E(Z_X \cap Z_Y) = E(Z_Y) + E(Z_X) = E_1(X) + E_2(Y)$, for all X and Y. If X and Y are bounded sets for which $O \neq E_1(X) \neq I$, and $O \neq E_2(Y) \neq I$, then, by the first line of equation (2.13), $E(Z_X \cap Z_Y) = O$. Therefore $E_1(X) \leq E(Z_X \cup Z_Y) - E_2(Y) \leq I - E_2(Y)$, which contradicts the second line of (2.13). Hence we have established the following.

M5. *Complementary observables do not admit any joint measurements. Neither do they have any order independent sequential measurements.*

Let $\mathcal{I}_1$ and $\mathcal{I}_2$ be state transformers compatible with E_1 and E_2, respectively. Assume that for some sets X and Y there is an operation Φ such that $\Phi \leq \mathcal{I}_{1,X}$ and $\Phi \leq \mathcal{I}_{2,Y}$. Such an operation is called a *test* of the effects $E_1(X)$ and $E_2(Y)$. It would allow one to construct a measurement (state transformer) which provides some probabilistic information on both $E_1(X)$ and $E_2(Y)$. We say that two state transformers are *mutually exclusive* if no such operation exists, that is, if

$$\mathcal{I}_{1,X} \wedge \mathcal{I}_{2,Y} \;=\; 0$$
$$\mathcal{I}_{1,X} \wedge \mathcal{I}_{2,\Omega_2 \setminus Y} \;=\; 0 \tag{2.15}$$
$$\mathcal{I}_{1,\Omega_1 \setminus X} \wedge \mathcal{I}_{2,Y} \;=\; 0$$

for all bounded sets X, Y for which neither $\mathcal{I}_{1,X}$ nor $\mathcal{I}_{2,Y}$ is a maximal operation. The following is then a measurement theoretical characterisation of the complementarity.

M6. *Two observables are complementary if and only if any of their associated state transformers are mutually exclusive.*

This result sharpens (M5):

M7. *Noncoexistent observables do not admit any joint measurements.*
Complementary observables do not admit any joint tests.

IV.2.4 Coupling properties of position and momentum. The canonically conjugate position and momentum observables Q and P, represented as a Schrödinger couple, are Fourier equivalent physical quantities [Eq. (III.2.29)]. This fundamental connection, which results from the Galilei covariance of the localisation observable, is the root of the many important coupling properties known for these observables. Some of them were already discussed in the previous sections and in Chapter III. In view of their relevance to the question of the joint measurability of position and momentum, we collect all these results here.

The basic coupling properties are, of course, the *commutation relation*

$$QP - PQ = iI \tag{2.16}$$

which holds on a dense domain, and the *uncertainty relation*

$$\Delta(Q, \varphi)\Delta(P, \varphi) \geq \tfrac{1}{2} \tag{2.17}$$

which holds for all unit vectors φ.

The Fourier relation $P = U_F^{-1} Q U_F$ extends also to the spectral measures, so that one has $F^P(Y) = U_F^{-1} E^Q(Y) U_F$ for all $Y \in \mathcal{B}(\mathbf{R})$. Thus if $E^Q(X)\varphi = \varphi$ and $F^P(Y)\varphi = \varphi$ for some vector φ, then, using the Schrödinger representation, the function φ vanishes (almost everywhere) in $\mathbf{R} \setminus X$ and its Fourier transform $\hat\varphi$ vanishes (almost everywhere) in $\mathbf{R} \setminus Y$. According to the identity theorem for analytic functions, the Fourier transform of a compactly supported function cannot vanish on any subset of $\mathbf{R}$ of positive measure unless it vanishes identically. Therefore the following relations for the spectral projections of a Schrödinger couple (Q,P) are obtained:

$$E^Q(X) \wedge F^P(Y) = O$$
$$E^Q(X) \wedge F^P(\mathbf{R} \setminus Y) = O \tag{2.18}$$
$$E^Q(\mathbf{R} \setminus X) \wedge F^P(Y) = O$$

for all bounded $X, Y \in \mathcal{B}(\mathbf{R})$. These relations show the *complementarity* of Q and P both in the sense of (2.12) and (2.13). It may be of interest to note that $E^Q(\mathbf{R} \setminus X) \wedge F^P(\mathbf{R} \setminus Y) \neq O$, for all bounded X and Y [4.11].

The uncertainty relations (2.17) as well as the complementarity (2.18) of Q and P imply their *total noncommutativity*:

$$\text{com}(Q,P) = \bigcap_{X,Y\in\mathcal{B}(\mathbf{R})} \text{com}\big(E^Q(X),F^P(Y)\big) = \{0\} \tag{2.19}$$

This is obvious from (2.18) but it follows also from (2.17) [4.12]. In spite of their strong noncommutativity, Q and P do have *commuting spectral projections*:

$$E^Q(X)\,F^P(Y) = F^P(Y)\,E^Q(X) \tag{2.20}$$

whenever X and Y are periodic (Borel) sets of type $X = X + 2\pi/a$ and $Y = Y + a$ [4.13]. This is another way of expressing the fact that Q and P admit coexistent coarse-grainings $f(Q)$ and $g(P)$ for periodic functions f and g, stated in (2.6).

IV.2.5 The Heisenberg interpretation of the uncertainty relations. Position and momentum are complementary observables and cannot be measured together. Nevertheless Werner Heisenberg (1927) maintained that these observables can be simultaneously determined provided that the measuring accuracies are in accordance with the uncertainty relations. This interpretation was often criticised on the grounds that the uncertainty relation is, in the first instance, a statistical scatter relation, showing only that in any state the measurement statistics of position and momentum are correlated according to (2.17). On the other hand the introduction of phase space observables (Section III.2.4) provides the basis for a conceptually sound formulation of Heisenberg's interpretation, which will in fact obtain a detailed operational justification in terms of a position-momentum measurement model (Section VI.2). Formally the key in solving the dilemma of measuring together complementary observables lies in the fact that the strict complementarity (2.13) can be broken by introducing a sufficient degree of measurement inaccuracy. The resulting unsharp observables may be coexistent. For position and momentum this is the case when the measuring inaccuracies satisfy the Heisenberg inequality. While such unsharp observables are no more complementary and can be measured together, they always remain probabilistically complementary so that their predictions are mutually exclusive.

A phase space observable (III.2.42) has the unsharp position E^e and the unsharp momentum F^f as marginal observables, which are thereby coexistent, so that the Fourier related confidence functions e and f satisfy the inequality

$$\Delta(e) \cdot \Delta(f) \geq \tfrac{1}{2} \tag{2.21}$$

Being coexistent observables, E^e and F^f cannot be complementary. This can also be seen directly since for any sets X and Y

$$\begin{aligned}
G_\xi(X \times Y) &\leq G_\xi(X \times \mathbf{R}) \leq E^e(X) \\
G_\xi(X \times Y) &\leq G_\xi(\mathbf{R} \times Y) \leq F^f(Y)
\end{aligned} \tag{2.22}$$

Assume that for some bounded sets X and Y the probabilities $\operatorname{tr}\left[TE^e(X)\right]$ and $\operatorname{tr}\left[TF^f(Y)\right]$ were both equal to one for some state T. This would imply that both $T(q,q)$ and $\widehat{T}(p,p)$ have bounded supports, which is excluded. Hence E^e and F^f are probabilistically complementary observables, whether coexistent or not.

IV.2.6 Coexistence of complementary spin observables. The unsharp spin (component) observables of a spin-$\frac{1}{2}$ system introduced in Section III.3.3 are generated by effects of the form (III.3.42),

$$F(\pm\mathbf{a}) \;=\; \tfrac{1}{2}\left(I \pm \mathbf{a}\cdot\sigma\right), \qquad \|\mathbf{a}\| \le 1 \tag{2.23}$$

Such observables arise as smeared versions of sharp spin observables. The degree of smearing is characterised by the parameter $1 - \lambda$ (with $\lambda = \|\mathbf{a}\|$) appearing in the nondiagonal elements of the stochastic matrix (III.3.43). The same parameter determines the degree of unsharpness $\varepsilon := 1 - \tfrac{1}{2}(1+\lambda) = \tfrac{1}{2}(1-\lambda)$, where the larger eigenvalue of $F(\mathbf{a})$, $\tfrac{1}{2}(1+\lambda)$, can be interpreted as the maximal available degree of reality of that effect.

Pairs of unsharp observables of the form (2.23) are probabilistically complementary but in general not complementary. We shall derive a simple geometric criterion for the coexistence of such observables by evaluating the coexistence conditions for pairs of effects (Section II.2.2). To ensure the coexistence of the effects $F(\mathbf{a}_1)$ and $F(\mathbf{a}_2)$, one needs to find an effect $G = \gamma F(\mathbf{c})$ such that

$$O \le \gamma F(\mathbf{c}), \qquad \gamma F(\mathbf{c}) \le F(\mathbf{a}_1), \qquad \gamma F(\mathbf{c}) \le F(\mathbf{a}_2)$$
$$F(\mathbf{a}_1) + F(\mathbf{a}_2) - \gamma F(\mathbf{c}) \le I \tag{2.24}$$

Taking into account that $\alpha F(\mathbf{a}) \le \beta F(\mathbf{b})$ is equivalent to $\|\beta\mathbf{b} - \alpha\mathbf{a}\| \le \beta - \alpha$, this system of inequalities is equivalent to the following one:

$$\|\gamma\mathbf{c}\| \le \gamma, \qquad \|\mathbf{a}_1 - \gamma\mathbf{c}\| \le 1 - \gamma, \qquad \|\mathbf{a}_2 - \gamma\mathbf{c}\| \le 1 - \gamma$$
$$\|\mathbf{a}_1 + \mathbf{a}_2 - \gamma\mathbf{c}\| \le \gamma \tag{2.25}$$

Let us denote by $S(\mathbf{a}, r)$ the closed ball with radius r and center point $\mathbf{a}$. Then (2.25) can be rewritten as follows:

$$\gamma\mathbf{c} \in S(\mathbf{a}_1, 1-\gamma) \cap S(\mathbf{a}_2, 1-\gamma) \cap S(\mathbf{a}_1 + \mathbf{a}_2, \gamma) \cap S(\mathbf{o}, \gamma) \tag{2.26}$$

The intersection of the first two balls is nonempty exactly when $\gamma \le 1 - \tfrac{1}{2}\|\mathbf{a}_1 - \mathbf{a}_2\|$, while the intersection of the last two balls is nonempty if and only if $\gamma \ge \tfrac{1}{2}\|\mathbf{a}_1 + \mathbf{a}_2\|$. Thus the coexistence of $F(\mathbf{a}_1)$ and $F(\mathbf{a}_2)$ implies the inequality $\tfrac{1}{2}\|\mathbf{a}_1 + \mathbf{a}_2\| \le 1 - \tfrac{1}{2}\|\mathbf{a}_1 - \mathbf{a}_2\|$. Conversely the validity of this relation entails the existence of some γ satisfying the preceding two inequalities. In turn these ensure that $\mathbf{c}_o := \tfrac{1}{2}(\mathbf{a}_1 + \mathbf{a}_2)$ is in the intersection of all four balls. Taking for γ the norm of this vector and defining $\mathbf{c} := \mathbf{c}_o / \|\mathbf{c}_o\|$, one has satisfied (2.26). We have thus established the following result.

M8. *The unsharp spin-$\frac{1}{2}$ properties $F(\mathbf{a}_1)$ and $F(\mathbf{a}_2)$ are coexistent if and only if*

$$\|\mathbf{a}_1 + \mathbf{a}_2\| + \|\mathbf{a}_1 - \mathbf{a}_2\| \leq 2 \tag{2.27}$$

If $\|\mathbf{a}_1\| = 1$, say, so that $F(\mathbf{a}_1)$ is a projection, then (2.27) is fulfilled exactly when $\mathbf{a}_2 = \pm \|\mathbf{a}_2\| \, \mathbf{a}_1$, that is, $F(\mathbf{a}_2)$ commutes with $F(\mathbf{a}_1)$. This confirms the general result that the coexistence of two observables amounts to their commutativity if one of them is a sharp observable. For the coexistence of two noncommuting unsharp observables a sufficient degree of unsharpness is required so that the value of $\|\mathbf{a}_i\|$, which determines the maximal degree of reality, must not be too close to unity.

A joint observable for a pair of coexistent effects $F(\mathbf{a}_1)$, $F(\mathbf{a}_2)$ can be easily constructed. For example, the following set of effects will do:

$$G_{ik} := \alpha_{ik} \, F\left(\frac{1}{\alpha_{ik}} \frac{1}{2}\left(\mathbf{a}_i + \mathbf{a}_k\right)\right) \tag{2.28}$$

where $\mathbf{a}_i$ is one of $\mathbf{a}_1$, $\mathbf{a}_{\bar{1}} = -\mathbf{a}_1$, and $\mathbf{a}_k$ is either $\mathbf{a}_2$ or $\mathbf{a}_{\bar{2}} = -\mathbf{a}_2$. The factor α_{ik} can be taken to be $\frac{1}{2}(1 + \mathbf{a}_1 \cdot \mathbf{a}_2)$. It is readily verified that (2.27) ensures the positivity of all G_{ik} and that e.g. $F(\mathbf{a}_1) = G_{12} + G_{1\bar{2}}$. Joint observables of this form can be realised in concrete experimental schemes such as the one described in Section VII.2.

Another type of joint observable is given by the covariant observable (III.3.32). Splitting the sphere S^2 into two hemispheres with the respective poles $\pm\mathbf{n}$, thus choosing $Z_\pm = S_\pm^2$, and taking into account the rotation invariance of μ and the normalisation $\mu(S^2) = 2$, one obtains a two-valued marginal observable,

$$M_s(S_\pm^2) = \frac{1}{2}\left[I \pm \frac{1}{2}\mathbf{n} \cdot \sigma\right] = F_\pm^{(\mathbf{n})} \tag{2.29}$$

This is an unsharp spin observable, with unsharpness parameter $\varepsilon = \frac{1}{4}$. In other words, the covariant POV measure M_s is a joint observable for the whole family of smeared spin observables characterised by this value of unsharpness.

IV.3 Measurements and conservation laws

The fundamental theorem of quantum measurement theory (Section II.3.3) ensures the measurability of every observable of a quantum system. This is an abstract result which furnishes the formal basis of measurement theory, but which must be qualified in view of further restrictions arising in concrete quantum theory due to the actual physical laws governing measurement processes. A particular instance of such constraints is given by the universal conservation laws. For example if the total momentum of an object-apparatus system is a conserved quantity during the measurement interaction, then a position measurement should not alter at all the total momentum distribution. However, due to the strong noncommutativity of the involved quantities, this is hardly possible, except in an approximate way. Such limitations on measurability were discovered by Wigner [4.14] and later put in the form of a theorem by Araki and Yanase [4.15]. More recent investigations into this subject can be found, for instance, in [2.13, 4.16–4.20]. The limitation in question is typically formulated as a no-go verdict, which is then circumvented on the basis of some more or less intuitive idea of approximate measurement. We shall show that the ensuing models can be described as measurements of some unsharp observables.

IV.3.1 Repeatable measurements and conservation laws. We present first the original result of Araki and Yanase [4.15] in a slightly modified form. This result applies only to discrete sharp observables, delimiting the feasibility of repeatable measurements. Let $E : i \mapsto E_i$ be a PV measure, with (φ_{ij}) being an orthonormal basis of $\mathcal{H}$ consisting of eigenvectors of E; $E_i = \sum_j P[\varphi_{ij}]$ for each i. Consider a measurement $\mathcal{M}$ of the observable E, the unitary measurement coupling U : $\mathcal{H} \otimes \mathcal{H}_\mathcal{A} \to \mathcal{H} \otimes \mathcal{H}_\mathcal{A}$ arising from the evolution of the composite system $\mathcal{S}+\mathcal{A}$. Further let the bounded self-adjoint operator $L := L_\mathcal{S} \otimes I_\mathcal{A} + I_\mathcal{S} \otimes L_\mathcal{A}$ correspond to a conserved quantity of the object-apparatus system so that $[L, U] = O$.

If the measurement $\mathcal{M}$ is repeatable, then $[L_\mathcal{S}, E_i] = O$ for all i. This is to say that $L_\mathcal{S}$ and E are coexistent. In fact for any pairs of indices (i, j) and (k, l),

$$\langle \varphi_{ij} \otimes \phi \,|\, L(\varphi_{kl} \otimes \phi) \rangle \;=\; \langle \varphi_{ij} \,|\, L_\mathcal{S}\varphi_{kl} \rangle + \delta_{ik}\,\delta_{jl}\, \langle \phi \,|\, L_\mathcal{A}\phi \rangle \qquad (3.1)$$

On the other hand using the relations $LU = UL$ and $U^*U = I$ one obtains

$$\langle \varphi_{ij} \otimes \phi \,|\, L(\varphi_{kl} \otimes \phi) \rangle \;=\; \langle U(\varphi_{ij} \otimes \phi) \,|\, LU(\varphi_{kl} \otimes \phi) \rangle \qquad (3.2)$$
$$=\; \langle U(\varphi_{ij} \otimes \phi) \,|\, L_\mathcal{S} \otimes I_\mathcal{A}U(\varphi_{kl} \otimes \phi) \rangle + \langle U(\varphi_{ij} \otimes \phi) \,|\, I_\mathcal{S} \otimes L_\mathcal{A}U(\varphi_{kl} \otimes \phi) \rangle$$

According to the calibration condition, the first term on the last line equals zero whenever $i \neq k$. Further if the measurement is repeatable, then also the second term vanishes for $i \neq k$. Therefore the above two equations lead to the following identity:

$$\langle \varphi_{ij} \,|\, L_\mathcal{S}\varphi_{kl} \rangle \;=\; \delta_{ik}\, \langle \varphi_{ij} \,|\, L_\mathcal{S}\varphi_{kl} \rangle \qquad (3.3)$$

But then, for any $n = 1, 2, \cdots$,

$$\langle \varphi_{ij} \,|\, (L_\mathcal{S}E_n - E_nL_\mathcal{S})\varphi_{kl} \rangle \;=\; (\delta_{nk} - \delta_{ni})\delta_{ik}\, \langle \varphi_{ij} \,|\, L_\mathcal{S}\varphi_{kl} \rangle \;=\; 0 \qquad (3.4)$$

which shows that E commutes with $L_\mathcal{S}$. We summarise this discussion in the following theorem.

M9. *If a sharp observable admits a repeatable measurement then it commutes with any bounded additive conserved observable of the object-apparatus system.*

This result, known as the Wigner-Araki-Yanase theorem, is a further restriction of the possibility of repeatable measurements. By now we have found that any sharp observable admits a repeatable measurement only if it is *discrete* and if it is *coexistent* with any bounded additive conserved quantity of the object-apparatus system. Since the work of Wigner [4.14], it is known that this verdict has important bearings, for instance, on spin measurements. However, it has largely remained an open question whether the corresponding limitation pertains also to the measurement of continuous observables or whether it holds in the case of unbounded conserved quantities. As a partial result it has been shown that theorem (M9) extends to unbounded additive conserved quantities whose spectra have finite degeneracy [4.16]. This covers the important case of angular momentum conservation. Another important test case for continuous observables is the measurement of position in view of momentum conservation to be studied in a model in Section IV.3.3. But first we shall revisit Wigner's analysis of spin measurements.

IV.3.2 Spin measurements and rotation invariance. Let us consider a Lüders measurement of the spin component $s_1 = \frac{1}{2}P[\varphi_+] - \frac{1}{2}P[\varphi_-]$ of a spin-$\frac{1}{2}$ object, given by a mapping

$$U_{vNL} : \varphi \otimes \phi = \sum_{k=+,-} c_k \varphi_k \otimes \phi \mapsto \sum_{k=+,-} c_k \varphi_k \otimes \phi_k \qquad (3.5)$$

for some $\phi, \phi_+, \phi_- \in \mathcal{H}_A$. This measurement does not respect the conservation of the third component of the angular momentum of the object-apparatus system (Sec. I.1.3). Therefore one may try to modify (3.5) as follows:

$$\begin{aligned} \varphi_+ \otimes \phi &\to \varphi_+ \otimes \chi_+ + \varphi_- \otimes \eta \\ \varphi_- \otimes \phi &\to \varphi_- \otimes \chi_- - \varphi_+ \otimes \eta \end{aligned} \qquad (3.6)$$

where $\langle \chi_+ | \chi_- \rangle = \langle \chi_+ | \eta \rangle = \langle \chi_- | \eta \rangle = 0$. Thus in addition to the spin-up and spin-down pointer states χ_+, χ_-, there is a third pointer state indicating an indeterminate spin, and this can be chosen so as to satisfy the conservation law. Denoting $\phi_\pm = \chi_\pm / \|\chi_\pm\|$, $\phi_o = \eta / \|\eta\|$, $\|\eta\|^2 = \varepsilon$, $\|\chi_\pm\|^2 = 1 - \varepsilon$, the above transformations can be extended by linearity into a unitary mapping:

$$U : \varphi \otimes \phi \mapsto \sqrt{1-\varepsilon} \sum_{k=+,-} c_k \varphi_k \otimes \phi_k + \sqrt{\varepsilon} \sum_{k=+,-} k c_k \varphi_{-k} \otimes \phi_o \qquad (3.7)$$

The spin effects F_k determined by this coupling and the three-valued pointer observable $k \mapsto P[\phi_k]$ are

$$F_+ = (1 - \varepsilon) P[\varphi_+], \quad F_- = (1 - \varepsilon) P[\varphi_-], \quad F_o = \varepsilon I \qquad (3.8)$$

The change of the spin state of the object caused by this measurement is given by the state transformer $k \mapsto \mathcal{I}_k$, with

$$\begin{aligned} \mathcal{I}_+(P[\varphi]) &= (1 - \varepsilon) |\langle \varphi_+ | \varphi \rangle|^2 P[\varphi_+] \\ \mathcal{I}_-(P[\varphi]) &= (1 - \varepsilon) |\langle \varphi_- | \varphi \rangle|^2 P[\varphi_-] \\ \mathcal{I}_o(P[\varphi]) &= \varepsilon \sigma_2 P[\varphi] \sigma_2 \end{aligned} \qquad (3.9)$$

These results show that the limiting case of a repeatable measurement can be approximated by making ε small. The following model demonstrates that a spin measurement may be achieved in conformance with the conservation of *all* components of the total angular momentum, and with arbitrarily small measurement inaccuracy.

In this model the spin of a particle is coupled with its orbital angular momentum via the rotation invariant spin-orbit coupling:

$$H = \lambda \mathbf{L} \cdot \mathbf{s} \tag{3.10}$$

Monitoring an appropriate orbital angular momentum quantity of the particle amounts to measuring a coarse-grained version of the corresponding spin component. For simplicity we therefore specify the 'apparatus' Hilbert space to be the angular part of the configuration space spanned by the orbital angular momentum eigenstates $\phi_{l,m}$ of $\mathbf{L}^2$ and L_3, $l = 0, 1, \ldots; m = -l, -l+1, \ldots, 0, \ldots, l$. If $\Psi(0) = (c_+\varphi_+ + c_-\varphi_-) \otimes \phi_{l,0}$ is the initial object-apparatus state, then its final state $\Psi(\tau) \equiv U_\tau \Psi(0) = e^{-i\tau H} \Psi(0)$ assumes the form

$$\Psi(\tau) = \alpha(\tau)\Psi(0) + \beta(\tau)\big[c_+\varphi_- \otimes \phi_{l,1} + c_-\varphi_+ \otimes \phi_{l,-1}\big]$$
$$\alpha(\tau) = \frac{l+1}{2l+1} e^{\frac{i}{2}\lambda l\tau} + \frac{l}{2l+1} e^{-\frac{i}{2}\lambda(l+1)\tau} \tag{3.11}$$
$$\beta(\tau) = \frac{\sqrt{l(l+1)}}{2l+1} \left\{ e^{\frac{i}{2}\lambda l\tau} - e^{-\frac{i}{2}\lambda(l+1)\tau} \right\}$$

The first part of $\Psi(\tau)$ comes essentially from an identity transformation, while the second part takes care of some angular momentum exchange.

Assume now that the pointer observable consists of the effects B which are diagonal in the $\phi_{l,m}$ basis:

$$B = \sum_m b_m P[\phi_{l,m}] \tag{3.12}$$

The corresponding measured spin effects F are then of the form

$$F = b_1 |\beta|^2 P[\varphi_+] + b_{-1} |\beta|^2 P[\varphi_-] + b_o |\alpha|^2 I \tag{3.13}$$

By considering the pointer observable as an unsharp observable, one can in principle circumvent the potential objection that Wigner's problem reappears on the level of the apparatus system: in view of angular momentum conservation, an angular momentum observable of the apparatus cannot be measured in a strict sense of a repeatable measurement. However, for the sake of simplicity we only consider the sharp L_3-observable as the pointer observable, thus choosing the pointer effects B from the collection of the spectral projections $P[\phi_{l,m}]$ of L_3 (for fixed l). The corresponding spin effects are thus

$$F_+ = |\beta|^2 P[\varphi_+], \quad F_- = |\beta|^2 P[\varphi_-], \quad F_o = |\alpha|^2 I \tag{3.14}$$

This observable is of the same form as the spin observable (3.8) derived from the approximate measurement scheme (3.7).

In order to make the measuring error small, one must require $|\alpha|^2 \ll 1$. But

$$|\alpha|^2 \; = \; 1 - |\beta|^2 \; = \; 1 - 2\frac{l(l+1)}{(2l+1)^2} \left[1 - \cos\left(\tfrac{1}{2}\lambda(2l + 1)\tau\right)\right] \tag{3.15}$$

which can be made small only by putting $\cos\left(\tfrac{1}{2}\lambda(2l + 1)\tau\right) = -1$; in that case,

$$|\alpha|^2 \; = \; (2l + 1)^{-2} \tag{3.16}$$

Thus the larger the initial angular momentum l of the apparatus, the better the accuracy of the spin measurement will be.

IV.3.3 Position measurements and momentum conservation.

The position of an object is a continuous quantity and does not therefore admit any repeatable measurements. In addition the momentum observable, which would have to be taken into account as a conserved quantity, is unbounded. Therefore the Wigner-Araki-Yanase theorem is not applicable to position measurements, nor is its proof easily adapted to this situation. In lack of a general theory we shall be content with illustrating the involved problems with the position measurement model studied in Sections II.3.4 and IV.1.

We consider again the coupling $U = e^{-i\lambda Q \otimes P_\mathcal{A}}$, which in conjunction with the pointer observable $Q_\mathcal{A}$ determines an unsharp position E^e. The resulting position measurement does not respect the momentum conservation. Indeed if the total momentum of the object-probe system, $P' := P \otimes I_\mathcal{A} + I \otimes P_\mathcal{A}$, were a conserved quantity, then it should commute with U; equivalently, the unitary operators $U_a = e^{-iaP'}$ should commute with U for any $a \in \mathbf{R}$. One finds instead:

$$U U_a \; = \; e^{-i\lambda a I \otimes P_\mathcal{A}} U_a U \tag{3.17}$$

Thus U cannot arise as part of any time evolution for the compound system satisfying the momentum conservation. However, an appropriate modification will do: instead of the operator Q in the above coupling one may introduce the relative coordinate operator $Q - Q_\mathcal{A}$ to obtain

$$\hat{U} \; = \; \exp\left(-i\tfrac{\lambda}{2}\left[(Q - Q_\mathcal{A})P_\mathcal{A} + P_\mathcal{A}(Q - Q_\mathcal{A})\right]\right) \tag{3.18}$$

Here, e.g., $(Q - Q_\mathcal{A})P$ is a short-hand notation for $(Q \otimes I_\mathcal{A} - I \otimes Q_\mathcal{A})I \otimes P_\mathcal{A}$. The conservation of the total momentum P' of the object-probe system is ensured due to the commutativity of P' with the relative coordinate:

$$\hat{U} U_a \; = \; U_a \hat{U} \qquad a \in \mathbf{R} \tag{3.19}$$

The actually measured observable E still is an unsharp position:

$$E(X) \; = \; E^e(X) \qquad X \in \mathcal{B}(\mathbf{R}) \qquad\qquad (3.20)$$

now with the confidence function

$$e(q) \; = \; \left(e^\lambda - 1\right) \left|\phi\big(-(e^\lambda - 1)q\big)\right|^2 \qquad x \in \mathbf{R} \qquad\qquad (3.21)$$

Here ϕ denotes the initial probe state, and the pointer function f is chosen such that $f^{-1}(X) = \left(1 - e^{-\lambda}\right)X$.

The above model may be questioned in view of the fact that the reading of the pointer observable Q_A requires again a measurement which should obey the momentum conservation. This objection can be avoided by replacing the pointer observable Q_A with an unsharp pointer observable $E^{Q_A,g}$. If one applies such a pointer observable instead of the sharp one, Q_A, the ensuing measured observable is again an unsharp position E^h, but then with the confidence function $h = e * g$.

V. Uncertainty

An observable establishes a connection between the set of states and the totality
of probability distributions for the possible outcomes of a given experiment. Accordingly there are two purposes a measurement can serve: to determine the value
of the measured observable; or to infer the object state by analysing the outcome
statistics. The first goal presupposes either that the observable to be measured
does have a definite value that can be determined; or it requires the measurement
to be repeatable so that it 'determines' the value in an active sense: the system
is forced to assume such a value. In other words satisfying the first goal amounts
to determining the system's future. If one does not want to limit oneself to this
rather specialised class of measurements, one is left with the latter goal of col-
lecting statistical information about the state before measurement, thereby trying
to determine the system's past. This demand has led to adopting the probability
reproducibility condition as the minimal requirement for a process to qualify as
a measurement of some observable (Chapter III). The interplay between the two
goals of a measurement allows one to elucidate the indeterminacy, or uncertainty,
that pervades the world of quantum phenomena in various senses. Indeed quantum
mechanics imposes fundamental limitations on both goals and their optimisation,
some of which shall be outlined in this chapter.

The problem of state determination from the measurement outcome statistics
entails some issues which deserve a more detailed discussion. First extending the
set of PV measures to all POV measures allows one to envisage a relation between
pairs of observables which can be rightfully characterised as a general relation of
coarse-graining (Section 1). One observable being coarser than another corresponds
to the different degrees of confidence in distinguishing between different states by
comparing the respective probability distributions. The optimal case of a one-to-
one association between the states and probability measures is realised whenever
an observable is informationally complete (Section 2). There are various ways of
characterising optimal state distinction procedures. It turns out that such measurements typically involve unsharp observables. This result calls for a reflection
on the degree of disturbance needed in order that a measurement provides any
information at all. The goals of optimising the measurement information and min-
imising the state disturbance are mutually exclusive. This furnishes a novel form
of complementarity: that between the determinations of a system's past and future
(Section 3).

The state inference problem has been extensively studied from a practical point of view [1.11, 1.12] and is currently under intense investigation in connection with quantum optical signal detection and quantum cryptography (see, e.g., [5.1], [5.2]). This recent interest in fundamental features of quantum mechanics coming from fairly practical needs has also established a link with foundational research into new formulations of uncertainty relations. Some novel measures of uncertainty, or confidence, are reviewed in Section 4.

V.1 Coarse-graining

From the operational point of view observables are representations of the totality of probability distributions for measurement outcomes. Thus any observable E, defined as a POV measure, induces a convex map V_E that associates with every state T a unique probability measure $V_E(T) = p_T^E$ on the outcome space $(\Omega, \mathcal{F})$ of the measurement in question. Moreover if $V_E = V_{E'}$, then $E = E'$. With regard to the perspective of probability theory we shall refer to V_E as the *classical embedding* of the quantum states induced by E.

On may alternatively base the definition of an observable on the notion of a classical embedding. Indeed let us define a classical embedding as a linear map

$$V : \mathcal{T}(\mathcal{H}) \to M(\Omega, \mathcal{F})$$

from the trace class operators into the space of σ-additive set functions (measures) on a measurable space $(\Omega, \mathcal{F})$ with the following properties:

$$\begin{aligned} T \geq 0 &\quad \Rightarrow \quad V(T) \geq 0 \\ V(T)(\Omega) &= \operatorname{tr}[T] \end{aligned} \tag{1.1}$$

Such a map sends states to probability measures. Moreover, for any $X \in \mathcal{F}$ the map $T \mapsto V(T)(X)$ is a bounded linear functional on $\mathcal{T}(\mathcal{H})$ and therefore can be written as $V(T)(X) = \operatorname{tr}[TE(X)]$ with some bounded operator $E(X)$. Conditions (1.1) imply that $E(X)$ is positive and bounded by I. This shows that any classical embedding yields a unique observable E. The properties of the map V correspond to spectral properties of E [5.3].

The classical embeddings V and the ensuing probability measures $V(T)$ provide *coarse-grained* descriptions of the quantum states T. It can be shown that this terminology is in full accordance with a proper generalisation of the conception of coarse-graining in the sense of partitionings [5.4]. In particular it turns out that the probability measures $V(T)$ and $V(T')$ associated with any two states T and T' are in general less distinct than the states themselves (if measured in the trace norm and total variation norm metrics, respectively). Furthermore it may happen that $V(T) = V(T')$ for $T \neq T'$.

Pairs of observables can be compared with respect to their state resolution power. Consider two observables E_i on $(\Omega_i, \mathcal{F}_i)$, $i = 1, 2$, with the associated classical embeddings $V_i : \mathcal{T}(\mathcal{H}) \to M(\Omega_i, \mathcal{F}_i)$. We say that observable E_2 is *coarser* than E_1, $E_2 \prec E_1$, if there exists a linear map $W : M_1 \to M_2$, where $M_i :=$ $V_i(\mathcal{T}(\mathcal{H}))$, such that probability measures are sent to probability measures and $V_2 = W \circ V_1$. This corresponds to the commutativity of the diagram shown in Figure 5.1. Sending probability measures to probability measures, the map W is contractive; hence

$$\left\| p_T^{E_2} - p_{T'}^{E_2} \right\|_1 \leq \left\| p_T^{E_1} - p_{T'}^{E_1} \right\|_1 \tag{1.2}$$

Observable E_1 leads thus to a better separation of states than E_2. The relation $\prec$ shall be called *(relative) coarse-graining*.

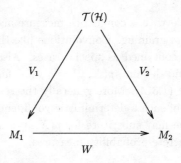

Figure 5.1. Relative coarse-graining

Examples of relative coarse-grainings are given by partitionings of the outcome space or smearings by means of confidence measures. In the first case, let $\Omega = \cup_i X_i$ be a partition of Ω into mutually disjoint sets $X_i \in \mathcal{F}$. The collection $\mathcal{P} = (X_i)$ generates a subalgebra $\mathcal{F}_\mathcal{P}$ of $\mathcal{F}$, and one may define a map

$$\begin{aligned} W_\mathcal{P} : M(\Omega, \mathcal{F}) &\to M(\Omega, \mathcal{F}_\mathcal{P}) \\ m &\mapsto W_\mathcal{P}(m) := m|_{\mathcal{F}_\mathcal{P}} \end{aligned} \tag{1.3}$$

This is a linear tranformation which sends probability measures to probability measures. If E_1 is a POV measure on $(\Omega, \mathcal{F})$, then the POV measure E_2 on $(\Omega, \mathcal{F}_\mathcal{P})$ associated with the classical embedding $W_\mathcal{P} \circ V_{E_1}$ is simply

$$E_2 = E_1|_{\mathcal{F}_\mathcal{P}} \tag{1.4}$$

An analogous procedure applies if instead of $\mathcal{F}_\mathcal{P}$ one takes an arbitrary sub-σ-algebra $\mathcal{F}'$ of $\mathcal{F}$. An important instance of that is given by a POV measure E which is a joint observable of two observables E_1 and E_2. The marginals E_1 and E_2 are now coarse-grained versions of their joint observable E since the algebras $\mathcal{F}_1$ and $\mathcal{F}_2$ can be identified with the subalgebras $\mathcal{F}_1' = \{X \times \Omega_2 \,|\, X \in \mathcal{F}_1\}$ and $\mathcal{F}_2' = \{\Omega_1 \times Y \,|\, Y \in \mathcal{F}_2\}$, respectively.

A conditional confidence measure $p(X, \omega)$, $\omega \in \Omega_1$, $X \in \mathcal{F}_2$, induces a map

$$W_p : M(\Omega_1, \mathcal{F}_1) \rightarrow M(\Omega_2, \mathcal{F}_2)$$
$$m \mapsto W_p(m) := \int_{\Omega_1} p(\cdot, \omega) \, dm(\omega) \tag{1.5}$$

Starting with an observable E_1 on $(\Omega_1, \mathcal{F}_1)$, the coarse-grained observable E_2 associated with the embedding $W_p \circ V_{E_1}$ is then given by

$$E_2(X) = \int_{\Omega_1} p(X, \omega) \, dE_1(\omega) \tag{1.6}$$

Smearing an observable with a confidence measure acquires therefore an interpretation as relative coarse-graining. Convolutions like those transforming sharp into unsharp positions are contained as special cases. Also taking $\Omega_1 = \Omega_2 = \Omega$, and $\mathcal{F}_1 = \mathcal{F}$, $\mathcal{F}_2 = \mathcal{F}_\mathcal{P}$, and defining $p(X, \omega) = \chi_X(\omega)$ for $X \in \mathcal{F}_\mathcal{P}$, one recovers the partitioning procedure (1.3–4). More generally the restriction to a subalgebra can be described in terms of such a deterministic confidence measure. For discrete sets $\Omega_1 = \Omega_2 = \mathbf{N}$, say, one may substitute for $p(X, \omega)$ a stochastic matrix $\Lambda := (\lambda_{kl})$ to obtain a map of discrete probability measures,

$$(W_\Lambda(m)_k) = \left(\sum_l \lambda_{kl} m_l\right) \tag{1.7}$$

The relation between the discrete observables E_1, E_2, with $V_{E_2} = W_\Lambda \circ V_{E_1}$, is

$$E_2(k) = \sum_l \lambda_{kl} E_1(l) \tag{1.8}$$

For instance, the map which effects the transition from the sharp photon number observable to an unsharp one obtained in the beam splitter experiment discussed in Section I.1.2 is of this form. The notion of relative coarse-graining of observables provides thus a unified description of the smearing procedures of observables. In addition it offers an information theoretical interpretation since the loss of information due to smearing is characterised by the decreasing dissimilarity of probability measures.

V.2 Informational completeness
An observable E is *informationally complete* if the classical embedding V_E is injective. The importance of such observables lies in the fact that their measurement outcome statistics entail a unique state determination: $p_T^E = p_{T'}^E$ if and only if $T = T'$. The corresponding measurements are called informationally complete as well. The notion of informational completeness can be extended to any collection of effects or observables in an obvious way.

Informationally complete observables represent maximal elements among the POV measures with respect to the relation $\prec$ in the following sense: if $E^{(2)}$ is coarser than $E^{(1)}$ and if $E^{(2)}$ is informationally complete, then so is $E^{(1)}$. It is a well-known fact of quantum mechanics that no sharp observable is informationally complete. More generally observables represented by commutative POV measures cannot be informationally complete. In fact for a commutative POV measure E one has $V_E(P[\varphi]) = V_E(P[e^{iE(X)}\varphi])$ for all φ and X; but, if $E(X) \neq \lambda I$, then $P[\varphi] \neq P[e^{iE(X)}\varphi]$ for some φ. Hence V_E is not injective (except in the trivial case of the Hilbert space being one-dimensional). Therefore any informationally complete observable E is necessarily totally noncommutative: $\mathrm{com}(E) = \{0\}$ [5.5].

There is an important limitation concerning the measurability of informationally complete observables. Consider an observable E that admits a repeatable measurement; then E is discrete, $E : i \mapsto E_i$, and each E_i has eigenvalue 1. Any eigenvector φ, $E_i\varphi = \varphi$, is in $\mathrm{com}(E)$ so that E cannot be informationally complete. In other words an informationally complete measurement cannot be repeatable. This is related to the fact that the injectivity of a classical embedding V_E excludes its being surjective, and vice versa [5.5]. In fact it can be shown that V_E can only be surjective if E is discrete and all $E(X)$ ($\notin \{O, I\}$) have both eigenvalues 1 and 0, so that repeatable measurements are possible. On the other hand if V_E is injective then no $E(X)$ can have both eigenvalues 1 and 0 together; this is to say that an informationally complete observable E is unsharp in a strong sense. In particular there can be no point measure in the range of V_E. It turns thus out that the two mutually exclusive possible properties of a classical embedding, the injectivity and surjectivity, reflect an important mode of complementarity manifesting itself in the competing aims of determining the past and the future of a physical system in quantum mechanics [5.6]. We summarise these facts as follows.

IC. *An informationally complete observable E has the following properties:*
(1) E is totally noncommutative and therefore not a sharp observable;
(2) E admits no repeatable measurement;
(3) no effect in the range of E can have both eigenvalues 0 and 1;
(4) for no state T can the probability measure p_T^E be a point measure;
(5) if E is coarser than an observable F, then F is informationally complete as well.

We present now our major physical examples of informationally complete spin quantities and joint position-momentum observables. This will also demonstrate that informational completeness is not merely an interesting theoretical possibility but on the contrary represents an important experimental option. Informationally complete measurements are feasible.

Example 1. Consider a spin-$\frac{1}{2}$ system, with its associated Hilbert space $\mathcal{H} = \mathbf{C}^2$. Let φ, ψ be orthogonal unit vectors in $\mathcal{H}$ and define $\eta_r := \frac{1}{\sqrt{2}}(\varphi + i^r \psi)$, $r = 0, 1, 2, 3$. The three pairs of projection operators $P[\varphi], P[\psi], P[\eta_0], P[\eta_2]$, and $P[\eta_1], P[\eta_3]$ can be viewed as the spectral projections of three mutually orthogonal spin components. Let w_1, w_2 and w_3 be positive numbers which sum up to 1. The mapping $i \mapsto E_i$ on $\Omega = \{1, 2, 3, 4, 5, 6\}$, with

$$
\begin{aligned}
E_1 &= w_1 P[\varphi] & E_2 &= w_1 P[\psi] \\
E_3 &= w_2 P[\eta_0] & E_4 &= w_2 P[\eta_2] \\
E_5 &= w_3 P[\eta_1] & E_6 &= w_3 P[\eta_3]
\end{aligned}
\tag{2.1}
$$

defines a POV measure E whose range spans the whole space $\mathcal{L}(\mathbf{C}^2)$, so that the mapping V_E is injective. Note that this observable has the same structure as the one corresponding to the compound spin measurement described in Section I.1.2.

A similar construction can be carried out in an infinite dimensional Hilbert space showing that also in this case there do exist discrete observables which are informationally complete [5.5].

Example 2. The informationally complete spin observable formulated above has a value space of six possible outcomes. Since a spin-$\frac{1}{2}$ state is uniquely specified by a set of three real parameters, the minimal number of outcomes for an informationally complete spin measurement must be four. It is not hard to construct such observables. Clearly no sharp spin observable $s_{\hat{\mathbf{a}}}$ is informationally complete. (We write $\hat{\mathbf{a}}$ for a unit vector, whereas $\mathbf{a}$ is a vector with the norm $\|\mathbf{a}\| \leq 1$). The measurement outcome statistics of $s_{\hat{\mathbf{a}}}$ determine, in general, only the component $\hat{\mathbf{a}} \cdot \mathbf{n}$ of the vector $\mathbf{n}$ specifying the state $T_{\mathbf{n}}$. Thus for a full determination of the spin state of the system one needs the measurement outcome statistics from three spin quantities $s_{\hat{\mathbf{a}}}$, $s_{\hat{\mathbf{b}}}$ and $s_{\hat{\mathbf{c}}}$ associated with linearly independent orientations $\mathbf{a}, \mathbf{b}, \mathbf{c}$. In particular no complementary spin pair $(s_{\hat{\mathbf{a}}}, s_{\hat{\mathbf{b}}})$ is informationally complete. However, one may replace the pair $(s_{\hat{\mathbf{a}}}, s_{\hat{\mathbf{b}}})$ with a pair of unsharp spin observables $F^{\mathbf{a}}$ and $F^{\mathbf{b}}$, with the generating effects $F_{\pm}^{\mathbf{a}}$ and $F_{\pm}^{\mathbf{b}}$ of the form (IV.2.23), which have a common informationally complete refinement G, a joint observable of $F^{\mathbf{a}}$ and $F^{\mathbf{b}}$. In fact choosing $\mathbf{a}$ and $\mathbf{b}$ such that $\|\mathbf{a}+\mathbf{b}\| + \|\mathbf{a}-\mathbf{b}\| < 2$, then $F^{\mathbf{a}}$ and $F^{\mathbf{b}}$ are coexistent [result (M8), Section IV.2.6], and any effect $C \equiv \gamma C(\mathbf{c})$, with $\frac{1}{2}\|\mathbf{a}+\mathbf{b}\| \leq \gamma \leq 1 - \frac{1}{2}\|\mathbf{a}-\mathbf{b}\|$, and $(\mathbf{a} \times \mathbf{b}) \cdot \mathbf{c} \neq 0$, generates an informationally complete joint observable G, with the range $\{O, C, F_+^{\mathbf{a}} - C, F_+^{\mathbf{b}} - C, I - F_+^{\mathbf{a}} - F_+^{\mathbf{b}} + C, I\}$ [5.6]. In Sections VII.2 and VII.4.2 we present measurement schemes which realise such observables.

Example 3. The covariant spin observable M on S^2,

$$
M : Z \mapsto \int_Z T_{\mathbf{z}} \, d\mu(\mathbf{z})
\tag{2.2}
$$

introduced in Section III.3.3 is informationally complete. This is easily seen as follows. Let $T_i = \frac{1}{2}(I + \mathbf{n}_i \cdot \sigma)$, $i = 1, 2$, be two states. Then $\operatorname{tr}\big[(T_2 - T_1)M(Z)\big] = 0$ holding for all sets Z on the sphere implies $(\mathbf{n}_2 - \mathbf{n}_1) \cdot \int_Z \mathbf{z} \, d\mu = 0$ for all Z. Picking $Z = Z_k$ to be polar caps around the three mutually orthogonal directions of $\mathbf{e}_k$ gives $\int_Z \mathbf{z} \, d\mu \sim \mathbf{e}_k$. Consequently, $\mathbf{n}_2 - \mathbf{n}_1 = 0$ and $T_1 = T_2$.

While it may not be evident how to devise a scheme for a direct measurement of this observable, it is possible in principle to perform the following experiment. Consider a collection of K pairs of points $\{\mathbf{z}_k, -\mathbf{z}_k\}$ uniformly distributed on the surface of the Poincaré sphere. To each of these pairs there corresponds a spin observable $s_{\mathbf{z}_k}$, with spectral projections $T_{\mathbf{z}_k}, T_{-\mathbf{z}_k}$. If each of these observables is measured on N particles in spin state $T_{\mathbf{n}}$, the expected average number of 'up' outcomes will be

$$N(\pm; k) \simeq \operatorname{tr}\left[T_{\mathbf{n}} T_{\pm \mathbf{z}_k}\right] N \equiv p(\pm; \mathbf{z}_k) N \tag{2.3}$$

Let Z be a Borel subset of the surface S^2 of the Poincaré sphere. Then the total number of counts corresponding to spin pointing in some direction within Z is

$$N_+(Z) = \sum_{\mathbf{z}_k \in Z} N(+; k) + \sum_{-\mathbf{z}_k \in Z} N(-; k) \tag{2.4}$$

Therefore the relative frequency $N_+(Z)/KN$ of obtaining an 'up' count 'within' Z approaches the probability

$$p_+(Z) = \sum_{\mathbf{z}_k \in Z} \frac{1}{K} p(+; \mathbf{z}_k) + \sum_{-\mathbf{z}_k \in S} \frac{1}{K} p(-; \mathbf{z}_k) \tag{2.5}$$

Noting that $p(-; \mathbf{z}_k) = p(+; -\mathbf{z}_k)$, one obtains in the limit $K \to \infty$ (applying the Riemann integral on the sphere S^2 with respect to μ, the uniform measure normalised to 2):

$$p_+(Z) \longrightarrow \int_Z d\mu(\mathbf{z})\, p(+; \mathbf{z}) = \int_Z d\mu(\mathbf{z}) \operatorname{tr}\left[T_{\mathbf{n}} T_{\mathbf{z}}\right] \tag{2.6}$$

This shows that the statistics of the continuous spin observable (2.2) emerge as a 'pattern' in the combined statistics of the present experiment.

Example 4. The canonically conjugate (Cartesian components of) position and momentum observables Q and P are both informationally incomplete. Even worse, as continuous observables neither of them is informationally complete with respect to any state, that is, for any state T there is another state $T' \neq T$ such that, for instance, $V_Q(T) = V_Q(T')$. Consequently neither the position measurement outcome statistics nor the momentum measurement outcome statistics ever determine the state of the system. Taken together position and momentum are totally noncommutative, $\operatorname{com}(Q, P) = \{0\}$. Therefore a necessary condition for their informational completeness is fulfilled. Still this pair is not informationally complete. It is an old question, known as the Pauli problem, under which conditions the position and momentum distributions do determine the state. For example it is not even clear whether there is an observable A such that the triple (Q, P, A) were informationally complete. If A is of the type $A = P^2 + v(Q)$, $v(Q) \geq O$, then the triple is demonstrably informationally incomplete [5.7]; for an overview of the Pauli problem and for further references the reader may wish to consult [5.6].

Consider next an unsharp position E^e and an unsharp momentum F^f defined by the confidence functions e and f, respectively, according to Eqs (III.2.33). Since, for instance, $E^e(X) = \int_{\mathbf{R}} (\chi_X * e)(q) E(dq)$, the unsharp position E^e is a coarse-graining of the sharp position E, the spectral measure of Q. Apart from that, E^e and E, as well as F^f and F, are *informationally equivalent*, that is, their state distinction properties are exactly the same: for any T, T', $V_{E^e}(T) = V_{E^e}(T')$ if and only if $V_E(T) = V_E(T')$ [5.8]. This implies that also the pairs (E^e, F^f) and (E, F) are informationally equivalent; the measurement outcome statistics of both of these pairs of observables have the same state distinction power and do not, in general, suffice to determine the state. There is an important difference between the two pairs. Contrary to the sharp pair (Q, P), the unsharp pair (E^e, F^f) can be coexistent; both E^e and F^f can be coarse-grainings of a single finer observable. Moreover their common refinement can be informationally complete. If E^e and F^f are Fourier-related, for instance $e = |\xi|^2$ and $f = |\hat{\xi}|^2$ for some unit vector ξ, then $G_\xi : Z \mapsto \frac{1}{2\pi} \int_Z P[W_{qp}\xi] \, dq \, dp$ is their joint observable, cf. Section III.2.4. The classical embedding V_{G_ξ} is injective whenever the vector ξ, from the common domain of Q and P, fulfills $\langle \xi \, | \, W_{qp}\xi \rangle \neq 0$ for almost all $(q, p) \in \mathbf{R}^2$ [5.9]. The Pauli problem is hereby solved in an unexpected way: the informationally incomplete pair (Q, P) is replaced by an informationally equivalent unsharp position-momentum pair (E^e, F^f) which has an informationally complete refinement G_ξ. In Sections VI.2. and VII.3.7 two different realisations of informationally complete measurement schemes are presented which lead to such phase space observables G_ξ.

V.3 Unsharpness

The quantum mechanical indeterminacy manifests itself as a limitation of both prediction and retrodiction. First, for a measurement on a system that is not in an eigenstate of the measured observable, the outcomes can only be predicted with some probability. Second, given a single measurement outcome and no prior knowledge about the state, there is no way to infer with certainty what the state was. Both features, the indeterminism of measurement outcomes and the uncertainty of retrodictions, acquire some new faces if unsharp observables are taken into account.

V.3.1 Quantum indeterminism and classicality conditions. The representation of states as positive trace-one operators is the most general one compatible with the probabilistic structure of quantum mechanics. This is essentially the content of Gleason's theorem. A a consequence, there is no map $v : \mathcal{E}(\mathcal{H}) \to \{0, 1\}$ from the set of effects onto the numbers $0, 1$ that would satisfy the additivity condition $v(\sum_k E_k) = \sum_k v(E_k)$ for any sequence of effects with $\sum_k E_k \leq I$. This fact

rules out to a large extent the possibility of embedding quantum mechanics into a hidden-variable theory. Thus there is no 'hidden' state description with respect to which every property would be a real one. In this respect the quantum and classical physical indeterminism are fundamentally different: while classical properties are always real though sometimes unknown, quantum mechanical properties are generically indeterminate.

A global reduction of quantum mechanics to a classical theory must therefore be regarded impossible. Nevertheless a classical description may be admissible with respect to a fixed state and for a restricted set of noncommuting observables: one may determine *classicality conditions* for the corresponding probability distributions which ensure the existence of joint distributions [5.10]. The investigation of this possibility has received much attention in connection with the Einstein-Podolsky-Rosen paradox and Bell's theorem [5.11]. It is of great importance for the problem of understanding, within quantum mechanics, the emergence of classical physical phenomena and properties. Rather than aiming at an exhaustive account we give a few examples illustrating the classicality conditions.

Any pair of discrete probability distributions $\mathbf{p} := \{p_k\}, \mathbf{q} := \{q_l\}$ possesses at least one joint distribution $\mathbf{r}$ such as, for instance, the 'product' $\mathbf{r} := \{p_k q_l\}$ which represents a pair of uncorrelated random variables. In quantum mechanics the distributions $\mathbf{p}, \mathbf{q}$ arise from a fixed state and refer to some observables: $\mathbf{p} = p_T^E, \mathbf{q} = p_T^F$. If a joint distribution is stipulated to exist for all states T and to derive from one common observable G such that $\mathbf{r} = p_T^G$, then this amounts to demanding the coexistence of E and F. As pointed out in Section II.2.2, one virtue of unsharp observables lies in the fact that their coexistence does not require commutativity.

A more complicated situation arises when triples or quadruples of observables are considered and besides the single distributions some pair distributions are also given. The classicality of such a system of probabilities means that it is embeddable into a joint probability distribution. Typically systems of quantum mechanical probabilities do not satisfy this classicality condition. To give a specific example, let $F^{(\mathbf{a})}, F^{(\mathbf{b})}, F^{(\mathbf{c})}$ be three observables of a spin-$\frac{1}{2}$ system, with the probability measures (for a fixed state T) denoted as $p(\pm\mathbf{a}), p(\pm\mathbf{b}), p(\pm\mathbf{c})$. Assuming that for each pair there are pair probabilities $p(\mathbf{x}, \mathbf{y})$, with $(\mathbf{x}, \mathbf{y}) \in \{(\pm\mathbf{a}, \pm\mathbf{b}), (\pm\mathbf{a}, \pm\mathbf{c}), (\pm\mathbf{b}, \pm\mathbf{c})\}$, then a necessary and sufficient condition for the existence of a triple joint probability is given by a set of four inequalities of the form

$$p(\mathbf{b}, -\mathbf{c}) \leq p(-\mathbf{a}, \mathbf{b}) + p(\mathbf{a}, -\mathbf{c}) \tag{3.1}$$

These are known as Bell's inequalities. If the three observables are coexistent, then there does exist a triple probability for any state and Bell's inequalities are fulfilled. As shown in Section IV.2.6, this will be the case whenever the degree of unsharpness in these observables is sufficiently large. The validity of (3.1) for only a single state is less restrictive and can be achieved also for noncoexistent quantities.

Consider next an EPR-correlated photon pair in the singlet state Ψ, with the polarisation observables $F^{(\mathbf{a})}, F^{(\mathbf{a}')}$ and $F^{(\mathbf{b})}, F^{(\mathbf{b}')}$ for the first and second photon, respectively. It is possible to measure jointly a pair of polarisation observables, one on each photon. Hence there are the single and joint probabilities which shall be denoted $p(\mathbf{x}), p(\mathbf{y}), p(\mathbf{x}, \mathbf{y})$, where $\mathbf{x} \in \{\pm\mathbf{a}, \pm\mathbf{a}'\}$, $\mathbf{y} \in \{\pm\mathbf{b}, \pm\mathbf{b}'\}$. A necessary and sufficient condition for the existence of a joint probability comprising all these entities is a set of four inequalities of the form

$$0 \leq p(\mathbf{a}) + p(\mathbf{b}') - p(\mathbf{a}, \mathbf{b}) - p(\mathbf{a}, \mathbf{b}') - p(\mathbf{a}', \mathbf{b}') + p(\mathbf{a}', \mathbf{b}) \leq 1 \qquad (3.2)$$

These are the Clauser-Horne-Shimony-Holt inequalities. Bell's inequalities are recovered in the case $\mathbf{a}' = \mathbf{b}' \equiv \mathbf{c}$ and for strict anticorrelation of this pair.

Assume that all the polarisation observables have identical unsharpness, i.e., $\|\mathbf{a}\| = \|\mathbf{a}'\| = \|\mathbf{b}\| = \|\mathbf{b}'\| = \lambda$. Denoting the corresponding unit vectors as $\hat{\mathbf{a}}$, etc., one then finds for the singlet state

$$p(\mathbf{a}, \mathbf{b}) = \tfrac{1}{4}\left(1 - \lambda^2 \hat{\mathbf{a}} \cdot \hat{\mathbf{b}}\right) \qquad (3.3)$$

Inserting this into (3.2) and writing $\hat{\mathbf{x}} \cdot \hat{\mathbf{y}} = \cos\theta_{\mathbf{x}, \mathbf{y}}$ yields

$$\left| \cos\theta_{\mathbf{a}, \mathbf{b}} + \cos\theta_{\mathbf{a}, \mathbf{b}'} + \cos\theta_{\mathbf{a}', \mathbf{b}'} - \cos\theta_{\mathbf{a}', \mathbf{b}} \right| \leq \frac{2}{1 - 2\varepsilon} \qquad (3.4)$$

Here $\varepsilon = \tfrac{1}{2}(1 - \lambda^2)$. The left hand side assumes its maximum $2\sqrt{2}$ when the three vectors are coplanar and satisfy $\theta_{\mathbf{a}, \mathbf{b}} = \theta_{\mathbf{a}, \mathbf{b}'} = \theta_{\mathbf{a}', \mathbf{b}'} = \tfrac{1}{3}\theta_{\mathbf{a}', \mathbf{b}} = 45^o$. This yields a minimal value of the unsharpness parameter ε such that (3.4) remains valid:

$$\varepsilon \geq \varepsilon_C = \frac{1}{2}\left(1 - \frac{1}{\sqrt{2}}\right), \qquad \lambda \leq \lambda_C = \left(\frac{1}{2}\right)^{1/4} \qquad (3.5)$$

Consider now the case $\mathbf{a}' = \mathbf{b}' \equiv \mathbf{c}$ which correpsonds to measuring polarisations of the same orientation on both photons. From (3.3) one obtains $p(\mathbf{c}, \mathbf{c}) = \tfrac{1}{4}(1 - \lambda^2) = \tfrac{1}{2}\varepsilon$ ($\neq 0$ unless $\varepsilon = 0$), which is to say that due to the unsharpness there is no strict anticorrelation. Consequently (3.2) reduces to the following:

$$p(\mathbf{b}, \mathbf{c}) \leq p(\mathbf{a}, \mathbf{b}) + p(\mathbf{a}, \mathbf{c}) + p(\mathbf{c}, \mathbf{c}) \qquad (3.6)$$

This is weaker than Bell's inequality (except in the case of strict anticorrelation) and can in fact be satisfied with a smaller degree of unsharpness than given in (3.5). To see this, rewrite (3.6) by using again (3.3):

$$\cos\theta_{\mathbf{a}, \mathbf{b}} + \cos\theta_{\mathbf{a}, \mathbf{c}} - \cos\theta_{\mathbf{b}, \mathbf{c}} \leq \frac{1 + 2\varepsilon}{1 - 2\varepsilon} \qquad (3.7)$$

The left hand side attains its maximum $\frac{3}{2}$ when the three vectors are coplanar and satisfy $\theta_{\mathbf{a},\mathbf{b}} = \theta_{\mathbf{a},\mathbf{c}} = \frac{1}{2}\theta_{\mathbf{b},\mathbf{c}} = 60^o$. In order that (3.6), (3.7) remain valid, there must therefore be a minimal value for the unsharpness parameter:

$$\varepsilon \geq \varepsilon_B = \frac{1}{10}, \qquad \lambda \leq \lambda_B = \frac{2}{\sqrt{5}} \tag{3.8}$$

It is evident that $\varepsilon_B < \varepsilon_C$ so that it is indeed easier to realise the classicality for a triple than for a quadruple of polarisation observables in the singlet state [5.12].

It must be emphasised that the fulfilment of the classicality conditions via the introduction of unsharpness does not lead back to a cryptodeterministic interpretation of quantum probabilities. But there is a gain: the existence of joint probabilities is a minimal condition for the possibility of simultanously ascribing values to noncommuting quantities. As a consequence the classical reasoning of Einstein, Podolsky and Rosen cannot be refuted along the lines of Bell's theorem since Bell's inequalities are no longer violated in quantum mechanics in the case of sufficiently unsharp observables. Hence there emerges a 'shadow of classicality' in the quantum world if unsharpness is taken into account.

V.3.2 Measurement: disturbance versus information gain. The quantum indeterminism of measurement outcomes entails that of the corresponding state changes which are described by the action of a state transformer. As pointed out in Section II.3.2, there is no state transformer associated with a nontrivial observable that would leave the system's state unchanged irrespectively of what the measurement outcome was. In other words if a measurement is to provide any statistical information about the state, there must be some 'disturbance' of that state. It follows that it is impossible to repeat the same measurement of an observable E many times on the same system originally in state T so that one could infer the probability measure p_T^E. A general proof that there is no individual state determination in quantum mechanics will be given in Section 3.4.

This state of affairs raises the question as to what extent one may control the state change due to a measurement and at the same time still ensure some information gain. One obvious form of minimal disturbance is given by the ideality condition: whenever a system is in an eigenstate associated with some value of the measured observable, then the measurement should not change that state. For a discrete sharp observable this amounts to stipulating the repeatability, and therefore also the first-kind property and value reproducibility (Sec. IV.1). However a repeatable measurement exerts drastic changes to all states that are not eigenstates. An apparently weaker form of minimal disturbance is afforded by the notion of value reproducibility. Again for sharp observables it is equivalent to repeatability. Hence it seems difficult to conceive of 'gentle' measurement procedures for sharp observables. This situation changes if unsharp observables are taken into consideration. In that case all the above notions have to be relaxed appropriately. Ideality is substituted with approximate ideality (Sec. II.4). The Lüders operation associated with an unsharp property is approximately ideal and approximately value-reproducible but no longer (approximately) repeatable. Thus the class of states which suffer only small disturbances in an unsharp measurement can be expected to increase with increasing unsharpness of the measured observable. This will be illustrated in the phase space measurement model of Chapter VI.

Instead of requiring minimal disturbance, one may consider the repeatability as a way of determining the system's future: it allows one to assert that after the measurement the measured observable has assumed a definite value. Such measurements cannot, however, provide optimal information gain in the sense of informational completeness. This is the complementarity of past and future determinations mentioned in Section 2. An informationally complete measurement is necessarily an unsharp one so that one can at most expect (ε, δ)-repeatability. The state transformer (IV.1.16) associated with an informationally complete phase space observable G_ξ does have this property. It will be further analysed in the next subsection.

V.3.3 Unsharpness in phase space inference.

The unsharpness of an observable shall be illustrated in its implications on prediction and retrodiction. Our example will represent a well tuned balance between the competing goals of controlling the future state and inferring the past state of the measured system.

In some experiments one is dealing with a situation where the state to be estimated is not completely arbitrary but is known to belong to a specified class of states. Therefore in order to find out the particular state at a given time, it may be sufficient to determine, or estimate, just a finite set of parameter values. Such is the case in the detection of weak signals by means of monitoring some quantum non-demolition observable. In such experiments the detector state is known to be an eigenstate of that observable for all times, its change indicating the presence of an external field and being detected by a repeatable measurement. Typically the measured observable is continuous so that there are no strict eigenstates and repeatable measurements. Moreover if one intends to monitor the position, then one has to be aware that fairly precise determination of the position goes along with high indeterminacy of the momentum and therefore large uncertainties in subsequent position measurements. A good compromise can be expected in a phase space measurement since this allows a control of both quantities simultaneously.

Consider a measurement of the phase space observable G_ξ modelled by the state transformer (IV.1.16):

$$\mathcal{I}_Z(T) = \frac{1}{2\pi} \int_Z P[\xi_{qp}] \operatorname{tr}\big[P[\xi_{qp}]T\big] \, dq \, dp \qquad (3.9)$$

Let the generating vector ξ be a Gaussian state with vanishing first moments of Q and P. This state transformer is *quasi-preparatory* in the following sense. If one confines the reading intervals to sets $Z = Z(q,p)$ centered at (q,p) and small compared to the variances of Q and P in the state ξ, then the system is left nearly in a vector state $P[\xi_{qp}]$ localised within that set. That is, to a good approximation,

$$\mathcal{I}_{Z(q,p)}(T) \simeq P[\xi_{qp}] \operatorname{tr}\big[P[\xi_{qp}]T\big] \, m\big(Z(q,p)\big) \tag{3.10}$$

Moreover assuming that there are only weak external forces coupling linearly to the system, it follows that the evolution between successive measurements drives the system through states $\xi_{q(t),p(t)}$ so that a measurement should just serve to determine the trajectory parameters $\big(q(t),p(t)\big)$. In this way the quantum non-demolition conditions are (approximately) realised for a phase space measurement [4.7]. We are now ready to discuss the following two problems.

(P) Given a state $P[\xi_{q_op_o}]$, what is the 'most likely' range of phase space points (q,p) to be registered in a measurement of G_ξ? That is, given a confidence level α such that

$$\operatorname{tr}\big[P[\xi_{q_op_o}]\, G_\xi(Z)\big] = 1 - \alpha \tag{3.11}$$

find the corresponding confidence region $Z = \mathcal{C}_\alpha(q_o,p_o)$, the smallest bounded set for which (3.11) holds.

(R) Given some registered 'small' set $Z(q,p)$ in a measurement of G_ξ, and given that the state T was one of the form $P[\xi_{q_op_o}]$, what is the 'most likely' region Z where the (q_o,p_o) came from?

Answer to (P): The sought set Z is a solid ellipse just large enough so that (3.11) is satisfied, the boundary curve of Z consisting of those points (q,p) for which the confidence function

$$f_{q_op_o}(q,p) := \operatorname{tr}\big[P[\xi_{q_op_o}]\, P[\xi_{qp}]\big] \tag{3.12}$$

assumes a constant value. It is evident that any deformation of this error ellipse preserving the confidence level condition (3.11) leads to a region of larger total area.

    The motivation for considering (3.11) as a formalisation of the confidence criterion is the following. Consider a partitioning of the phase space into 'small' sets $Z_k = Z(q_k,p_k)$ such that (3.10) can be used. Given Z, the probability of finding some $Z_k \subseteq Z$ (that is, some 'point' $(q_k,p_k) \in Z$,) is

$$\operatorname{prob}_{\xi_{q_op_o}}(\text{some } Z_k \subseteq Z) = \sum_{(q_k,p_k)\in Z} \operatorname{tr}\big[P[\xi_{qp}]\, P[\xi_{q_op_o}]\big]\, m(Z_k) \tag{3.13}$$

and this approaches $\operatorname{tr}\big[G_\xi(Z)P[\xi_{q_op_o}]\big]$ under refinement of the partitioning.

Answer to (R): Let α be some confidence level, then the sought region Z satisfies

$$\operatorname{tr}\big[G_\xi(Z)P[\xi_{qp}]\big] = 1 - \alpha \tag{3.14}$$

so it is the same error ellipse as above, $Z = \mathcal{C}_\alpha(q,p)$.

To justify this answer, observe that if $Z(q,p)$ was registered, one would assume that the state T was some $P[\xi_{q_o p_o}]$ with $(q_o, p_o) \in Z$. The chance that this guess is wrong is required to be $\leq \alpha$. A wrong guess means: $(q_o, p_o) \notin Z$. Consider a partitioning of the phase space into 'small' sets $Z_k = Z(q_k, p_k)$, $m(Z_k) = m$. Assume that $T = P[\xi_{q_o p_o}]$ was one such T_{z_k}, with $z_k = (q_k, p_k)$. The question now is: 'how often' does one get a registration $Z(z)$, $z = (q,p)$, from some $T = P[\xi_{z_k}]$ with $z_k \notin Z$? The probability for this is:

$$\text{prob}\,(z|z_k) \;=\; \text{tr}\big[P[\xi_{z_k}]\,G_\xi\big(Z(z)\big)\big] \;\simeq\; \text{tr}\big[P[\xi_{z_k}]\,P[\xi_z]\big]\,m\big(Z(z)\big) \tag{3.15}$$

Summing over all $z_k \notin Z$ and noting that $m\big(Z(z)\big) = m$ yields:

$$\begin{aligned}
\text{prob}\,(z|\text{error}) \;=\; \sum_{z_k \notin Z} \text{prob}\,(z|z_k) \;&\simeq\; \sum_{z_k \notin Z} \text{tr}\big[P[\xi_{z_k}]P[\xi_z]\big]\,m \\
\longrightarrow\; \int_{\mathbf{R}^2 \setminus Z} \text{tr}\big[P[\xi_{z'}]P[\xi_z]\big]\,dm(z') \;&=\; 1 - \text{tr}\big[G_\xi(Z)P[\xi_{qp}]\big] \;=\; \alpha
\end{aligned} \tag{3.16}$$

(the limit being understood in the sense of a Riemann sum under refinement of the partition, i.e., $m \to 0$).

In summary the error ellipse $C_\alpha(q, p)$, or the associated confidence function $f_{q_o p_o}$ of (3.12), admits the following interpretations with reference to quasi-preparatory phase space measurements:

(P) A prediction '$z = (q, p)$', given a prepared (unsharp) phase space point $z_o = (q_o, p_o)$, can only be made within the uncertainty determined by the error ellipse $Z = C_\alpha(z_o)$ specified by the prepared distribution $f_{z_o}(z)$ of Eq. (3.12).

(R) A retrodiction of a state $P[\xi_{z_o}]$ prior to a measurement with outcome $z = (q, p)$ is subject to the uncertainty represented by the error ellipse $Z = C_\alpha(z)$, or by the corresponding confidence function $f_z(z_o)$.

In the present model the two basic goals that may be pursued with a measurement, prediction and retrodiction, are simultaneously optimised: the distribution (3.12), which describes the degree of unsharpness inherent in the phase space observable, derives from a minimal uncertainty state. This situation represents in a sense the best approximation to a repeatable measurement of a discrete sharp observable. If the system is in an unknown eigenstate, then the outcome allows one to conclude with certainty to which eigenspace the state belonged. The repeatability ensures that the state has not changed at all during the measurement. Hence it is with respect to the family of (approximate) eigenstates of an observable that retrodiction and prediction can be simultaneously optimised.

V.3.4 Impossibility of individual state determination. Why is it impossible to determine the state of an individual system in quantum mechanics? Would it be bad if one could? The latter question must be answered in the positive. Individual state determination would allow one to lift the statistical indistinguishability of different ways of preparing mixed states; for instance, one could find out whether a system is isolated or entangled with some environment, by performing a single measurement on that system alone. This would render the quantum mechanical state description incomplete. Moreover it would lead to the possibility of superluminal signalling by exploiting EPR correlations. In fact the signal could consist of performing either one of two selected polarisation measurements on one photon of a pair prepared in a singlet state. The receiver may perform his measurement on the second photon at a spacelike separation and would find out which of the two polarisations was being measured on the first photon, in plain contradiction with relativistic causality.

Fortunately quantum mechanics protects itself against such disastrous consequences. The identification of a state by means of a single measurement would require the existence of an observable E such that every state T would give probability one to some outcome X_T while for any other state T' the probability of obtaining X_T would be zero. For two vector states φ, ψ this would entail $E(X_\varphi)\varphi = \varphi$, $E(X_\psi)\psi = \psi$, and $E(X_\varphi)\psi = O$. But then $\langle \psi \mid \varphi \rangle = \langle \psi \mid E(X_\varphi)\varphi \rangle = 0$. Hence any two different states would have to be mutually orthogonal which is wrong. In other words, no pair of overlapping states can be uniquely separated by means of a single measurement.

There are different ways of optimising the separation of nonorthogonal state pairs, some of which shall be briefly reviewed [5.13]. For simplicity we consider only the case of a two-dimensional Hilbert space. First one may try to maximise the degree of certainty with which a given pair of nonorthogonal states φ, ψ can be separated. To this end one needs to find an observable such that for two effects $E_1, E_2 = I - E_1$ in its range one would guess from the occurrence of E_1 or E_2 that the state was φ or ψ, respectively. The task is to find E_1 such that on the average the probability of a correct guess is maximal. Suppose that the apriori probabilities for φ, ψ are r, s, respectively (where $r + s = 1$). The total probability of success is

$$p = r \operatorname{tr}[E_1 P[\varphi]] + s \operatorname{tr}[E_2 P[\psi]] = s + \operatorname{tr}[E_1(rP[\varphi] - sP[\psi])] \qquad (3.17)$$

Writing the operator $D := rP[\varphi] - sP[\psi]$ in its spectral decomposition, $D = uP[\xi] - vP[\eta]$ ($u, v \geq 0$), shows immediately that the sought effect is the projection $E_1 = P[\xi]$. The positive eigenvalue u is easily determined and one obtains

$$p_{\max} = s + u = \tfrac{1}{2}\left(1 + \sqrt{1 - 4rs|\langle \varphi \mid \psi \rangle|^2}\right) \qquad (3.18)$$

This probability equals 1 exactly when φ and ψ are orthogonal, and for $r = s = \tfrac{1}{2}$, it is close to $\tfrac{1}{2}$ when the states are very similar.

Interestingly there are measurement procedures which allow correct inferences with certainty for some of their outcomes. The idea is to perform two measurements in succession, where the first one serves to turn the states φ, ψ into a known pair of mutually orthogonal states. Then a suitably chosen second measurement can uniquely distinguish between these and thus one may conclude whether the original state was φ or ψ.

To sketch the most general scheme for the first measurement, choose a probe system and fix its initial state ϕ and a unitary coupling U. Then for any initial state ξ of the original system the final state of the compound system is $U(\xi \otimes \phi)$. The first measurement is completed as soon as some pointer observable has been measured. Thus there should be a projection Q associated with the probe such that the states $I \otimes QU(\varphi \otimes \phi)$ and $I \otimes QU(\psi \otimes \phi)$ are orthogonal:

$$\langle U(\psi \otimes \phi) | I \otimes QU(\varphi \otimes \phi) \rangle = 0 \qquad (3.19)$$

Then after the registration of a pointer reading represented by Q, it is possible to distinguish uniquely between these two states. Note that given (3.19), a corresponding orthogonality relation cannot hold for Q replaced with $I - Q$, since if it did, then the states $U(\varphi \otimes \phi)$ and $U(\psi \otimes \phi)$, and therefore (by unitarity) φ and ψ would have to be orthogonal. Hence the task is to maximise the probability p for obtaining Q. Assuming that both states φ and ψ have equal apriori probabilities, one obtains for p

$$p = \tfrac{1}{2} \langle U(\varphi \otimes \phi) | I \otimes QU(\varphi \otimes \phi) \rangle + \tfrac{1}{2} \langle U(\psi \otimes \phi) | I \otimes QU(\psi \otimes \phi) \rangle \qquad (3.20)$$

We now make use of the fact that the map $p(\xi, \eta) := \langle U(\xi \otimes \phi) | I \otimes QU(\eta \otimes \phi) \rangle$ is a positive semidefinite bilinear form of the vectors of $\mathcal{H}$. This ensures the existence of an effect E acting in $\mathcal{H}$ such that $p(\xi, \eta) = \langle \xi | E\eta \rangle$. Therefore

$$p = \tfrac{1}{2} \langle \varphi | E\varphi \rangle + \tfrac{1}{2} \langle \psi | E\psi \rangle \qquad (3.21)$$

This is the probability to be maximised by a proper choice of the effect E and under the orthogonality constraint (3.19), which now reads:

$$\langle \psi | E\varphi \rangle = 0 \qquad (3.22)$$

Thus starting with an operational description of the problem to be solved, one has obtained a formulation in terms of the first Hilbert space alone. Before giving the solution it is instructive to show how the unitary map can be realised for any given effect E that satisfies the constraint (3.22): define for any $\xi \in \mathcal{H}$

$$\begin{aligned} U(\xi \otimes \phi) &:= A_1 \xi \otimes \phi_1 + A_2 \xi \otimes \phi_2 \\ A_1 &:= E^{1/2}, \quad A_2 := (I - E)^{1/2} \end{aligned} \qquad (3.23)$$

Here ϕ_1, ϕ_2 are some mutually orthogonal (normalised) pointer states. One can easily verify that unitarity is ensured by virtue of the relation $A_1^2 + A_2^2 = I$. Furthermore it is obvious that (3.19) boils down to (3.22) if one chooses $Q = P[\phi_1]$.

The maximum value of p turns out to be

$$p_{\max} = 1 - |\langle \varphi | \psi \rangle| \tag{3.24}$$

One can show that there is only one effect that leads to this value, and it can be written as

$$E = \frac{1}{1 + |\langle \varphi | \psi \rangle|} \left[P[\varphi]^\perp + P[\psi]^\perp \right] \tag{3.25}$$

This is indeed a positive operator of norm 1 so that its spectral decomposition has the form $E = P[\xi] + eP[\eta]$ $(e > 0)$, while the complement effect is a multiple of a projection, $I - E = (1 - e)P[\eta]$. It follows that the solution of the optimisation problem is unique and involves necessarily unsharp measurements.

The preceding example is based on the possibility of turning a nonorthogonal state pair into an orthogonal one by a suitable coupling with a measurement device. This allows one to envisage another new feature of measurements brought about by unsharp observables: the state changes associated with the various outcomes may be invertible so that the conditional final states of the measured system contain a complete memory of the unknown initial state [5.14]. A measurement shall be called *reversible* if its state transformer is composed of invertible operations. As an example consider again the model (3.23) whose state transformer is the Lüders transformer. If E is taken to be a projection, then the state change $\varphi \mapsto E\varphi$ is certainly not invertible. But any unsharp property E in $\mathcal{H} = \mathbf{C}^2$ is an invertible operator. Now one can conceive of a sequence of measurements where the second measurement is chosen so as to allow a reversal of the state changes brought about by the first measurement. This seems to suggest that state changes can be undone so that a measurement can be repeated many times on a single system until sufficient statistics have been collected. We shall see that this does not undermine the impossibility of individual state determination demonstrated above.

Let A_1, A_2 from (3.23) be diagonal matrices with positive eigenvalues,

$$A_1 = \begin{pmatrix} \alpha & 0 \\ 0 & \sqrt{1 - \alpha^2} \end{pmatrix} \qquad A_2 = \sqrt{I - A_1^2} = \begin{pmatrix} \sqrt{1 - \alpha^2} & 0 \\ 0 & \alpha \end{pmatrix} \tag{3.26}$$

The number α should satisfy $1 > \alpha^2 > \frac{1}{2} > 1 - \alpha^2$, which ensures the invertibility of A_1, A_2. Suppose that a measurement of the type (3.23) has led to an outcome '1' so that the initial state φ has been changed into $A_1\varphi$. Then one can perform a second measurement, with new operators $A_1' := A_2$, $A_2' := A_1$. For the outcome '1' the final state is $A_1' A_1 \varphi = \alpha\sqrt{1 - \alpha^2}\, \varphi$. Hence there is a chance to recover knowingly the unknown initial state. If in the second measurement the outcome was '2', then no reversal has taken place but the resulting state change, $\varphi \to A'_2 A_1 \varphi$, is still invertible so that a third measurement could be carried out with a chance of a successful reversal. A similar reasoning applies to the second outcome, '2', of the first measurement. One can therefore perform a sequence of measurements, each chosen according to the previous outcome, and such that after each successful reversal the original measurement is repeated on the original state. In this way one collects a random sequence of N_1 outcomes '1' and N_2 outcomes '2' of $N = N_1 + N_2$ repetitions of the original measurement, each performed on the state φ; but it is easily seen that this sequence carries as much information about the state φ as a single measurement outcome does. In fact the probability for any specific sequence of measurements ending with a successful reversal is given by a product of eigenvalues of the corresponding A-matrices and does in no way depend on the state φ. Therefore the 'correct' frequencies $N_1 \simeq N \langle \varphi \, | \, E\varphi \rangle$, $N_2 \simeq N \langle \varphi \, | \, (I - E)\varphi \rangle$ are in no way preferred over any other combination of values of N_1, N_2.

V.4 Uncertainty relations

The measurement outcome probability measures p_T^E constitute the physical basis of quantum mechanics. As pointed out in the previous section, they have some features that distinguish quantum from classical probability theory. In particular these probability measures generally take values other than 0 or 1 and they cannot be expressed as (σ-)convex combinations of a common family of Dirac measures. This feature is intimately related to the noncommutative nature of the set of experimental propositions as represented by the set of effects. As a consequence one is facing a fundamental indeterminacy, or randomness, on the level of individual (measurement) events which calls for a principal reconsideration of the meaning of probability that goes beyond making reference merely to a statistical interpretation. Closely related to the irreducibly probabilistic nature of quantum theory is the fact that for some observables, such as position and momentum, the measurement outcome distributions are mutually dependent. These quantum correlations have been related to probabilistic complementarity as discussed in Chapter IV, and they have been formally represented in terms of various kinds of uncertainty relations. Some of the measures of uncertainty introduced for this purpose shall be reviewed here.

V.4.1 Variance. The *variance* of (the measurement outcome distribution of) an observable E in the state T, $\mathrm{Var}(E, T)$, is defined as the variance of the probability measure p_T^E on the real line,

$$\mathrm{Var}(E, T) \; := \; \int x^2 \, p_T^E(dx) - \left[\int x \, p_T^E(dx) \right]^2 \tag{4.1}$$

Its square root $\Delta(E, T) := \sqrt{\mathrm{Var}(E, T)}$ is the *standard deviation* of (the measurement outcomes of) E in the state T.

Both $\mathrm{Var}(E,T)$ as well as $\Delta(E,T)$ are defined, and are finite, only when the integrals in (4.1) exist. If E is the spectral measure of a self-adjoint operator A, then $\mathrm{Var}(E,T) \equiv \mathrm{Var}(A,T) = \mathrm{Exp}(A^2,T) - \mathrm{Exp}(A,T)^2$, whenever the *first moment* (*expectation value*) $\mathrm{Exp}(A,T) = \mathrm{tr}[TA]$ and *second moment* $\mathrm{Exp}(A^2,T) = \mathrm{tr}[TA^2]$ are finite. For a vector state $T = P[\varphi]$ this is the case exactly when φ is in the domain of A, $D(A)$, whereas for mixed states T some additional convergence requirements are to be posed [5.15]. For observables represented by POV measures, conditions ensuring the existence of the variance are less easily established, and therefore the whole concept is less useful. If E is associated with a symmetric operator A (in the way specified in Section II.2.5), then $\mathrm{Var}(E,\varphi)$ is again defined for all $\varphi \in D(A)$. But there are also POV measures E for which the second moment $\int x^2 p_\varphi^E(dx)$ converges for no vector states φ [2.10]. In such a case $\mathrm{Var}(E,T)$ is never defined. We conclude from these observations that the concept of variance is to be applied with some care.

According to Eq. (4.1) the variance $\mathrm{Var}(E,T)$ is determined by the first and the second moments of the probability measure p_T^E. As pointed out in Section II.2.5, there is a one-to-one correspondence between maximal symmetric or self-adjoint operators and POV measures for which the set

$$D = \left\{ \varphi \in \mathcal{H} \mid \int x^2 d\,\langle \varphi \mid E(x)\varphi \rangle < \infty \right\} \tag{4.2}$$

is dense. This fact can be rephrased by saying that in those cases the totality of all first and second moments determine the measured observable. It may well occur that in terms of expectation values alone, two different observables appear to be the same. For instance the position Q and an unsharp position E^e have the same first moments whenever the first moment of the confidence function e vanishes (III.2.34). But the variance of an unsharp position is always greater than that of the sharp position (III.2.35). In such cases the variance can be used to characterise the sharp observable in question. Consider a self-adjoint operator A with the spectral measure E^A. Let $\mathcal{O}_A$ denote the family of all the POV measures E on the real line for which the set D from (4.2) is dense and which agree with A on the first moment,

$$E^{(1)} = \int x\,E(dx) = A \tag{4.3}$$

Clearly $E^A \in \mathcal{O}_A$, and it is uniquely determined by the property $E^{(n)} = \int x^n E(dx) = A^n$, which is to hold for all n. Consequently E^A has the smallest dispersion, for all $E \in \mathcal{O}_A$,

$$\mathrm{Var}(E^A,T) \leq \mathrm{Var}(E,T) \tag{4.4}$$

and equality holds exactly for $E = E^A$ [2.9, 2.7]. The inequality (4.4) can be rephrased in the form of the statement that the operator

$$R := E^{(2)} - A^2 \tag{4.5}$$

is positive. Taking into account the relation (4.3), one finds

$$\text{Var}(E, T) = \text{Var}(A, T) + \text{Exp}(R, T) \qquad (4.6)$$

Hence the operator R accounts for the additional scatter in the statistics of an unsharp measurement of the observable A. It is therefore called *noise operator*.

The relation between unsharp and sharp positions E^e and Q is a particular instance of relative coarse-graining, $E^e \prec E^Q$, and the variance of an unsharp position is greater than that of the sharp position, $\text{Var}(E^e, T) = \text{Var}(Q, T) + \text{Var}(e)$, the noise operator being $R = \text{Var}(e) \, I$ in this case. One might therefore ask whether it holds true in general that a finer observable has a smaller dispersion than a coarser one. The answer is in the negative. Consider a discretised position $f(Q) = \sum f_k E^Q(X_k)$ associated with a partition (X_k). This is a coarse-graining of the position. But $\text{Var}(Q, \varphi) \neq 0$ for all φ, whereas $\text{Var}(f(Q), \varphi) = 0$ for all eigenstates of $f(Q)$. Hence the variance may not always reflect the increase of uncertainty brought about by coarse-graining. This may be taken as another weakness of the variance as a probabilistic measure of uncertainty.

We shall consider next the interrelations between any two observables E_1 and E_2 in terms of the product of their variances, $\text{Var}(E_1, T) \cdot \text{Var}(E_2, T)$. For spectral measures the familiar uncertainty relations, the Heisenberg inequalities, give an estimate for a lower bound of this number, see, e.g., [2.1, 2.2]. As in the case of position and momentum, it may happen that the uncertainty product $\text{Var}(E_1, T) \cdot \text{Var}(E_2, T)$ has a positive lower bound h, that is, for each T,

$$\text{Var}(E_1, T) \cdot \text{Var}(E_2, T) \geq h > 0 \qquad (4.7)$$

Such a relation shows, first of all, that whatever the prepared state is, the measurement outcome distributions of the two observables are correlated in such a way that if one variance is 'small', then the other is necessarily 'large'. For observables represented by PV measures, this result should be confronted with the fact that the variance of one of the quantities E_1 and E_2 can be made arbitrarily small by a suitable choice of the state T. In addition for such observables the relation (4.7) implies their total noncommutativity, $\text{com}(E_1, E_2) = \{0\}$ [4.12]. For position and momentum observables the minimum of their 'uncertainty products' $\text{Var}(Q, T) \cdot \text{Var}(P, T)$ is $\frac{1}{4}$, and indeed, $\text{com}(Q, P) = \{0\}$ (Sec. IV.2.4). Again observables represented by POV measures behave differently. They may be coexistent and still be correlated according to (4.7). The prototypical example is the Fourier-related unsharp position-momentum pair (E^e, F^f). These are coexistent observables, hence they can be measured together, but their measurement outcomes are correlated such that for all states $\text{Var}(E^e, T) \cdot \text{Var}(E^f, T) \geq 1$. The lower bound 1 instead of the usual $\frac{1}{4}$ can be interpreted by saying that complementarity is lifted by trading extra measurement noise for coexistence.

V.4.2 Alternative measures of uncertainty. The notion of variance is restricted in its applicability by the requirement that the first and second momenta of the statistics must be finite. Also, the 'uncertainty product' $\text{Var}(E_1, T) \cdot \text{Var}(E_2, T)$ of two quantities may not always be a good measure of their mutual probabilistic dependence. For instance for the photon number and phase, the product of the variances does not behave according to (4.7). Other, more suitable 'measures of uncertainty', have therefore been investigated. The literature on this topic is rather vast. We shall only illustrate a few of these alternative concepts.

Consider the phase observable M of Eq. (III.5.12) which is canonically conjugate to the number observable in the sense of the covariance condition (III.5.9). The first and second moments of M are just the self-adjoint operators $M^{(1)}, M^{(2)}$ of Eqs. (III.5.20,21). Clearly $M^{(2)} \neq \left(M^{(1)}\right)^2$. These moments can be used to determine the variance $\text{Var}(M, T)$ of M in any state T. For the number states $T = |n\rangle\langle n|$, one finds $\text{Var}(M, |n\rangle\langle n|) = \pi^2/3$, so that $\text{Var}(M, |n\rangle\langle n|) \cdot \text{Var}(N, |n\rangle\langle n|) = 0$ for those states. Therefore the phase-number 'uncertainty product' $\text{Var}(M, T) \cdot \text{Var}(N, T)$ does not have the expected property that, for instance, a small number uncertainty goes along with a large phase uncertainty.

It has been argued [1.12, 5.16] that a certain function of the expectation value of the number shift operator $V = \int_0^{2\pi} e^{i\phi} M(d\phi)$ is a more appropriate measure of phase uncertainty. This function is the following one:

$$\widetilde{\text{Var}}(M, T) := \frac{1 - |\text{Exp}(V, T)|^2}{|\text{Exp}(V, T)|^2} \tag{4.8}$$

Since $0 \leq |\text{Exp}(V, T)| \leq 1$, one has $0 \leq \widetilde{\text{Var}}(M, T) \leq \infty$. It can then be shown that the following phase-number uncertainty relation holds

$$\widetilde{\text{Var}}(M, T) \cdot \text{Var}(N, T) > \tfrac{1}{4} \tag{4.9}$$

This inequality has the desired property that a small variance of one of the quantities implies a big variance for the conjugate entity.

The above type of 'uncertainty relation' can also be obtained for other observables, such as position and momentum. For instance one may consider the expectation value $\text{Exp}(e^{i\alpha Q}, T)$ of the unitary operator $e^{i\alpha Q}, \alpha \in \mathbf{R}$, and define

$$\widetilde{\text{Var}}_\alpha(Q, T) := \frac{1 - |\text{Exp}(e^{i\alpha Q}, T)|^2}{\alpha^2 |\text{Exp}(e^{i\alpha Q}, T)|^2} \tag{4.10}$$

to obtain

$$\widetilde{\text{Var}}_\alpha(Q, T) \cdot \text{Var}(P, T) \geq \tfrac{1}{4} \tag{4.11}$$

as a 'scaled' position-momentum uncertainty relation. The advantages of such a formulation is that the quantity (4.10) is defined for all states T, and therefore the left-hand side of (4.11) is finite, in particular, for all vector states φ in $D(P)$. Furthermore $\lim_{\alpha \to 0} \widetilde{\mathrm{Var}}_\alpha(Q, T) = \mathrm{Var}(Q, T)$, so that in the limit $\alpha \to 0$ the relation (4.9) leads back to the ordinary uncertainty relation.

Apart from some formal advantages the modified variance of (4.8) or (4.10), and the ensuing 'uncertainty relations' (4.9), (4.11) may offer when compared with ordinary formulations, it is to be noted that the physical relevance of these quantities is less evident. Yet there are some further notions which are rather well understood in their operational and probabilistic meaning.

Perhaps the most natural and straightforward characterisation of the 'width' of a probability distribution $p(x)$ on $\mathbf{R}$ is offered by the *overall width*, defined as the length of the smallest interval which yields a given level α of total probability $(0 < \alpha < 1)$. Thus the overall width $W^Q(\varphi; \alpha)$ of the probability distribution p_φ^Q of the position in a vector state φ associated with some level α is

$$W^Q(\varphi; \alpha) := \min \left\{ a > 0 \ \Big| \int_{x_o - a/2}^{x_o + a/2} |\varphi(x)|^2 \, dx = \alpha \text{ for some } x_o \in \mathbf{R} \right\} \quad (4.12)$$

This concept was considered long ago in signal theory [5.17], but it took some time until it was recognised in a wider context. Denoting the overall width of the momentum distribution with level α' as $W^P(\varphi; \alpha')$, one can prove that for $\alpha, \alpha' < 1$ there exists a finite positive number $C(\alpha, \alpha')$ such that for all vector states φ the following uncertainty relation holds [5.18]:

$$W^Q(\varphi; \alpha) \cdot W^P(\varphi; \alpha') \geq C(\alpha, \alpha') \quad (4.13)$$

Furthermore, this relation is known to entail the ordinary one for the variances. It is has the advantage that it applies to all states (the generalisation of (4.13) to mixed states being straightforward).

There is still another notion of width that accounts for the fine structure of a wave function (in an L^2-representation): this is the *mean peak width*, defined (for momentum) as the minimal distance $w^P(\varphi; \beta)$ over which the autocorrelation function does not drop below a certain magnitude β:

$$w^P(\varphi; \beta) := \min \left\{ |a| : \left| \int_{-\infty}^{+\infty} \hat{\varphi}(p)^* \hat{\varphi}(p + a) \, dx \right| = \beta \right\} \quad (4.14)$$

This quantity gives a measure of the width of oscillations in a 'wave function'; hence it can be used to state an aspect of complementarity in the sense that the total width of the position distribution $|\varphi(x)|^2$ and the fine structure of the momentum wave function $\hat{\varphi}(p)$ cannot be arbitrarily small in one and the same state:

$$W^Q(\varphi; \alpha) \cdot w^P(\varphi; \beta) \geq D(\alpha, \beta) \quad (4.15)$$

for some state-independent number $D(\alpha, \beta)$ [5.18]. It can be shown that it is only this relation, and not the variance uncertainty relation, which allows one to confirm Bohr's argument against the possibility of determining the path taken by a particle in a double-slit experiment without destroying the interference pattern.

The notion of overall width (4.12) can be extended in an obvious way to apply to more general observables E on $\mathcal{B}(\mathbf{R})$. We denote it $W^E(\varphi; \alpha)$. Now the overall width of a smeared position observable E^e is never less than the overall width of the sharp position Q,

$$W^{E^e}(\varphi, \alpha) \geq W^Q(\varphi; \alpha) \tag{4.16}$$

For $X = [x_o - \frac{a}{2}, x_o + \frac{a}{2}]$, consider the quantity $I_a^{E^e} := \int_X \int |\varphi(y)|^2 e(y - x) \, dy \, dx = \int dy \, e(y) \, I_a^Q(y)$, where $I_a^Q(y) = \int_{X+y} |\varphi(x)|^2 \, dx$. Choose the minimal a such that $I_a^{E^e} = \alpha$ for some suitable x_o. Hence $a = W^{E^e}(\varphi; \alpha)$. Suppose $W^Q(\varphi; \alpha) < \alpha$, which is to say that $I_a^Q(y) < \alpha$ for all $x_o + y$. But then $I_a^{E^e}$, the average of I_a^Q over the distribution e, must be less than α itself, which is false. This proves (4.16).

One may also generalise the mean peak width by casting its definition in a more abstract form: whenever an observable E is covariant with respect to a group of translations on its outcome space, with the associated unitary representation $a \mapsto V_a$, it is possible to define the autocorrelation $A(a; \varphi) := \langle \varphi \, | \, V_a \varphi \rangle$ and the mean peak width $w^E(\varphi; \beta)$ in the same way as given in (4.14). This can be applied to obtain, for instance, energy-time uncertainty relations [5.18].

Finally we give an indication how similar procedures of formulating uncertainty relations may be worked out for higher-dimensional outcome spaces. In the case of a phase space observable G, (III.2.42), one would introduce the area in phase space as a measure of uncertainty. Let $dm(q, p) = (2\pi)^{-1} dq \, dp$, then as a simple illustration we note that $p_T^G(Z) = \int_Z \text{tr}[S_{qp}T] \, dm \leq \int_Z dm = m(Z)$, hence one has

$$\min\{m(Z) \, | \, p_T^G(Z) = 1 - \varepsilon\} \geq 1 - \varepsilon \tag{4.17}$$

Although this is only a crude estimate, it shows that localising a particle in phase space with a level of confidence close to unity is impossible unless the allowed area is of the order of Planck's unit, $2\pi\hbar$.

V.4.3 Entropy. The notion of entropy can be used to characterise the mixing degree of a state in terms of the von Neumann entropy of a state, or the average lack of information for predicting a particular measurement outcome in a given state of the system, the Shannon entropy of the outcome probability distribution. It is the latter concept which is of interest here. But also the former can be recovered as an instance of the Shannon concept. In fact the *entropy of a state* T is defined as the (non-negative) number

$$S(T) := -\text{tr}[T \ln T] \tag{4.18}$$

which assumes the value 0 if and only if the state is a vector state. Evaluating the trace in (4.18) with respect to an orthonormal eigenbasis $\{\varphi_k\}$ of T, one may write $T = \sum_k t_k P[\varphi_k]$ to get $S(T) = H(\{t_k\})$, where $H(\{w_i\}) := -\sum t_k \ln t_k$ is the Shannon entropy of the discrete probability distribution (t_k). If $T = \sum w_i T_i$ is any decomposition of T into some other states T_i with the sequence of weights (w_i), $w_i \geq 0, \sum w_i = 1$, then, due to the concavity and the subadditivity of the entropy functional,

$$\sum w_i S(T_i) \leq S(\sum w_i T_i) \leq \sum w_i S(T_i) + H(\{w_i\}) \qquad (4.19)$$

with $H(\{w_i\}) := -\sum w_i \ln w_i$. The first inequality is an equality if and only if all T_i are equal, whereas the second inequality is an equality exactly when the states T_i are mutually orthogonal, $T_i T_j = O$ for all $i \neq j$ [5.19]. This shows clearly that $S(T)$ measures the degree of mixing in T. But in view of the nonunique decomposability of mixed states, the number $S(T)$ cannot in general be interpreted as the lack of information about the actual state T_k, say, of the system, if the system is known to be in the state $T = \sum w_i T_i$. Such an ignorance interpretation of the mixed state as a mixture of states is certainly inapplicable whenever the system is an entangled part of a larger compound system [1.1, 2.1].

Given an observable E and a state T, the Shannon entropy of the probability measure p_T^E describes the lack of information for predicting the outcomes of an E-measurement when performed on the system in state T. In order to formulate this concept properly, we assume that E is either discrete or continuous. In the first case the probability measures p_T^E are discrete, $i \mapsto p_T^E(i)$, while in the second case they have the distribution functions $\omega \mapsto p_T^E(\omega)$, normalised with respect to some reference measure μ. In either case the *Shannon entropy* is defined as the entropy of the probability distribution:

$$
\begin{aligned}
H(E,T) &:= -\sum p_T^E(i) \ln p_T^E(i) & (E \text{ discrete}) \\
H(E,T) &:= -\int_\Omega p_T^E(\omega) \ln p_T^E(\omega) \, d\mu(\omega) & (E \text{ continuous})
\end{aligned}
\qquad (4.20)
$$

For an arbitrary observable E, one can define its Shannon entropy in a state T with respect to a partition $\mathcal{P} = (X_i)$ of its value space Ω as the Shannon entropy of the coarse-grained observable $E|_{\mathcal{F}_\mathcal{P}}$ [Eq. (V.1.4)] in that state:

$$H(E,T|\mathcal{P}) := H(E|_{\mathcal{F}_\mathcal{P}}, T) \qquad (4.21)$$

Like the state-entropy, the functional $p_T^E \mapsto H(E,T)$ is both concave and subadditive. Moreover the concavity property guarantees that if $\mathcal{P}'$ is a finer partition than $\mathcal{P}$, so that $E|_{\mathcal{F}_\mathcal{P}} \prec E|_{\mathcal{F}_{\mathcal{P}'}}$, then also $H(E|_{\mathcal{F}_\mathcal{P}}, T) \leq H(E|_{\mathcal{F}_{\mathcal{P}'}}, T)$.

Apart from partitionings the monotonicity of the Shannon entropy is obtained also under the other discrete or continuous coarse-graining operations discussed earlier, where the actually measured observables appear as smeared versions of the sharp observables one intends to measure. For instance the unsharp number observable E^ε of (I.1.11), or (VII.3.11), resulting from non-ideal photocounting with quantum efficiency ε, or the unsharp position E^e of (II.3.29) obtained from a 'non-demolishing' measurement coupling $e^{-i\lambda Q \otimes P_A}$, were found to have the following probability distributions:

$$p_T^{E^\varepsilon}(n) = \sum_{m=n}^{\infty} p(n|m,\varepsilon)\, p_T^N(m) \tag{4.22}$$

$$p_T^{E^e}(x) = \int_{\mathbf{R}} e(y - x)\, p_T^Q(y)\, dy \tag{4.23}$$

which gives [5.20]

$$H(N,T) \leq H(E^\varepsilon, T) \tag{4.24}$$

$$H(Q,T) \leq H(E^e, T) \tag{4.25}$$

Given two observables E_1 and E_2, let $E_1|_{\mathcal{P}_1}$ and $E_2|_{\mathcal{P}_2}$ be any of their coarse-grainings defined by the partitions $\mathcal{P}_1$ and $\mathcal{P}_2$. The mutual dependence of the (discrete) probability distributions $i \mapsto p_T^{E_1}(X_i)$ and $j \mapsto p_T^{E_2}(Y_j)$ can now be expressed in terms of the entropies (4.21), as the *entropic uncertainty relations*, either with a state independent lower bound

$$H(E_1, T|\mathcal{P}_1) + H(E_2, T|\mathcal{P}_2) \geq -2\ln\left(\tfrac{1}{2}\sup_{ij} \left\| E_1(X_i) + E_2(Y_j) \right\|\right) \tag{4.26}$$

or with a sharper state dependent lower bound

$$H(E_1, T|\mathcal{P}_1) + H(E_2, T|\mathcal{P}_2) \geq -2\ln\left(\tfrac{1}{2}\sup_{ij} \left[p_T^{E_1}(X_i) + p_T^{E_2}(Y_j) \right]\right) \tag{4.27}$$

(see [5.21]). For probabilistically complementary quantities, such as position and momentum, the lower bound of (4.27) is strictly positive for all states with respect to any partitions into bounded sets.

Relations of the form (4.26) and (4.27) are particularly interesting if the observables in question are coexistent. In this case the entropies are characteristics of the joint outcome distributions and refer to one single experiment rather than to two independent ones. The entropic uncertainty relations reflect then limitations of joint measurements of noncommuting observables, including the extra unsharpness, or noise, to be introduced for achieving coexistence. Typically the relations obtained are stronger than the corresponding relations for variances. We shall illustrate this for the canonical pairs of the photon number and phase (N, M), position and momentum (Q, P), and the phase space observable G.

The entropy of the number observable $N = \sum n|n\rangle\langle n|$ in a state T is a non-negative number given by the expression

$$H(N,T) = -\sum_0^\infty p_T^N(n) \ln p_T^N(n) \qquad (4.28)$$

Using the representation (III.5.15) of the phase observable M, one can write

$$H(M,\psi) = -(2\pi)^{-1} \int_0^{2\pi} |\psi(\phi)|^2 \ln |\psi(\phi)|^2 \, d\phi \qquad (4.29)$$

for all vector states $\psi \in \mathcal{H}^2$. This number can also be negative. In fact this occurs for $\psi = \frac{1}{\sqrt{2}}(|\,0\rangle + |\,1\rangle)$, since $H(M,\psi) = \ln 2 - 1 < 0$. However, it can be shown that [5.22]

$$H(M,\psi) + H(N,\psi) \geq 0 \qquad (4.30)$$

The lower bound is attained exactly for the number eigenstates. Hence the entropic uncertainty relation respects the conjugacy of number and phase more truly than the standard uncertainty relation, expressed in terms of the variances.

For position and momentum the entropic uncertainty relation is known to be stronger than the standard one in the sense that it implies the Heisenberg inequality. To see this, consider the phase space observable G of (III.2.42), taken here for one degree of freedom. Then

$$H(G,T) = -(2\pi)^{-1} \int_{\mathbf{R}^2} \mathrm{tr}\big[TS_{qp}\big] \ln \mathrm{tr}\big[TS_{qp}\big] \, dqdp \geq 0 \qquad (4.31)$$

It can be proven [5.23, 5.24] that for each generating state S_0, the entropy (4.31) has a state independent lower bound

$$H(G,T) \geq \ln(2\pi e) \qquad (4.32)$$

By the subadditivity of H this gives a lower bound to the sum of the marginal distributions of $p_T^G(q,p)$,

$$\begin{aligned}
\rho_1(q) &:= (2\pi)^{-1} \int_{-\infty}^\infty \mathrm{tr}\big[TS_{qp}\big] \, dp = p_T^{E^e}(q) \\
\rho_2(p) &:= (2\pi)^{-1} \int_{-\infty}^\infty \mathrm{tr}\big[TS_{qp}\big] \, dq = p_T^{F^f}(p)
\end{aligned} \qquad (4.33)$$

namely,

$$H(G,T) \leq H(E^e,T) + H(F^f,T) \qquad (4.34)$$

Thus one also has

$$H(E^e,T) + H(F^f,T) \geq \ln(2\pi e) \qquad (4.35)$$

This inequality implies, first of all, that

$$\text{Var}(E^e, T) \cdot \text{Var}(F^f, T) \geq 1 \qquad (4.36)$$

and also

$$\text{Var}(Q, T) \cdot \text{Var}(P, T) \geq \frac{1}{4} \qquad (4.37)$$

The proof [5.24] of the last two inequalities is based on the fact that the sum of entropies $H(E^e, T) + H(F^f, T)$ attains its minimum value when S_0 and T are Gaussian minimal uncertainty states. In this case the distributions $\rho_1(q)$ and $\rho_2(p)$ are Gaussians. For such distributions one directly computes, for instance, $H(E^e, T) \equiv H(\rho_1^{Gauss}) = \frac{1}{2} \ln(\pi e \text{Var}(\rho_1^{Gauss}))$, showing that (4.35) implies (4.36). Choosing $S_0 = T$ and using the result (III.2.35), one finally obtains (4.37) from (4.36).

VI. Phase Space

The phrase 'quantum mechanics on phase space' [1.14] epitomises the fundamental question of the relationship between quantum mechanics and classical mechanics. The deep structural differences between the two theories make it difficult to see how to justify the common claim that the latter theory emerges as an approximation to the former in circumstances under which Planck's constant can be regarded as small. On the side of the states one would have to explain why superpositions of (vector) states representing macroscopically distinct properties of large systems are practically never observed. We do not enter into this difficult issue here.

On the part of the classical limit problem which refers to observables, we shall argue that a promising step towards a satisfactory answer may be reached with using phase space observables. Phase space concepts had been introduced quite early in the history of quantum mechanics, namely, the Wigner quasi-distribution, the Glauber-Sudarshan P-function, or the Husimi Q-function, and these turned out to provide both a valuable guide of intuition as well as powerful technical tools. Furthermore we have seen how to construct well defined probability distributions for unsharp joint measurements of position and momentum on the basis of phase space observables, and indeed the Husimi distribution is one example of these. Section 1 offers a brief survey and comparison of the various phase space representations of quantum states. In Section 2 we shall give a model description of a phase space measurement which shows that the fulfilment of the Heisenberg indeterminacy relation for individual measurement inaccuracies is a necessary precondition for the feasibility of such measurements. Moreover this model provides a basis for formulating and realising operational criteria for a quasi-classical measurement situation, which will be described in Section 3.

VI.1 Representations of states as phase space functions
In Chapter V informationally complete observables were found to induce classical embeddings of the quantum states, representing them as probability distributions via injective convex mappings. We shall now develop the idea of classical representations in greater detail for the two dimensional phase space $\Gamma = \mathbf{R}^2$.

The most widely used methods of casting quantum physical statements into a classical phase space language are due to Weyl [3.1], Wigner [6.1], and Moyal [6.2]. The task consists in finding associations $T \leftrightarrow \rho$, $A \leftrightarrow f$ between states and distribution functions, and observables and phase space functions, such that the quantum mechanical expectation values can be expressed as classical averages:

$$\mathrm{tr}[TA] = \int_\Gamma \rho(q,p)\, f(q,p)\, dq\, dp \tag{1.1}$$

The answer to this question depends on the way it is turned into a precise mathematical form. One may begin with requiring a mapping $T \mapsto \rho_T$ from the space of trace class operators to some space of functions. Then one will have to stipulate further requirements such as the linearity and boundedness of the sought mapping. These properties are equivalent to the condition that ρ_T assumes the usual quantum mechanical form:

$$\rho_T(q,p) = \mathrm{tr}\big[T\,\mathcal{P}_{qp}\big] \tag{1.2}$$

where $\mathcal{P}_{qp}$ are appropriate bounded operators associated with the phase space points (q, p). In addition one could demand that ρ_T have the usual position and momentum densities as its marginals,

$$\int_{\mathbf{R}} \rho_T(q,p)\, dp = T(q,q), \qquad \int_{\mathbf{R}} \rho_T(q,p)\, dq = \widehat{T}(p,p) \tag{1.3}$$

where $T(q, q')$ and $\widehat{T}(p, p')$ denote the matrix elements of T in the position and momentum representations, respectively. As is well known, conditions (1.2) and (1.3) cannot be reconciled with the positivity:

$$\rho_T(q,p) \geq 0 \quad \text{a.e. whenever} \quad T \geq O \tag{1.4}$$

Indeed satisfying all these requirements would amount to establishing a POV measure on phase space that has the sharp position and momentum observables as its marginals, which is impossible.

Giving up the positivity, one can satisfy (1.2) and (1.3) by introducing the Wigner quasi-distribution:

$$\rho_T(q,p) := \pi^{-1} \int_{\mathbf{R}} T(q+x, q-x)\, e^{-i2px}\, dx = \mathrm{tr}\big[T\mathcal{P}_{qp}\big]$$
$$\mathcal{P}_{qp} := \pi^{-1} W_{qp}\, \mathcal{P}\, W_{qp}^{-1} \tag{1.5}$$

Here $\mathcal{P}$ is the space inversion operator which satisfies

$$\mathcal{P}Q\mathcal{P}^{-1} = -Q, \quad \mathcal{P}P\mathcal{P}^{-1} = -P, \quad \mathcal{P} = \mathcal{P}^{-1} = \mathcal{P}^* \tag{1.6}$$

and the W_{qp} are Weyl operators,

$$W_{qp} = \exp\!\big(-iqP + ipQ\big) = e^{iqp/2}\, U_q V_p \tag{1.7}$$

From (1.5), (1.6) it is evident that the $\mathcal{P}_{qp}$ are self-adjoint but not positive operators since the spectrum of $\mathcal{P}$ consists of the eigenvalues ± 1. The only vector states $T = P[\psi]$ yielding a positive function ρ_T are known to be the Gaussian coherent states.

One can easily confirm the marginals (1.3), either by direct integration or by verifying the following operator relations:

$$\int_{X \times \mathbf{R}} \mathcal{P}_{qp} \, dq \, dp \; = \; E^Q(X), \quad \int_{\mathbf{R} \times Y} \mathcal{P}_{qp} \, dq \, dp \; = \; E^P(Y) \qquad (1.8)$$

It should be noted that (1.5) is well defined for any trace-class operator. The following consideration shows that the mapping $T \mapsto \rho_T$ can be extended to a somewhat larger domain. Taking any two trace class operators S, T, one computes

$$\begin{aligned}
\langle \rho_S \,|\, \rho_T \rangle_{L^2(\Gamma)} &\equiv \int_\Gamma \overline{\rho_S(q,p)} \, \rho_T(q,p) \, dq \, dp \\
&= \pi^{-2} \int_\Gamma \iint dx \, dy \, \overline{S(q+x, q-x)} \, T(q+y, q-y) \, e^{2ip(x-y)} \, dq \, dp \\
&= \pi^{-1} \int dq \int dx \, S^*(q-x, q+x) \, T(q+x, q-x) \\
&= (2\pi)^{-1} \operatorname{tr}[S^*T] \; \equiv \; (2\pi)^{-1} \langle S \,|\, T \rangle_{HS} \qquad (1.9)
\end{aligned}$$

The last expression denotes the Hilbert-Schmidt inner product of the two operators. If $S = P[\psi]$ and $T = P[\varphi]$, and we write $\rho_S = \rho_\psi$, $\rho_T = \rho_\varphi$, then (1.9) reads

$$2\pi \int_\Gamma \overline{\rho_\psi(q,p)} \, \rho_\varphi(q,p) \, dq \, dp \; = \; |\langle \psi \,|\, \varphi \rangle|^2 \qquad (1.10)$$

so that the 'transition probability' between two vector states is given by the L^2-inner product of their Wigner functions.

The identity (1.9) has several important applications. First of all it tells that $T \mapsto \rho_T$ establishes a Hilbert space isomorphism of the Hilbert-Schmidt class $\mathcal{B}_2(\mathcal{H})$ onto (a closed subspace of) $L^2(\Gamma, 2\pi dq \, dp)$. (Note that any trace class operator is also in the Hilbert-Schmidt class.) Hence there will be both an inverse mapping and dual mapping, and it turns out that these coincide and are all onto. Let us introduce the mappings

$$\begin{aligned}
\mathcal{W} &: \mathcal{B}_2(\mathcal{H}) \to L^2(\Gamma), \quad T \mapsto \mathcal{W}(T) := \rho_T \\
W &: L^2(\Gamma) \to \mathcal{B}_2(\mathcal{H}), \quad f \mapsto W(f) := \int_\Gamma f(q,p) \, \mathcal{P}_{qp} \, dq \, dp
\end{aligned} \qquad (1.11)$$

The mappings $\mathcal{W}$ and W are dual to each other:

$$\operatorname{tr}[T W(f)] \; = \; \int_\Gamma \mathcal{W}(T)(q,p) \, f(q,p) \, dq \, dp \qquad (1.12)$$

In this way one has established a correspondence between operators and functions which provides a representation of quantum mechanical expectations formally as classical expectation functionals in the sense of (1.1). As a consequence of (1.9), one finds that $\langle S\,|\,T\rangle_{HS} = \langle S\,|\,W(\rho_T)\rangle_{HS}$ for all operators S. Therefore

$$T \;=\; W(\rho_T) \;=\; \int_\Gamma \rho_T(q,p)\,\mathcal{P}_{qp}\,dq\,dp \tag{1.13}$$

so that W is indeed the inverse of $\mathcal{W}$. In other words the operator valued measure given by the density $(q,p) \mapsto \mathcal{P}_{qp}$ is informationally complete.

A straightforward calculation for carrying out the Fourier transform of the functions ρ_T gives the following useful connection:

$$\int_\Gamma \rho_T(q',p')\,e^{ipq'-iqp'}\,dq'\,dp' \;=\; \mathrm{tr}\big[TW_{qp}\big]$$
$$\int_\Gamma \mathcal{P}_{qp}\,e^{ipq'-iqp'}\,dq'\,dp' \;=\; W_{qp} \tag{1.14}$$

This is to say that the operator functions $\mathcal{P}_{qp}$ and W_{qp} are phase space Fourier transforms of each other.

Inserting $f(q,p) = (2\pi)^{-1}\int \exp\big(iqp' - ipq'\big)\hat{f}(p',q')dq'dp'$ in (1.11) and using (1.14), one recovers that W is the Weyl correspondence:

$$W(f) \;=\; (2\pi)^{-1}\int_\Gamma \hat{f}(p,q)\,W_{qp}\,dq\,dp \tag{1.15}$$

(Further results concerning the properties of W can be found, for example, in [6.3].)

Another implication of (1.9) is the existence of the convolution operation for Wigner functions. Let S, T be two state operators. One can write $\rho_S(q-q',p-p') = \mathrm{tr}\big[S_{qp}\mathcal{P}_{q'p'}\big]$, where $S_{qp} = W_{qp}S_0W_{qp}^{-1}$ and $S_0 = \mathcal{P}S\mathcal{P}^{-1}$. This yields

$$(\rho_S * \rho_T)(q,p) \;=\; (2\pi)^{-1}\mathrm{tr}\big[S_{qp}T\big] \;=\; p_T^G(q,p) \tag{1.16}$$

where G is the phase space observable generated by S_0. It follows that any such observable can be obtained via convoluting the respective Wigner functions. It is this 'smearing' procedure which re-establishes the lacking positivity. This may be taken as a further illustration of the fact that joint position-momentum measurements become possible if and only if an intrinsic measurement inaccuracy is taken into account so as to fulfil the uncertainty relations.

The Fourier transform of a convolution of two functions is equal to the product of their Fourier transforms. This then helps to recover the Wigner function of T from the phase space probability distribution p_T^G under a certain condition. The Fourier transform of p_T^G is:

$$\int_\Gamma e^{ipq'-iqp'}\,p_T^G(q',p')\,dq'\,dp' \;=\; \mathrm{tr}\big[SW_{qp}\big]\,\mathrm{tr}\big[TW_{qp}\big] \tag{1.17}$$

If one could divide out the first factor of the right-hand-side, one would find an explicit expression for the Fourier transform of the Wigner function ρ_T, cf. (1.14). Thus G is informationally complete exactly when $\text{tr}[S_0 W_{qp}] \neq 0$ for almost all (q, p).

The classical embedding induced by a phase space observable G,

$$V_G : \mathcal{T}(\mathcal{H}) \rightarrow L^1(\Gamma), \qquad T \mapsto V_G(T) := p_T^G \tag{1.18}$$

is positive and has the expectation value form required in (1.2); but rather than giving the usual marginals, Eq. (1.3), it leads to smeared position and momentum observables, cf. Eq. (III.2.48). This is another way out of the dilemma posed by the postulates (1.2–4), which acknowledges the fact that joint position-momentum measurements inevitably involve some unsharpness. V_G has a dual mapping that sends L^∞-functions to bounded operators:

$$W_G : L^\infty(\Gamma) \rightarrow \mathcal{L}(\mathcal{H}), \quad f \mapsto W_G(f) := (2\pi)^{-1} \int_\Gamma f(q, p)\, S_{qp}\, dq\, dp \tag{1.19}$$

The range of W_G is not all of $\mathcal{L}(\mathcal{H})$, but it is (ultraweakly) dense in $\mathcal{L}(\mathcal{H})$ exactly when G is informationally complete [5.5]. In this case all observables can be approximated (with respect to expectation values) by operators of the form (1.19).

We have so far presented two possible solutions to the classical representation problem posed in Eq. (1.1): the Wigner-Weyl correspondence, $(\mathcal{W}, W)$, and the classical embedding induced by an informationally complete phase space observable, (V_G, W_G). Eqs. (1.16), (1.17) show how closely these approaches are related to each other and, in fact, how one can be turned into the other one. In order to offer a cursory exposition of the remaining phase space approaches based on the Q- and P-functions, it is useful to switch to the oscillator notation introduced in Section III.5.1. For instance we put $a = \frac{1}{\sqrt{2}}(Q + iP)$, $z = \frac{1}{\sqrt{2}}(q + ip)$.

The (Husimi) Q-function [6.4] is a particular instance of the mapping V_G, with the generating operator S_0 being an oscillator ground state:

$$T \mapsto Q_T, \qquad Q_T(z) \equiv Q(z) := \pi^{-1}\langle z | T | z \rangle \tag{1.20}$$

It allows one to compute the expectation values of anti-normally ordered operators in a particularly simple way:

$$\langle a^r (a^*)^s \rangle_T = \int z^r \bar{z}^s\, Q(z)\, d^2 z \tag{1.21}$$

Moreover, the inverse of $T \mapsto Q_T$ can be given explicitly: noting that $Q(z) = \exp(-|z|^2) \sum_{n,m} z^m \bar{z}^n \langle n | T | m \rangle / (\pi\sqrt{n!m!}) = \sum_{n,m} z^m \bar{z}^n Q_{nm}$, one finds an expansion of the state operators in terms of normally ordered operators:

$$T = \pi \sum_{n,m} (a^*)^n a^m Q_{nm} \tag{1.22}$$

The (Glauber-Sudarshan) P-function [3.23, 6.5] results from the concern to find a practicable representation of (free field) states that display some classical features. Since the coherent states are those states which optimally follow the classical oscillator trajectories, it is natural to consider states of the form

$$T = \int P_T(z) |z\rangle\langle z| \, d^2z \qquad (1.23)$$

Providing that P_T is essentially bounded, this formulation can be seen as a particular instance of the dual mapping W_G, Eq. (1.19), of V_G, with V_G leading to the Q-functions. As the range of W_G is not all of $\mathcal{L}(\mathcal{H})$, it cannot be expected that all state operators T admit such a representation. For this, one must in fact allow for singular distributions. As an example note that $T = |z\rangle\langle z|$ corresponds to $P_T(z') = \delta(z' - z)$. Furthermore while P_T is normalised for any state operator, it will not always be positive and does not therefore admit a probabilistic interpretation. Nevertheless expectation values of normally ordered operators can be neatly computed:

$$\langle (a^*)^r a^s \rangle_T = \int \bar{z}^r z^s \, P_T(z) \, d^2z \qquad (1.24)$$

In analogy to (1.22), any state T associated with a P-function can be expanded into a power series of antinormally ordered operator products. Finally it may be remarked that also the Wigner function leads to expectation values of the form (1.24) if symmetric operator products are taken into consideration.

VI.2 Joint position-momentum measurement

In Section II.3.5 a coupling of the form $\exp(-i\lambda Q \otimes P_1)$ between an object system $\mathcal{S}$ and an apparatus $\mathcal{A}_1$ was found to afford a measurement of an unsharp position E^{e_o} of the object system. If such a measurement is followed sequentially by any (sharp) momentum measurement, one obtains the sequential joint observable of Eq. (IV.2.9), the marginals of which are the unsharp position E^{e_o} and the unsharp momentum F^{f_o}, with the Fourier related confidence functions $e_o(q) = \lambda|\phi_1(-\lambda q)|^2$ and $f_o(p) = \frac{1}{\lambda}|\hat{\phi}_1(-\frac{p}{\lambda})|^2$. Similarly one may perform first an unsharp momentum measurement, given by a coupling $\exp(i\mu P \otimes Q_2)$, and let it be followed by any (sharp) position measurement. Such a measurement sequence defines again a phase space observable with Fourier related unsharp momentum and unsharp position observables as its marginals. We investigate next a measurement that results from the joint application of the above standard-model couplings for unsharp position and momentum measurements. This model was first investigated by Arthurs and Kelly [6.6] for the minimal uncertainty case (see Section 2.4 below), whereas in [6.7] it was used to explore the role of the indeterminacy relation as a condition for the joint measurability of position and momentum.

VI.2.1 The model. We consider a measuring apparatus consisting of two probe systems, $\mathcal{A} = \mathcal{A}_1 + \mathcal{A}_2$, with initial state $\phi_1 \otimes \phi_2$. It will be coupled to the object system $\mathcal{S}$, originally in state φ, by means of the interaction

$$U := \exp\left(-\tfrac{i}{\hbar}\lambda Q \otimes P_1 \otimes I_2 + \tfrac{i}{\hbar}\mu P \otimes I_1 \otimes Q_2\right) \tag{2.1}$$

(In this and the next section we let Planck's constant $\hbar$ explicitly appear in the formulas.) The Baker-Cambell-Hausdorff decomposition of U

$$\begin{aligned}
U = \;&\exp\left(-\tfrac{1}{2\hbar^2}\lambda\mu\,[Q,P]\otimes P_1 \otimes Q_2\right) \cdot \\
&\exp\left(-\tfrac{i}{\hbar}\lambda Q \otimes P_1 \otimes I_2\right)\exp\left(\tfrac{i}{\hbar}\mu P \otimes I_1 \otimes Q_2\right)
\end{aligned} \tag{2.2}$$

shows that the position and momentum measurement probes become intertwined via their interactions with the object. A natural pointer observable is $Z = Z_1 \otimes Z_2 = E^{Q_1} \otimes E^{P_2}$, whereas a convenient pointer function is $(x,y) \mapsto (\lambda^{-1}x, \mu^{-1}y)$.

The coupling (2.1) changes the state of the object-apparatus system $\Psi_o \equiv \varphi \otimes \phi_1 \otimes \phi_2$ into $\Psi = U\Psi_o$ which has the position representation

$$\Psi(q,\xi_1,\xi_2) = \varphi(q + \mu\xi_2)\,\phi_1(\xi_1 - \lambda q - \tfrac{\lambda\mu}{2}\xi_2)\,\phi_2(\xi_2) \tag{2.3}$$

Since the second pointer observable is momentum, it is useful to Fourier transform this wave function with respect to the last variable:

$$\hat{\Psi}(q,\xi_1,\pi_2) = \tfrac{1}{\sqrt{2\pi\hbar}}\int e^{-(i/\hbar)\xi_2\pi_2}\,\Psi(q,\xi_1,\xi_2)\,d\xi_2 \tag{2.4}$$

The measured observable G as well as the state transformer $\mathcal{I}$, both of which are measures on $\mathcal{B}(\mathbf{R} \times \mathbf{R})$, are determined from the basic conditions

$$\langle\varphi\,|\,G(X \times Y)\,\varphi\rangle := \langle\Psi\,|\,I \otimes E^{Q_1}(\lambda X) \otimes E^{P_2}(\mu Y)\,\Psi\rangle \tag{2.5}$$

$$\mathrm{tr}\big[\mathcal{I}_{X \times Y}\big(P[\varphi]\big)\,A\big] := \langle\Psi\,|\,A \otimes E^{Q_1}(\lambda X) \otimes E^{P_2}(\mu Y)\,\Psi\rangle \tag{2.6}$$

which are to hold for all initial object states φ, for all outcome sets X, Y, and for all self-adjoint operators A acting in $\mathcal{H}$. One obtains:

$$G(X \times Y) = \int_{X \times Y} K_{qp}^* K_{qp}\,dq\,dp \equiv \tfrac{1}{2\pi\hbar}\int_{X \times Y} S_{qp}\,dq\,dp \tag{2.7}$$

$$\mathcal{I}_{X \times Y}\big(P[\varphi]\big) = \int_{X \times Y} K_{qp}\,P[\varphi]\,K_{qp}^*\,dq\,dp \tag{2.8}$$

$$K_{qp}(x,x') = \tfrac{1}{\sqrt{2\pi\hbar}}\,e^{\frac{i}{\hbar}p(x-x')}\,\phi_1^{(\lambda)}\big(q - \tfrac{1}{2}(x+x')\big)\,\phi_2^{(\mu)}(x - x') \tag{2.9}$$

Here we have introduced the scaled functions

$$\phi_1^{(\lambda)}(\xi_1) := \sqrt{\lambda}\,\phi_1(\lambda\xi_1), \qquad \phi_2^{(\mu)}(\xi_2) := \tfrac{1}{\sqrt{\mu}}\,\phi_1\big(\tfrac{1}{\mu}\xi_2\big) \tag{2.10}$$

At this stage we formulate some conditions to ensure that (2.9) and the subsequent operator relations are well-defined. We shall assume that the functions $\phi_1(\xi_1)$, $\phi_2(\xi_2)$, as well as their Fourier transforms, are continuous and bounded, and that they all have vanishing first and finite second moments. If seen in the position representation, the operators K_{qp} in (2.9) are integral operators with a kernel which is a function from $L^2(\mathbf{R}^2)$. This is to say that the K_{qp} are Hilbert-Schmidt operators and consequently, that the operators S_{qp} in (2.7) are positive trace class, or state operators. Since the normalisation of the POV measure G is guaranteed a priori by virtue of (2.5), one immediately has $\mathrm{tr}[S_{qp}] = 1$. A little further analysis of the K_{qp} shows that they form a phase-space translation covariant family:

$$K_{qp} = W_{qp} K_{oo} W_{qp}^{-1} \tag{2.11}$$

It is evident that G inherits this covariance and therefore is a phase space observable; one obtains

$$K_{oo} = \int dx \mid x \rangle \langle \psi_x \mid, \qquad \overline{\psi_x(x')} = K_{oo}(x', x)$$

$$S_0 = S_{oo} = 2\pi\hbar \int dx \, |\psi_x \rangle \langle \psi_x| \tag{2.12}$$

Thus S_0 is a mixed state in general. It can be a pure state only when $K_{oo}(x, x')$ factorises, which is possible only for certain Gaussian functions ϕ_1, ϕ_2.

VI.2.2 Indeterminacy relations. The marginals of the phase space observable (2.7) are unsharp position and momentum observables:

$$G(X \times \mathbf{R}) = E^e(X) = \chi_X * e(Q), \quad G(\mathbf{R} \times Y) = F^f(Y) = \chi_Y * f(P) \tag{2.13}$$

It is straightforward to determine the explicit forms of the confidence functions e, f:

$$e(q) = \int dq' \left|\phi_1^{(\lambda)}(\tfrac{1}{2}q' - q)\right|^2 \left|\phi_2^{(\mu)}(q')\right|^2 = e_o * \left|\phi_2^{(\frac{\mu}{2})}\right|^2(q)$$

$$f(p) = \int dp' \left|\hat{\phi}_2^{(\mu)}(\tfrac{1}{2}p' - p)\right|^2 \left|\hat{\phi}_1^{(\lambda)}(p')\right|^2 = f_o * \left|\hat{\phi}_1^{(\frac{\lambda}{2})}\right|^2(p) \tag{2.14}$$

Here e_o and f_o are the confidence functions of the original single measurements which can be recovered from the present joint measurement model by switching off one ($\mu = 0$) or the other ($\lambda = 0$) coupling. One thus obtains a full specification of all features of a phase space observable in operational terms. As indicated by the convolution structure, the original undisturbed inaccuracies are each changed due to the presence of the other device. In other words the simultaneous application of the measuring devices for E^{e_o} and F^{f_o} is a joint measurement of coarse-grained versions E^e and F^f of these observables.

The mutual influence of the two measurements being carried out simultaneously becomes manifest in the variances of e and f:

$$\text{Var}(e) = \frac{1}{\lambda^2}\,\text{Var}(Q_1,\phi_1) + \frac{\mu^2}{4}\,\text{Var}(Q_2,\phi_2)$$
$$\text{Var}(f) = \frac{1}{\mu^2}\,\text{Var}(P_2,\phi_2) + \frac{\lambda^2}{4}\,\text{Var}(P_1,\phi_1)$$

(2.15)

There are two ways to make the 'undisturbed' variances (the first terms) small: either by choosing large coupling constants or by preparing 'pointer' states having sharply peaked distributions $|\phi_1|^2$, $|\hat\phi_2|^2$. Both options have the same consequence: they produce large contributions to the other quantity's unsharpness (the second terms). Thus there is no way of getting both quantities $\text{Var}(e)$ and $\text{Var}(f)$ small in one and the same experiment. In fact it is apriori clear that they fulfil the uncertainty relation (III.2.50) since E^e and F^f are a Fourier couple. But we are now in a position to see the dynamical mechanism at work that ensures the phase space measurement inaccuracies to be in accord with this relation. Let us evaluate the product of the variances,

$$\text{Var}(e)\cdot\text{Var}(f) = \mathcal{Q} + \mathcal{D}$$
$$\mathcal{Q} := \frac{1}{4}\,\text{Var}(Q_1,\phi_1)\,\text{Var}(P_1,\phi_1) + \frac{1}{4}\,\text{Var}(Q_2,\phi_2)\,\text{Var}(P_2,\phi_2)$$

(2.16)

$$\mathcal{D} := \frac{1}{\lambda^2\mu^2}\,\text{Var}(Q_1,\phi_1)\,\text{Var}(P_2,\phi_2) + \frac{\lambda^2\mu^2}{16}\,\text{Var}(Q_2,\phi_2)\,\text{Var}(P_1,\phi_1)$$

Making use of the uncertainty relations $\text{Var}(Q_k,\phi_k)\,\text{Var}(P_k,\phi_k) \geq \hbar^2/4$ for the two probe systems, we find that both terms $\mathcal{Q},\mathcal{D}$ can be estimated from below. Putting $x := 16\,\text{Var}(Q_1,\phi_1)\,\text{Var}(P_2,\phi_2)/(\lambda\mu\hbar)^2$, we obtain:

$$\mathcal{Q} \geq \frac{1}{4}\left(\frac{\hbar^2}{4} + \frac{\hbar^2}{4}\right) = \frac{\hbar^2}{8}$$
$$\mathcal{D} \geq \frac{\hbar^2}{16}\left(x + \frac{1}{x}\right) \geq \frac{\hbar^2}{8}$$

(2.17)

This shows finally that

$$\text{Var}(e)\cdot\text{Var}(f) = \mathcal{Q} + \mathcal{D} \geq \frac{\hbar^2}{8} + \frac{\hbar^2}{8} = \frac{\hbar^2}{4}$$

(2.18)

It is remarkable that either one of the terms $\mathcal{Q}$ and $\mathcal{D}$ suffices to provide an absolute lower bound for the uncertainty product. Hence there are two sources of inaccuracy that give rise to an uncertainty relation. Neglecting $\mathcal{D}$ it would be simply the uncertainty relations for the two parts of the apparatus which forbids making the term $\mathcal{Q}$ arbitrarily small. This is in the spirit of Bohr's argument according to which it is the quantum nature of part of the measuring device that makes it impossible to escape the uncertainty relation. Note that the two terms occuring in $\mathcal{Q}$ each refer to one of the probe systems, and they contribute independently to the lower bound for $\mathcal{Q}$; furthermore no coupling parameters appear in $\mathcal{Q}$. There is no trace of a mutual influence between the two measurements being carried out simultaneously. On the other hand neglecting the term $\mathcal{Q}$, one would still be left with the two contributions collected in $\mathcal{D}$, the combination of which has again a lower bound. The terms in $\mathcal{D}$ are products of variances and coupling terms associated with the two probe systems, showing that $\mathcal{D}$ reflects the mutual disturbance of the two measurements. This is in accord with Heisenberg's illustrations of the uncertainty relation. For example if a particle is measured so as to have a rather well-defined momentum, then a subsequent measurement of position by means of a slit influences the effect of the preceding momentum measurement to the extent required by the uncertainty relation.

Finally we should like to emphasise that the nature of the measurement 'inaccuracy', or unsharpness, is determined by the preparations of the apparatus. Insofar as the pointer observables are indeterminate and not merely subjectively unknown, this interpretation applies to the measurement uncertainties as well: each individual measurement outcome is intrinsically unsharp, reflecting thereby a genuine quantum noise inherent in the measurement process, so that the inequality (2.18) should be properly called an indeterminacy relation. This interpretation will be further substantiated by analysing the way the measurement affects the object's states. To this end we shall have to investigate the state transformer (2.8).

VI.2.3 Mutual disturbance. The measurement coupling (2.1) is not exactly equivalent to the evolution operator describing a sequential position-momentum measurement. This is evident from the first factor appearing in the decomposition (2.2) of U, which contains a coupling between the two probe systems and which does not reduce to a unit operator due to the nonvanishing commutator of Q and P. One may ask whether there would be a way to compensate for this induced coupling term by introducing an extra interaction between the probe systems. The obvious modification of (2.1) would be the following one:

$$U^{(\kappa)} := \exp\left(-\tfrac{i}{\hbar}\lambda Q \otimes P_1 \otimes I_2 + \tfrac{i}{\hbar}\mu P \otimes I_1 \otimes Q_2 - \kappa\tfrac{i}{2\hbar}\lambda\mu I \otimes P_1 \otimes Q_2\right) \quad (2.19)$$

Here κ is some real parameter. The Baker-Campbell-Hausdorff decomposition of this coupling reads

$$\begin{aligned} U^{(\kappa)} = \exp\left(-(\kappa+1)\tfrac{i}{2\hbar}\lambda\mu I \otimes P_1 \otimes Q_2\right) \cdot \\ \exp\left(-\tfrac{i}{\hbar}\lambda Q \otimes P_1 \otimes I_2\right) \exp\left(\tfrac{i}{\hbar}\mu P \otimes I_1 \otimes Q_2\right) \end{aligned} \quad (2.20)$$

The measured observable is again a phase space observable which now depends on the new parameter κ. For our purposes it suffices to look at the variances of the confidence functions e_κ, f_κ of the new marginal position and momentum observables:

$$
\begin{aligned}
\mathrm{Var}(e_\kappa) &= \frac{1}{\lambda^2}\,\mathrm{Var}(Q_1, \phi_1) + (\kappa - 1)^2\,\frac{\mu^2}{4}\,\mathrm{Var}(Q_2, \phi_2) \\
\mathrm{Var}(f_\kappa) &= \frac{1}{\mu^2}\,\mathrm{Var}(P_2, \phi_2) + (\kappa + 1)^2\,\frac{\lambda^2}{4}\,\mathrm{Var}(P_1, \phi_1)
\end{aligned}
\tag{2.21}
$$

These variances do still satisfy the uncertainty relation, but this time the contributions corresponding to $\mathcal{Q}$ and $\mathcal{D}$ from (2.16) will both depend on the coupling between the probe systems unless $\kappa = 0$. Hence it does not help to make the extra factor in (2.20) disappear by choosing $\kappa = -1$. This would make only the second one of the variances (2.21) equal to its undisturbed value, while in the first variance the disturbing term is appropriately increased. This consideration explains why also in the sequential measurement model the validity of the uncertainty relation is ensured. In general there is no way to eliminate the mutual influence between the two measuring systems.

VI.2.4 Repeatability features. It is straightforward to realise minimal measurement uncertainty within the present model. Equality in (2.18) is equivalent to the state operator S_0 from Eq. (2.12) being a Gaussian state. The inequalities for $\mathcal{Q}, \mathcal{D}$ in (2.17) show how to reach this situation. Indeed the limiting case of equality in the first of them is obtained exactly when both ϕ_1 and ϕ_2 are Gaussian states so that $\mathrm{Var}(Q_k, \phi_k)\mathrm{Var}(P_k, \phi_k) = \hbar^2/4$. Using this in the second inequality, and noting that equalities there can only hold for $x = 1$, it follows that the two pointer variances must be correlated. This condition can be written in two equivalent ways:

$$
\begin{aligned}
\frac{1}{\lambda^2}\,\mathrm{Var}(Q_1, \phi_1) &= \frac{\mu^2}{4}\,\mathrm{Var}(Q_2, \phi_2) \\
\frac{1}{\mu^2}\,\mathrm{Var}(P_2, \phi_2) &= \frac{\lambda^2}{4}\,\mathrm{Var}(P_1, \phi_1)
\end{aligned}
\tag{2.22}
$$

This shows that minimal total noise can be achieved only if for both observables measured the additional noise due to the joint coupling equals the original undisturbed unsharpness. With (2.22) applied to Gaussian functions ϕ_1, ϕ_2 the kernel $K_{qp}(x, x')$ of Eq. (2.9) is readily seen to factorise, and one obtains $K_{oo} = (2\pi\hbar)^{-1/2}|\psi^{Gauss}\rangle\langle\psi^{Gauss}|$ and therefore $S_0 = |\psi^{Gauss}\rangle\langle\psi^{Gauss}|$.

In this case the state transformer (2.8) assumes a particularly simple form:

$$
\mathcal{I}_{X \times Y}(T) = (2\pi\hbar)^{-1}\int_{X \times Y}\langle\psi_{qp}\,|\,T\psi_{qp}\rangle\,P[\psi_{qp}]\,dq\,dp
\tag{2.23}
$$

This is a mixture of coherent states which depends on the initial state T only via the weights, cf. Eqs. (IV.1.16) and (V.3.9). Consider 'point-like' readings, that is, intervals $X = [q_o - \delta, q_o + \delta]$, $Y = [p_o - \gamma, p_o + \gamma]$ of lengths small compared to the respective variances of ψ_{qp}. One may expect that for a large class of states T the functions $\langle \psi_{qp} | T \psi_{qp} \rangle$ are slowly varying within such intervals. (Indeed this can be achieved for any state by adapting the interval lengths, cf. Section V.3.3.) For such intervals one can approximate the integration to obtain

$$\mathcal{I}_{X \times Y}(T) \simeq P[\psi_{q_o,p_o}] \langle \psi_{q_o,p_o} | T \psi_{q_o,p_o} \rangle \frac{4\delta\gamma}{2\pi\hbar} \qquad (2.24)$$

After normalisation no mark is left of the original object state. On the contrary, the characteristics of the measurement scheme are ideally imprinted into the system: its final state is localised at the 'point' (q_o, p_o) indicated by the reading, and the variances are exactly given by the measurement inaccuracies (2.15). This means that the measurement is quasi-preparatory in the sense of Section V.3.3. Moreover as noted in Section IV.1, this measurement is (ε, δ)-repeatable.

The last property does not pertain to the more general non-Gaussian case of our model. Still the measurement is quasi-preparatory for point-like readings and for states which are 'slowly varying' in phase space in a sense described symbolically by $\mathrm{Var}(Q, \varphi) \gg \mathrm{Var}(e)$, $\mathrm{Var}(P, \varphi) \gg \mathrm{Var}(f)$. Thus in a situation where the measurement can be regarded as highly accurate, the system is brought into states which reflect the outcomes in both the first and second moments of the corresponding confidence functions. In this way the outcomes refer to the object as its unsharp and only approximately real properties: one can predict with any level of confidence the result of a subsequent measurement to lie within an interval around the previous point reading provided the interval is made sufficiently large. The remaining uncertainty corresponds to the indeterminateness of the position and momentum observables that is irreducibly left in the final state of the object.

VI.3 Classical limit

One would expect that being able to perform phase space measurements it should also be possible to observe trajectories of microscopic particles. Thus one may hope to achieve a detailed quantum mechanical understanding of the formation of cloud or bubble chamber tracks. We shall show here that within the present model necessary conditions for such quasi-classical measurement behaviour are good localisation of the object and macroscopically large inaccuracies contributed by the device.

VI.3.1 Classical measurement situation.
In contrast to the quasi-preparatory measurement described in the preceding section, a classical measurement situation is characterised among others by the possibility of observing a particle without necessarily influencing it. The above phase space measurement model allows one to formalise this and some further classicality conditions and to demonstrate their realisability. We shall explicate four such requirements.

First, it should be admissible to think of the particle having 'extremely sharp' values of position and momentum. This cannot be meant in an absolute sense but only relative to the scale defined by the resolution of the means of measurement.
(C1) *Near value determinateness.*

$$\text{Var}(Q,\varphi) \ll \text{Var}(e), \qquad \text{Var}(P,\varphi) \ll \text{Var}(f) \tag{3.1}$$

This will be taken to imply that the functions $\phi_1^{(\lambda)}$ and $\hat{\phi}_2^{(\mu)}$ are slowly varying over the lengths within which $\varphi(q)$ and $\hat{\varphi}(p)$ are appreciably different from zero, respectively. Such states shall be referred to as 'localised' (in phase space).

Next, the position and momentum measurements should not disturb each other when performed jointly. This can be controlled by the variances (2.15) in terms of the condition that the additional noise terms should remain negligible:
(C2) *Small mutual disturbance.* $\text{Var}(e) \simeq \text{Var}(e_o)$ and $\text{Var}(f) \simeq \text{Var}(f_o)$; therefore

$$\begin{aligned}
\frac{1}{\lambda^2}\,\text{Var}(Q_1,\phi_1) &\gg \frac{\mu^2}{4}\,\text{Var}(Q_2,\phi_2) \\
\frac{1}{\mu^2}\,\text{Var}(P_2,\phi_2) &\gg \frac{\lambda^2}{4}\,\text{Var}(P_1,\phi_1)
\end{aligned} \tag{3.2}$$

This will again be taken to mean that the functions $\phi_1^{(\lambda)}$ and $\hat{\phi}_2^{(\mu)}$ should be slowly varying over lengths in which $\phi_2^{(\mu)}$ and $\hat{\phi}_1^{(\lambda)}$ are noticeably different from zero.

Third, in view of the uncertainty relation (2.18) for the measurement inaccuracies it should be kept in mind that in a classical measurement the imprecisions seem to be so large that no indication of Planck's constant can ever be observed.
(C3) *No limit of accuracy.* The position and momentum measurement inaccuracies can be made arbitrarily small:

$$\text{Var}(e)\,\text{Var}(f) \gg \frac{\hbar^2}{4} \tag{3.3}$$

Finally, since the properties to be measured are practically determinate, one should expect that a measurement will not necessarily disturb the system but will merely register the corresponding values. This is to say that the measurement should be approximately ideal and a fortiori approximately value reproducible.
(C4) *Approximate ideality.* Localised states [cf. (C1)] should not be disturbed much in a joint position-momentum measurement.

$$\mathcal{I}_{X\times Y}\big(P[\varphi]\big) \simeq P[\varphi]\,\langle\,\varphi\,|\,G(X\times Y)\,\varphi\,\rangle \qquad \text{(for localised } \varphi) \tag{3.4}$$

These features are not mutually independent. It is evident that (C1) implies (C3). The second property, (C2), gives somewhat more:

$$\text{Var}(e)\,\text{Var}(f) \geq \text{Var}(e_o)\,\text{Var}(f_o) \gg \frac{\hbar^2}{4} \tag{3.5}$$

This shows that mutual nondisturbance can only be achieved if from the outset one starts with highly unsharp measurements. If the slow variation properties expressed in (C1) and (C2) are taken together, they can be proven to entail (C4); we shall not go into details here (cf. [6.8]) but mention only that the sign $\simeq$ in (3.4) is to be understood as an approximation with respect to the trace norm metric. The basic approximations are found to be in the following steps:

$$
\begin{aligned}
K_{qp}\varphi(x) &= \int \frac{dx'}{\sqrt{2\pi\hbar}}\, \phi_1^{(\lambda)}\big(q - \tfrac{1}{2}(x + x')\big)\, \phi_2^{(\mu)}(x' - x)\, e^{\frac{i}{\hbar}p(x-x')}\, \varphi(x) \\
&\simeq \phi_1^{(\lambda)}(q - x) \int \frac{dx'}{\sqrt{2\pi\hbar}}\, \phi_2^{(\mu)}(x' - x)\, e^{\frac{i}{\hbar}p(x-x')}\, \varphi(x) \\
&= \phi_1^{(\lambda)}(q - x) \int \frac{dp'}{\sqrt{2\pi\hbar}}\, \hat{\phi}_2^{(\mu)}(p - p')\, \hat{\varphi}(p')\, e^{\frac{i}{\hbar}xp'} \\
&\simeq \phi_1^{(\lambda)}(q - x)\, \hat{\phi}_2^{(\mu)}\big(p - \langle P\rangle_\varphi\big)\, \varphi(x) \\
&\simeq \phi_1^{(\lambda)}\big(q - \langle Q\rangle_\varphi\big)\, \hat{\phi}_2^{(\mu)}\big(p - \langle P\rangle_\varphi\big)\, \varphi(x)
\end{aligned}
\tag{3.6}
$$

The first approximation is due to the slow variation of $\phi_1^{(\lambda)}$ against $\phi_2^{(\mu)}$ (C2), the second and third use the slow variations of $\hat{\phi}_2^{(\mu)}$ and $\phi_1^{(\lambda)}$ against $\hat{\varphi}$ and φ, respectively (C1). This shows, in particular, that the probability distribution depends, in this limit, only on the first moments of position and momentum:

$$
p_\varphi(q, p) = (2\pi\hbar)^{-1}\, \langle \varphi \,|\, S_{qp}\varphi \rangle \simeq \left| \phi_1^{(\lambda)}\big(q - \langle Q\rangle_\varphi\big) \right|^2 \left| \hat{\phi}_2^{(\mu)}\big(p - \langle P\rangle_\varphi\big) \right|^2
\tag{3.7}
$$

The same sequence of approximations can be applied to determine the total 'wave function' (2.4) of the compound system to be

$$
\hat{\Psi}(x, \xi_1, \pi_2) \simeq \phi_1\big(\xi_1 - \lambda\langle Q\rangle_\varphi\big)\, \hat{\phi}_2\big(\pi_2 - \mu\langle P\rangle_\varphi\big)\, \varphi(x)
\tag{3.8}
$$

This seems to suggest that one simply would have to read off the shift of the pointers in order to measure the expectation values of position and momentum of a single particle, without even changing its state. But it is quite obvious from the assumptions underlying the approximations made that the variances of the pointer observables are so large that one obtains only very low confidence estimates of the peak shifts of ϕ_1, $\hat{\phi}_2$. Still one can give (not too small) confidence intervals, in terms of a suitable phase space cell partition, such that a localised state φ will be found, with high probability, within a cell containing the point $(\langle Q\rangle_\varphi, \langle P\rangle_\varphi)$. This confirms the value-reproducibility of the quasi-classical measurement.

In the classical measurement situation described here, one is facing two kinds of uncertainty which have to be interpreted quite differently. If one starts with a localised though otherwise unknown state, then it is a matter of *subjective ignorance* what the 'true values' $(q_o, p_o) = (\langle Q\rangle_\varphi, \langle P\rangle_\varphi)$ of position and momentum are.

The measurement will give some (point-like) outcome (q, p) most likely in a region around (q_o, p_o), in which the probabilities given by (3.7) are non-negligible: $|q-q_o| \leq n\sqrt{\text{Var}(e)}$, $|p - p_o| \leq n\sqrt{\text{Var}(f)}$, with n of the order of unity. *Which* result will come out is objectively undecided as the unsharpnesses originate from the pointer indeterminacies. Hence with respect to the state inference problem one is dealing with subjective uncertainties, while preditions of future measurement outcomes are objectively indeterminate.

The above considerations reveal the decisive role of Planck's constant $\hbar$ for the classical limit of quantum mechanics in a new sense. Only with respect to measuring instruments yielding macroscopic inaccuracies, Eq. (3.5), is it possible to neglect the quantum mechanical restrictions and to make approximate use of the classical physical language as laid down in (C1–4). It is remarkable that the indeterminacy product $\text{Var}(Q, \varphi)\text{Var}(P, \varphi)$ of the object and, as will become apparent, the products $\text{Var}(Q_k, \phi_k)\text{Var}(P_k, \phi_k)$ of the probe systems need not at all be small for classical measurements. Thus one can conceive of measuring classical trajectories for microscopic particles. This is in accord with the situation encountered in the Wilson cloud chamber where elementary particles propagate along visible macroscopic paths. The 'collapse' of wave packets caused by the ionisation of scattering molecules – the microscopic probe systems – becomes negligible if the particles travel fast enough to avoid any appreciable spreading of the packets. In addition the ionisation will not influence too much the particle as long as its energy is high enough.

VI.3.2 Approximately ideal measurements. It remains to be shown that the conditions (C1), (C2), and thereby (C3) and (C4), can be fulfilled by suitable choices of the probe states ϕ_1, ϕ_2. This follows easily from a special scaling property of our model. As the computations in Section 2 have shown, the state transformer and measured observable will be the same in two models which are specified by the following choices of coupling parameters, apparatus state, and pointer scales: (λ, μ), (ϕ_1, ϕ_2), $\left(E^{Q_1}(\lambda X), E^{P_2}(\mu Y)\right)$, or $(1, 1)$, $\left(\phi_1^{(\lambda)}, \phi_2^{(\mu)}\right)$, $\left(E^{Q_1}(X), E^{P_2}(Y)\right)$. In other words a change of the couplings with a corresponding rescaling of the pointers has the same effect as simply rescaling the apparatus state with appropriate parameters. This can be verified in an elegant way by interpreting the scale transformations $\lambda \to \lambda'$, $\mu \to \mu'$ applied to Eq. (2.3) as unitary mappings changing P_1 into $(\lambda'/\lambda)P_1$ and Q_2 into $(\mu'/\mu)Q_2$. It is due to the particular form of the coupling operator that these two physically different rescaling operations cannot be distinguished on the level of the object system.

The dependence of the state transformer and the observable on the coupling parameters is explicitly determined in Eqs. (2.5) and (2.6). With this at hand, one is in a position to confirm that the approximating conditions (C1) and (C2) are self-consistent, can be realised, and do indeed lead to (C4). This can be most easily illustrated at one stroke by considering a simple special case. Let $\phi_1^{(\lambda)}$, $\phi_2^{(\mu)}$

be Gaussian states,

$$
\begin{aligned}
\phi_1^{(\lambda)}(\xi_1) &= \left(\frac{\lambda^2}{\pi\hbar}\right)^{\frac{1}{4}} \exp\left(-\frac{1}{2\hbar}\lambda^2\xi^2\right) \\
\hat{\phi}_2^{(\mu)}(\pi_2) &= \left(\frac{\mu^2}{\pi\hbar}\right)^{\frac{1}{4}} \exp\left(-\frac{1}{2\hbar}\mu^2\pi^2\right)
\end{aligned}
\tag{3.9}
$$

By making λ, μ sufficiently large one can manage $\phi_1^{(\lambda)}$ and $\hat{\phi}_2^{(\mu)}$ to be slowly varying against $\phi_2^{(\mu)}$ and $\hat{\phi}_1^{(\lambda)}$, respectively. This ensures small mutual distubance, (C2), and, in particular, Eq. (3.2). At the same time it becomes evident that there will be a large family of object states φ which are localised in the sense of comparably slow variations of $\phi_1^{(\lambda)}$ against $\varphi(q)$ and $\hat{\phi}_2^{(\mu)}$ against $\hat{\varphi}(p)$. Then the approximations indicated in (3.6–8) can be actually carried out.

The confidence functions e, f are now also Gaussian functions. As can be seen from the variances (2.15), an increase of λ, μ leads to larger values of the undisturbed variances $\mathrm{Var}(e_o), \mathrm{Var}(f_o)$ and to a decrease of the additional noise terms. For sufficiently large λ, μ any further enlargement will always increase the variances of e, f, and in this case the increase can be interpreted as being due to a convolution operation with some suitable Gaussian distributions. This is to say that the process of going to larger values of λ, μ may be effected by coarse-graining the marginal unsharp position and momentum observables.

Finally we illustrate the weakly disturbing or almost ideal character of the resulting measurement by a simulation for a Gaussian object state φ [6.8],

$$
\varphi(q) = (\pi\hbar\sigma)^{-\frac{1}{4}} \exp\left(-\frac{1}{2\hbar}\frac{1}{\sigma^2}(q-q_o)^2\right)
\tag{3.10}
$$

with variances $\mathrm{Var}(Q,\varphi) = (\hbar/2)\sigma^2$, $\mathrm{Var}(P,\varphi) = \hbar/(2\sigma^2)$. Noting that the variances of the pointer states (3.9) are $\mathrm{Var}(Q_1, \phi_1^{(\lambda)}) = \hbar/(2\lambda^2)$, $\mathrm{Var}(P_2, \phi_2^{(\mu)}) = \hbar/(2\mu^2)$, one may realise both extreme cases investigated in this hapter, that of a quasi-preparatory measurement by letting $\sigma^2\lambda^2 \gg 1$, $\mu^2/\sigma^2 \gg 1$ (a highly spread-out object state), and the weakly disturbing measurement with the opposite choice $\sigma^2\lambda^2 \ll 1$; $\mu^2/\sigma^2 \ll 1$ (a localised object state).

The state transformations in both cases can be illustrated graphically by plotting the Q-functions of the (normalised) states before ($P[\varphi]$) and after ($T_{X\times Y} := \mathcal{I}_{X\times Y}(T)/\mathrm{tr}[\mathcal{I}_{X\times Y}(T)]$) measurement. The plots displayed in Figures 6.1–3 are based on the specifications

$$
\hbar = 1, \qquad \sigma = 1, \qquad q_o = 4
\tag{3.11}
$$

The pointer unsharpnesses are indicated by means of 'error ellipses' centered around the coordinate origin. The object state is represented as a grey disc with center point $(4,0)$. The rectangular box corresponds to the reading intervals

$$
X = [2,6], \qquad Y = [-6,-2]
\tag{3.12}
$$

The plots provide a comparison of the initial and final object states for different values of the coupling constants λ, μ. Figure 6.1 shows a reference case, $\lambda = \mu = 0.5$, where the error ellipse, the object disc, and the reading box are of comparable sizes. The initial Q-function is smeared and shifted towards the box. In Figure 6.2 one has $\lambda = \mu = 3$, which corresponds to a strongly disturbing measurement: the Q-function is strongly smeared and its peak is shifted close to the center of the box. This shows the preparatory features of the measurement in this limit. Finally, choosing $\lambda = \mu = 0.1$, as indicated in Figure 6.3, one obtains a weakly disturbing measurement. The shifts and smearing are quite small in this case so that the state is indeed left nearly unchanged.

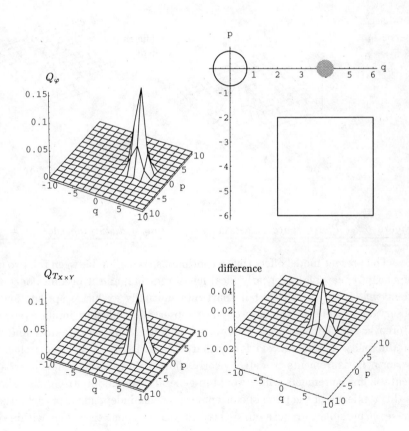

Figure 6.1. Phase space measurement – reference case $(\lambda = \mu = 0.5)$

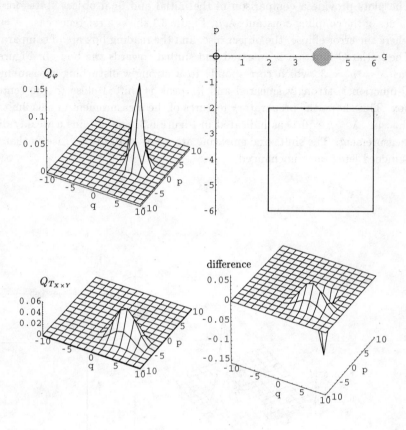

Figure 6.2. Phase space measurement – strongly disturbing case $(\lambda = \mu = 3)$

The present model offers thus a continuous transition between the two extremes of nearly repeatable and nearly ideal measurements, and it becomes once more evident that these two ideals fall apart into mutually exclusive options for unsharp observables, while in the case of sharp observables ideality implies repeatability. Moreover the tight connection between the concepts of repeatability and value reproducibility is also relaxed for unsharp observables. This means that one can perform measurements to detect nearly sharp values without having to apply repeatable measurements which would in general bring about strong disturbances. It is this tendency of the pairs of concepts to become independent of each other which makes it possible to realise quasi-classical measurement situations within quantum mechanics.

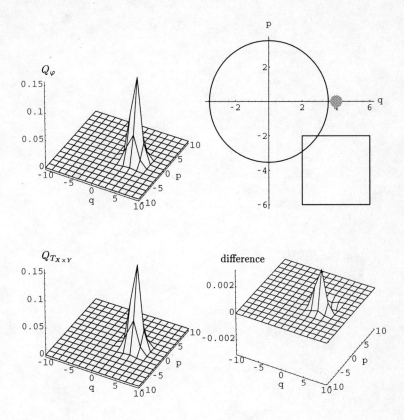

Figure 6.3. Phase space measurement – weakly disturbing case $(\lambda = \mu = 0.1)$

VII. Experiments

In the preceding chapters the representation of quantum observables as POV measures has been elucidated mostly in theoretical and general operational respects. We shall now investigate some concrete experimental setups, showing how each of them determines a POV measure as the measured observable. Thus it becomes clear that even under the most idealised circumstances the observable actually measured is, as a rule, a smeared version of the sharp observable that one perhaps intends to measure. We start with a closer look at the classic Stern-Gerlach experiment and conclude that it constitutes a nonideal measurement of an atomic spin component. The second example exhibits an informationally complete photon polarisation measurement. In Section 3 we discuss a variety of quantum optical experiments. First the number statistics of a single-mode electromagnetic field in a photodetector of quantum efficiency ε are determined; then we investigate some first kind measurements of the photon number via a two-mode coupling produced by the optical Kerr effect. These examples serve to illustrate the dependence of the actually measured 'number' observable on the applied readout observable. Next the photon-atom interaction in a cavity field is used in two ways, either for determining the photon number in a microwave cavity, or as a position measurement of an atom passing through a standing light wave. The questions of measuring the phase and quadrature components of a single-mode radiation field are addressed in Sections 3.4 through 3.7. The final section of this chapter is devoted to the wave-particle duality for photons. Applying the Mach-Zehnder interferometer for a photon split-beam experiment, it will be demonstrated on various levels of description that photons may simultaneously though unsharply display wave and particle behaviour.

VII.1 Stern-Gerlach experiment

The Stern-Gerlach experiment furnishes a school example of a quantum mechanical measurement process. The coupling of the spin degrees of freedom of an atom with its spatial degrees of freedom produced by a Stern-Gerlach magnet is well understood, and the resulting splitting of the atomic beam is usually considered to afford the prototype of an ideal spin measurement. Still we shall see that even under quite idealised conditions the actually measured spin quantity of an atom is rather an unsharp than a sharp observable. Studying the behaviour of the spin state of an atom in a Stern-Gerlach device, one can exhibit the lack of ideality of the resulting spin measurement. In Section IV.3.2 the spin-orbit coupling was found to afford another unsharp spin measurement scheme.

Let a beam of silver atoms be sent into a Stern-Gerlach device, cf. Figure 1.2. The beam is split into two parts, the atoms found in the upper section of the screen have spin up and those in the lower part spin down. This is the simplified textbook picture of the experiment. We shall now investigate what actually happens to the spin state and in which sense the device, when equipped with an appropriate screen observable, constitutes a spin measurement of the atom.

An appropriate realisation of the Hilbert space of the atom is $L^2(\mathbf{R}^3) \otimes \mathbf{C}^2$, with $L^2(\mathbf{R}^3)$ and $\mathbf{C}^2$ accounting for its spatial and spin degrees of freedom, respectively. Let $\mathbf{B}$ be the magnetic field produced by the Stern-Gerlach device. It will be treated classically. The field acts on the atom via the interaction

$$H_{SG} = \mu \cdot \mathbf{B}, \quad \mu = \mu_o \sigma, \quad \mu_o = \frac{g_s e}{2mc} \qquad (1.1)$$

where $\mu \cdot \mathbf{B} = \sum_{k=1}^{3} B_k \otimes \mu_k$, and it correlates the spin and the space degrees of freedom of the atom. Assume that the field strength and gradient are so strong that changes due to the 'free' evolution of the atom are negligible in comparison to the effect of the interaction as long as the particle is in the interaction region. In other words we apply the 'impulsive measurement' approximation. Suppose that the atom entering the device is initially in a state $\phi \otimes \varphi$, with the spatial part ϕ fairly well localised relative to the extension of the magnetic field region. Assume also that the interaction between the device and the atom is confined to a finite region of the space, the location of the magnetic field.

The initial state of the atom is transformed in the interaction according to

$$\phi \otimes \varphi \mapsto U(\phi \otimes \varphi) =: \Psi_\tau, \quad U = e^{-i\tau\mu \cdot \mathbf{B}} \qquad (1.2)$$

with τ denoting the duration of the interaction. Monitoring some spatial observable of the atom, say A, with the spectral measure E^A, the measured spin observable $E \equiv E^{U,\phi,A}$ is determined by the condition

$$\langle \varphi | E(X)\varphi \rangle := \langle U(\phi \otimes \varphi) | E^A(X) \otimes I\, U(\phi \otimes \varphi) \rangle \qquad (1.3)$$

which is stipulated to hold for all initial spin states φ and for all possible value sets X of the monitoring observable A. Note that an appropriate pointer function can be introduced without any complications when needed.

The question at issue is whether the Stern-Gerlach device can realise a measurement of a component s_3, say, of the atom's spin, that is, whether

$$E^{s_3} = E^{U,\phi,A} \qquad (1.4)$$

for some U, ϕ, and A. Since the components of the spin are given by discrete self-adjoint operators, like $s_3 = \frac{1}{2}P[\varphi_+] - \frac{1}{2}P[\varphi_-]$, the ideality of a spin measurement already implies its equivalence to a von Neumann-Lüders measurement

of this quantity, cf. Sec. II.3.5. In particular, the state transformations associated
with an ideal measurement of s_3 are given by the Lüders transformer

$$P[\varphi] \;\mapsto\; P[\varphi_\pm]\,P[\varphi]\,P[\varphi_\pm] \;=\; |\langle \varphi_\pm | \varphi \rangle|^2\, P[\varphi_\pm] \;\equiv\; |c_\pm|^2\, P[\varphi_\pm] \qquad (1.5)$$

with $c_\pm = \langle \varphi_\pm | \varphi \rangle$. The spin state of the atom, after passing the device, but before
reading the result, would therefore be the Lüders mixture

$$|c_+|^2\, P[\varphi_+] \;+\; |c_-|^2\, P[\varphi_-] \qquad (1.6)$$

We study first to what extent the properties (1.4)–(1.6) can be realised by a Stern-
Gerlach device.

VII.1.1 Ideal field. Consider the simple case (which is actually inconsistent with
Maxwell's equations):

$$\mathbf{B} \;=\; (B_o - bz)\,\mathbf{e}_3 \qquad (1.7)$$

The state of the atom after passing the field region is

$$\begin{aligned}
\Psi_\tau &= e^{-i\tau\mu_o B_3 \otimes \sigma_3}\, \phi \otimes \varphi \\
&= c_+\phi_+ \otimes \varphi_+ \;+\; c_-\phi_- \otimes \varphi_-, \qquad c_\pm = \langle \varphi_\pm | \varphi \rangle
\end{aligned} \qquad (1.8)$$

where $\varphi_\pm$ are spin eigenstates of $s_3 = \tfrac{1}{2}\sigma_3$, and the deflected wave functions $\phi_\pm$ are

$$\phi_\pm(\mathbf{q}) \;=\; e^{\mp i\tau\mu_o(B_o - bz)}\, \phi(\mathbf{q}) \qquad (1.9)$$

Writing these functions in the momentum representation

$$\widetilde{\phi}_\pm(\mathbf{p}) \;=\; e^{\mp i\tau\mu_o B_o}\, \widetilde{\phi}(\mathbf{p} \mp \tau\mu_o b\mathbf{e}_3) \qquad (1.10)$$

shows that the inhomogeneous part of the magnetic field produces shifts of mag-
nitudes $\mp\tau\mu_o b$ in the third component of the (center-of-mass) momentum of the
atom. Therefore it appears as if the two components of the state separate, causing
two distinguishable spots on the screen. In order to have a strict spatial separation
of the component wave functions ϕ_+ and ϕ_-, and thus an ideal spin determination,
one should have $\langle \phi_+ | \phi_- \rangle = 0$. Given such a 'pointer state' ϕ, one may take any
'pointer observable' A, with $A(X)\phi_+ = \phi_+$ and $A(Y)\phi_- = \phi_-$ for some disjoint
sets X and Y, to obtain an ideal spin measurement. The most direct way would be
to monitor the z-component of the atom's momentum, $A = P_3$, and to prepare the
atom initially such that, for instance, $E^{P_3}(\mathbf{R}_+)\phi_+ = \phi_+$ and $E^{P_3}(\mathbf{R}_-)\phi_- = \phi_-$.
The obvious difficulty with this is that one cannot prepare the atom in a state ϕ
which is localised both in configuration and momentum space. It appears therefore
that the present scheme cannot constitute an ideal measurement of s_3 in a strict
sense. In order to tackle the separation problem of the component wave functions

$\phi_\pm$ and the related spin correlations in greater detail, we need to study further the general structure of the measurement.

First of all one must introduce some 'screen observable' representing the registration of atoms impinging on the screen. Suppose that the screen absorption has 100% efficiency, independently of the momentum and the spin. Then localisation in regions of the screen yields all available information so that the use of a localisation observable instead of a proper screen observable will be sufficient. We consider still an arbitrary monitoring observable A. The actually measured spin observable E, determined by (1.3), is now

$$E(X) = \left\langle \phi_+ \,|\, E^A(X)\phi_+ \right\rangle P[\varphi_+] + \left\langle \phi_- \,|\, E^A(X)\phi_- \right\rangle P[\varphi_-] \qquad (1.11)$$

showing that $E(X)$ is a projection operator only if the numbers $\left\langle \phi_+ \,|\, E^A(X)\phi_+ \right\rangle$ and $\left\langle \phi_- \,|\, E^A(X)\phi_- \right\rangle$ are either 0 or 1. Otherwise the measured observable E is a smeared version of the sharp spin observable s_3.

The moment operators of the measured observable are easily obtained from (1.11). For instance its first moment is

$$E^{(1)} := \int_\Omega x E(dx) = \left\langle \phi_+ \,|\, A\phi_+ \right\rangle P[\varphi_+] + \left\langle \phi_- \,|\, A\phi_- \right\rangle P[\varphi_-] \qquad (1.12)$$

(provided that the vectors $\phi_\pm$ are in the domain of A). This shows that on the statistical level of first moments, the measured observable E appears as a function of s_3,

$$E^{(1)} = f(s_3), \qquad f(\pm\tfrac{1}{2}) := \left\langle \phi_\pm \,|\, A\phi_\pm \right\rangle \qquad (1.13)$$

Similarly the other moments of E are

$$E^{(k)} = f^{(k)}(s_3), \qquad f^{(k)}(\pm\tfrac{1}{2}) = \left\langle \phi_\pm \,|\, A^k\phi_\pm \right\rangle \qquad (1.14)$$

(with the assumption that $\phi_\pm$ are in the domain of A^k). It follows that in general $E^{(k)} \neq (E^{(1)})^k$, which is just another way of saying that the measured observable is represented by a POV measure, and not by a PV measure.

The general structure of the ensuing state transformer $\mathcal{I}$ can be directly computed, too:

$$\mathcal{I}_X\big(P[\varphi]\big) = \sum_{k,l=+,-} p_{lk}(X) \, P[\varphi_k] \, P[\varphi] \, P[\varphi_l]$$

$$p_{kl}(X) = \left\langle \phi_k \,|\, A(X)\,\phi_l \right\rangle, \quad k,l \in \{+,-\} \qquad (1.15)$$

This shows, first of all, that the spin states $P[\varphi_\pm]$ remain essentially unchanged in the course of the measurement

$$\mathcal{I}_X\big(P[\varphi_\pm]\big) = \left\langle \phi_\pm \,|\, E^A(X)\phi_\pm \right\rangle P[\varphi_\pm] \qquad (1.16)$$

the factor $\langle \phi_\pm \, | \, E^A(X)\phi_\pm \rangle$ exhibiting its deviation from an ideal s_3-measurement, (1.5). Second, the spin state of the atom after it has passed the Stern-Gerlach device is

$$
\begin{aligned}
\mathcal{I}_\Omega\big(P[\varphi]\big) = \; & |c_+|^2 P[\varphi_+] + |c_-|^2 P[\varphi_-] \\
& + \{\langle \phi_+ | \phi_- \rangle \, P[\varphi_-]\, P[\varphi]\, P[\varphi_+] + h.c.\}
\end{aligned}
\tag{1.17}
$$

This is the Lüders-mixture (1.6) only if $\langle \phi_+ | \phi_- \rangle = 0$. However, as pointed out above, such a choice is not feasible. Apart from this, for any ϕ and A, the measurement is a first kind measurement of the actually measured unsharp spin observable. Indeed the outcome probabilities for E are the same both before and after the measurement: for each X and for each initial spin state $P[\varphi]$,

$$
\langle \varphi \, | \, E(X)\varphi \rangle \; = \; \mathrm{tr}\big[\mathcal{I}_\Omega\big(P[\varphi]\big)\, E(X)\big]
\tag{1.18}
$$

It may be remarked that the present scheme preserves also the outcome probabilities of s_3, even though it is not strictly speaking a measurement of this quantity.

As evident from (1.11), the measured spin observable E is a smeared version of the sharp spin s_3. In order to make the unsharpness of E more transparent, we write the spin effects $E(X)$ in the form

$$
E(X) \; = \; \tfrac{1}{2}\,\varepsilon(X)\,\big(I + \mathbf{e}(X)\cdot\sigma\big)
\tag{1.19}
$$

with

$$
\begin{aligned}
\varepsilon(X) \; &= \; p_{++}(X) + p_{--}(X) \\
\varepsilon(X)\,\mathbf{e}(X) \; &= \; \big(p_{++}(X) - p_{--}(X)\big)\,\mathbf{e}_3
\end{aligned}
\tag{1.20}
$$

Similarly writing the initial spin state as $P[\varphi] = \tfrac{1}{2}(I + \mathbf{t}\cdot\sigma)$, we obtain

$$
\mathcal{I}_X\big(P[\varphi]\big) \; = \; \frac{1}{2}\alpha(X)\big(I + \mathbf{n}_{\mathrm{SG}}(X)\cdot\sigma\big)
\tag{1.21}
$$

where

$$
\begin{aligned}
\alpha(X) \; &= \; \langle \varphi \, | \, E(X)\varphi \rangle \; = \; \tfrac{1}{2}\varepsilon(X)\,(1 + \mathbf{t}\cdot\mathbf{e}(X)) \\
\mathbf{n}_{\mathrm{SG}}(X) \; &= \; [2\alpha(X)]^{-1}\bigg\{ \mathbf{e}_3 \Big[\big(p_{++}(X) - p_{--}(X)\big) \\
& \qquad\qquad + \big(p_{++}(X) + p_{--}(X) - 2\,\mathrm{Re}\,p_{+-}(X)\big)\mathbf{t}\cdot\mathbf{e}_3\Big] \\
& \qquad + \mathbf{t}\Big[2\,\mathrm{Re}\,p_{+-}(X)\Big] + \mathbf{t}\times\mathbf{e}_3\Big[2\,\mathrm{Im}\,p_{+-}(X)\Big] \bigg\}
\end{aligned}
\tag{1.22}
$$

A direct computation shows that the norm of $\mathbf{n}_{\mathrm{SG}}(X)$ equals one exactly when the incoming spin state $P[\varphi]$ is either $P[\varphi_+]$ or $P[\varphi]_-$; that is, $\mathcal{I}_X(P[\varphi])$ (when normalised) is a pure state if and only if the incoming spin state is a spin-up or a spin-down state in the field direction.

The measurement defined by U, ϕ, and A is not a measurement of s_3, and the ensuing state transformer $\mathcal{I}$ is not a Lüders transformer of s_3. For exhibiting the lack of ideality it is instructive to compare the state transformations $\mathcal{I}_X$ with the Lüders transformer of the effects $E(X)$,

$$\mathcal{I}_{L,X}\left(P[\varphi]\right) = E(X)^{1/2} P[\varphi] E(X)^{1/2} = \tfrac{1}{2}\alpha(X)\left(I + \mathbf{n}_L(X)\cdot\sigma\right)$$

$$\mathbf{n}_L(X) = [2\alpha(X)]^{-1}\left\{\mathbf{e}_3\left[\left(p_{++}(X) - p_{--}(X)\right)\right.\right.$$

$$+ \left(p_{++}(X) + p_{--}(X) - 2\left[p_{++}(X)p_{--}(X)\right]^{1/2}\right)\mathbf{t}\cdot\mathbf{e}_3\right]$$

$$\left. + \mathbf{t}\left[2[p_{++}(X)p_{--}(X)]^{1/2}\right]\right\} \tag{1.23}$$

This is of the same form as $\mathcal{I}_X\left(P[\varphi]\right)$, the only difference being that $p_{+-}(X)$ is here replaced by $\left[p_{++}(X)p_{--}(X)\right]^{1/2}$. It is this difference which guarantees that $\mathcal{I}_{L,X}(P[\varphi])$ is always a pure state (modulo normalisation). The vector $\mathbf{n}_L(X)$ is of the same form as $\mathbf{n}_{SG}(X)$, but its norm is one.

To get an idea of the distribution of outgoing (unsharp) spin directions, consider the case of a real Gaussian wave function ϕ, which gives rise to real $p_{+-}(X)$. Then let us have a look at the following cases:

(a) If X is a narrow stripe centered at and symmetric under reflection at the line $z = 0$, then $p_{++}(X) = p_{--}(X) \simeq p_{+-}(X)$, and therefore

$$\mathbf{n}_{SG}(X) \simeq \mathbf{e}_3\left[1 - \frac{p_{+-}(X)}{p_{++}(X)}\right](\mathbf{t}\cdot\mathbf{e}_3) + \mathbf{t}\frac{p_{+-}(X)}{p_{++}(X)} \simeq \mathbf{t}$$

$$E(X) \simeq p_{++}(X)\,I \tag{1.24}$$

Thus one obtains no information about the spin and the spin state is practically left unchanged. For comparison we note that under the same circumstances $\mathbf{n}_L(X) \simeq \mathbf{t}$.

(b) If X is a stripe located at the upper peak, then typically $p_{--}(X) \ll p_{+-}(X) \ll p_{++}(X)$. Assuming that $\mathbf{t}\cdot\mathbf{e}_3 = 0$, one obtains

$$\mathbf{n}_{SG}(X) \simeq \mathbf{e}_3\left[1 - 2\frac{p_{--}(X)}{p_{++}(X)}\right] + \mathbf{t}\,2\frac{p_{--}(X)}{p_{++}(X)} \simeq \mathbf{e}_3$$

$$E(X) \simeq P[\varphi_+] \tag{1.25}$$

Also $\mathbf{n}_L(X) \simeq \mathbf{e}_3$. This is a good approximation to the desired result. Similarly at the lower peak one is preparing 'spin down' $(\mathbf{n}_{SG}(X) \simeq -\mathbf{e}_3 \simeq \mathbf{n}_L(X))$ to a good approximation. In the regions between the peaks the outgoing spin is varying in the plane spanned by $\mathbf{e}_3$ and $\mathbf{t}$.

(c) If the incoming spin state is the degenerate mixture $T = \frac{1}{2} I$, so that $\mathbf{t} = 0$, then the outgoing states $\mathcal{I}_X(T)$ have vectors

$$\mathbf{n}_{\mathrm{SG}}(X) = \mathbf{n}_{\mathrm{L}}(X) = \mathbf{e}(X), \text{ and } \mathcal{I}_X(T) = \frac{1}{2} E(X) \qquad (1.26)$$

In this case the final spin state happens to coincide (up to a factor) with the measured effect.

In order to come closer to a realistic 'read out process' corresponding to a screen position observable, one has to take into account the fact that the reading does not take place immediately after the atom has passed the Stern-Gerlach device. Assuming that the extension of the atomic wave packet in the direction of propagation towards the screen is small, there is a fairly well defined time $\tau + t$ at which the atom reaches the screen and the localisation measurement takes place. Thus the state to be analysed on the screen is

$$\Psi_t := e^{-it\mathbf{P}^2/2M} \Psi_\tau = c_+ \phi_{+,t} \otimes \varphi_+ + c_- \phi_{-,t} \otimes \varphi_- \qquad (1.27)$$

Since the free evolution does not affect the spin part of the state, the structure of the measured observable and the state transformer are the same as those in Eqs. (1.11) and (1.15), but now with the time dependent coefficients

$$\begin{aligned} p_{kl}(X) &= \langle \phi_{k,t} \mid E^A(X) \phi_{l,t} \rangle, \quad k, l \in \{+, -\} \\ \phi_{k,t} &= e^{-it\mathbf{P}^2/2M} \phi_k \end{aligned} \qquad (1.28)$$

Taking $A = Q_3$ it is clear that due to the spreading of the wave function one cannot have $\phi_{\pm,t}$ localised in any finite region of the screen, nor in its upper or lower half planes. Therefore the actually measured observable E is an unsharp observable and there is no partition of the screen such that the associated discretised observable were a sharp one. The degree of unsharpness can, however, be made arbitrarily small, as indicated by the above item (b).

VII.1.2 Realistic magnetic field. We consider next a more realistic description of the Stern-Gerlach magnetic field that is consistent with Maxwell's equations. The simplest choice is

$$\mathbf{B} = bx\mathbf{e}_1 + (B_o - bz)\mathbf{e}_3 \equiv (B_1, B_2, B_3) \qquad (1.29)$$

Applying again the impulsive measurement approximation, the state of the atom leaving the magnetic field region is easily computed. Denoting $\omega := \mu_o|\mathbf{B}| = \omega(\mathbf{q})$ we find

$$\begin{aligned} \Psi_\tau(\mathbf{q}) = &\left\{ c_+ \left[\frac{1}{2} \left(1 + \frac{B_3}{|\mathbf{B}|} \right) e^{-i\omega\tau} + \frac{1}{2} \left(1 - \frac{B_3}{|\mathbf{B}|} \right) e^{i\omega\tau} \right] \right. \\ &\left. + c_- \left[-i \frac{B_1}{|\mathbf{B}|} \right] \sin(\omega\tau) \right\} \phi(\mathbf{q})\varphi_+ \\ &+ \left\{ c_- \left[\frac{1}{2} \left(1 + \frac{B_3}{|\mathbf{B}|} \right) e^{i\omega\tau} + \frac{1}{2} \left(1 - \frac{B_3}{|\mathbf{B}|} \right) e^{-i\omega\tau} \right] \right. \\ &\left. + c_+ \left[-i \frac{B_1}{|\mathbf{B}|} \right] \sin(\omega\tau) \right\} \phi(\mathbf{q})\varphi_- \end{aligned} \qquad (1.30)$$

As a consistency check one may observe that putting $B_1 = 0$ leads back to the previous result (1.8–9). Assuming that $|bx|/B_o, |bz|/B_o \ll 1$ and putting $\omega_o = \mu_o B_o$ we have

$$
\begin{aligned}
\Psi_\tau \simeq &\left[c_+ e^{-i(\omega_o - \mu_o bz)\tau} + ic_+ \tfrac{bx}{B_o} \sin(\omega_o \tau) \right] \phi \otimes \varphi_+ \\
&+ \left[c_- e^{i(\omega_o - \mu_o bz)\tau} + ic_- \tfrac{bx}{B_o} \sin(\omega_o \tau) \right] \phi \otimes \varphi_-
\end{aligned}
\tag{1.31}
$$

Comparing this state with (1.8–9), we see that if the inhomogeneous part of the field is weak, the 'disturbing' terms ($\sim \tfrac{bx}{B_o}$) are small. In this case the measured observable assumes the form

$$
E'(X) = \sum_{k,l} p'_{kl}(X) \, |\varphi_k\rangle\langle \varphi_l|
\tag{1.32}
$$

where

$$
\begin{aligned}
p'_{++}(X) &= \langle \phi_+ \mid E^A(X)\phi_+ \rangle + \sin^2(\omega_o \tau)\langle \phi_1 \mid E^A(X)\phi_1 \rangle \\
p'_{--}(X) &= \langle \phi_- \mid E^A(X)\phi_- \rangle + \sin^2(\omega_o \tau)\langle \phi_1 \mid E^A(X)\phi_1 \rangle \\
p'_{+-}(X) &= e^{i\omega_o \tau} \sin(\omega_o \tau) \left[\langle \phi_1 \mid E^A(X)\phi_- \rangle + \langle \phi_+ \mid E^A(X)\phi_1 \rangle \right] \\
p'_{-+}(X) &= \overline{(p'_{+-}(X))}
\end{aligned}
\tag{1.33}
$$

and

$$
\phi_1(\mathbf{q}) = i\tfrac{bx}{B_o} \, \phi(\mathbf{q})
\tag{1.34}
$$

A straightforward computation shows that the distance $\|E'(X) - E(X)\|$ between the effects (1.32) and (1.11) is of the order of $\langle \phi \mid \left(\tfrac{bx}{B_o}\right)^2 \phi \rangle$, which can be made arbitrarily small. Thus the statistics of E' can be approximated by that of E within an error of the order of this quantity. In this way we have reestablished the familiar result that the x-component of the magnetic field, when weak, does not significantly disturb the measurement of the z-component of the spin.

In order to further specify the influence of the B_1-component of the Stern-Gerlach field, we now assume that $\langle \phi \mid x\phi \rangle = 0$ and we restrict our attention to strip-shaped sets X on the screen. Then it follows that $\langle \phi_1 \mid E^A(X)\phi_\pm \rangle = 0$ and therefore $p'_{+-}(X) = 0$ so that $E'(X)$ becomes diagonal. Writing $E'(X) = \tfrac{1}{2}\alpha'(X)\left(I + \mathbf{e}'(X) \cdot \sigma\right)$, we find that

$$
\begin{aligned}
\alpha'(X)\mathbf{e}'(X) &= \left(p_{++}(X) - p_{--}(X)\right) \mathbf{e}_3 \\
\alpha'(X) &= p_{++}(X) + p_{--}(X) + \sin^2(\omega_o \tau) \langle \phi_1 \mid A(X)\phi_1 \rangle
\end{aligned}
\tag{1.35}
$$

The lengths of the vectors $\mathbf{e}_X$ and $\mathbf{e}'_X$ are measures of the unsharpness of the spin properties $E(X)$ and $E'(X)$. Thus we see that the x-component of the magnetic field only produces an increase of unsharpness in the measured spin z-component by an amount proportional to the quantity $\langle \phi_1 \mid E^A(X)\phi_1 \rangle \sim \langle \phi \mid \left(\tfrac{bx}{B_o}\right)^2 \phi \rangle$.

We could also study the state change due to the atom's passage through a slit of shape X at the screen; to this end one must determine the outgoing spin state $\mathcal{I}'_X(P[\varphi])$ as above. Leaving the study of the general case to an interested reader, we consider here only the case $\varphi = \varphi_+$, and instead of a slit we take again the stripes discussed above. With the ensuing simplifications one obtains

$$
\begin{aligned}
\mathcal{I}'_X(P[\varphi]) &= p_{++}(X)\,P[\varphi_+] + \big(p'_{++}(X) - p_{++}(X)\big)\,P[\varphi_-] \\
&= \tfrac{1}{2}\alpha'\left(I + \mathbf{n}'(X)\cdot\sigma\right) \\
\mathbf{n}'(X) &= \left[2\frac{p_{++}(X)}{p'_{++}(X)} - 1\right]\mathbf{e}_3
\end{aligned}
\tag{1.36}
$$

This shows that even if the particle originally had spin up, after leaving the magnetic field the spin state will be a mixture of spin-up and spin-down components, since $\|\mathbf{n}'(X)\| < 1$ due to the presence of B_1. Moreover there is a nonzero probability of registering some X in the lower half of the screen, indicating that the incoming particle should have had spin down. In such cases the outgoing particle even might have almost spin down (namely, if it happens that $p_{++}(X) \ll p'_{++}(X)$). Yet for the 'expected' outcomes (X close to the location of the peak of ϕ_+) one has $p'_{++}(X) \simeq p_{++}(X)$ and therefore $\mathbf{n}'(X) \simeq \mathbf{e}_3$; in this case the measurement is almost ideal, a property which is essential for interpreting appropriate multiples of the effects $E'(X)$ as nearly sharp spin properties.

One could easily extend the above analysis even to more realistic Stern-Gerlach fields $\mathbf{B}$. But the study of the ideal cases (1.7) and (1.29) already indicates that any step towards a more realistic description will only increase the degree of unsharpness in the measured spin quantity, meaning that the actually measured observable $E^{U,\phi,A}$ resembles less and less the expected one, s_3, say. In the next subsection we shall briefly discuss more realistic screen observables, a step which introduces yet another source of unsharpness in the operational definition of spin in terms of the Stern-Gerlach device.

VII.1.3 Proper screen observables. Screen observables take into account the fact that detectors, or screens, do not constitute an instantaneous space localisation measurement but rather are activated during some period of time, thus capturing appreciable fractions of the impinging atoms [7.1, 3.10]. In the Stern-Gerlach experiment we may assume the screen to be located in the plane $y = 0$; the wave function $\Psi_\tau = \phi_+ \otimes \varphi_+ + \phi_- \otimes \varphi_-$ leaving the magnetic field region undergoes free evolution until it arrives at the screen. For simplicity we assume that the instant τ lies in the 'remote past' so that the following time integrals can be extended from $-\infty$ to $+\infty$. Further assuming the simple magnetic field (1.29) and an initial preparation of the form $\phi(x, y, z, t_o) = \phi_{13}(x, z, t_o)\phi_2(y, t_o)$, the wave packet maintains this form for all times, especially for $t \geq \tau$. Then the probability distribution for registering a particle in the region X of the screen within the time interval $[t_1, t_2] = \theta$ is given as the expectation value of some screen effect $F(X;\theta)$ defined by the following:

$$\langle\,\Psi_\tau\,|\,F(X;\theta)\otimes I\,\Psi_\tau\,\rangle = \langle\,\phi_+\,|\,F(X;\theta)\phi_+\,\rangle + \langle\,\phi_-\,|\,F(X;\theta)\phi_-\,\rangle$$

$$\langle\,\phi_\pm\,|\,F(X;\theta)\phi_\pm\,\rangle = (2\pi)^{-3}\int_\theta f(t)\int_X |\phi_{13}^\pm(x,z,t)|^2\,dx\,dz\,dt$$

$$f(t) = \tfrac{1}{2\pi M}\left|\int_{-\infty}^{+\infty}|p_2|^{1/2}\,\phi_2^\pm(p_2)\,\exp\left[-i\tfrac{1}{2M}p_2^2 t\right]dp_2\right|^2 \tag{1.37}$$

$$|\phi_{13}^\pm(x,z,t)|^2 = (2\pi)^{-2}\left|\iint_{\mathbf{R}^2}\phi_{13}^\pm(p_1,p_3)\exp\left[-i(p_1 x + p_2 y)\right]\right.$$
$$\left.\times\exp\left[-i\tfrac{1}{2M}(p_1^2 + p_3^2)t\right]dp_1\,dp_3\right|^2$$

With these equations the operator $F(X;\theta)$ is implicitly defined and it is seen to be positive. Further one can show that its expectation values always lie between 0 and 1 so that it is an effect. What is most important in the present context is the fact that the first of the above equations gives rise to a spin POV measure E'' on the screen (via $\langle\,\varphi\,|\,E''(X)\varphi\,\rangle := \langle\,\Psi_\tau\,|\,F(X;\mathbf{R})\otimes I\,\Psi_\tau\,\rangle$), which is precisely of the form of the $E'(X)$ obtained above. The only difference is that the monitoring observable A is now replaced by a POV measure $X\mapsto F(X;\mathbf{R})$. Thus the simplified description of the screen measurement given earlier shows already the essential features of the registration process in the Stern-Gerlach experiment. However, it should be emphasised that in addition to the dynamical contributions to the measurement noise already incorporated in the discussions of subsections 1.1 and 1.2, the screen observable takes into account further unsharpness arising from the extension in time of the registration process.

VII.2 Informationally complete polarisation measurement

We study next a simple photon polarisation measurement based on a polarisation-dependent beam splitter followed by two analysers placed in the reflected and transmitted beam paths, respectively (Figure 7.1). The beam splitter will be characterised by complex transmission and reflection coefficients (amplitudes) t_h, t_v, r_h, r_v, depending on the direction of the incident plane polarisation. These coefficients satisfy the usual normalisation conditions:

$$|t_h|^2 + |r_h|^2 = 1, \quad |t_v|^2 + |r_v|^2 = 1 \tag{2.1}$$

The thickness of the plate and the wavelength of the photons can be chosen such that one has the following reality conditions for the amplitudes:

$$r_h,\ r_v \text{ real, and } t_h,\ t_v \text{ purely imaginary.} \tag{2.2}$$

We shall restrict ourselves to this case. Also we will substitute t_h, t_v with it_h, it_v and treat t_h, t_v as real numbers.

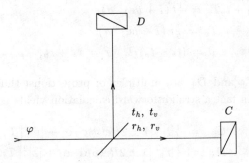

Figure 7.1. Photon polarisation experiment

The description of the photon is based on the Hilbert space $\mathcal{H} = \mathcal{H}_{pol} \otimes \mathcal{H}_{prop}$ where $\mathcal{H}_{pol} = \mathbf{C}^2$ describes the photon polarisation modes and $\mathcal{H}_{prop}$ represents its modes of propagation. This 'particle' picture for the photon derives from the one-particle section of the Fock space. In the present experiment it is enough to take into account only three modes of propagation $\mathbf{k}_i, \mathbf{k}_t, \mathbf{k}_r$, the incident, transmitted, and reflected modes, and the respective normalised vector states $|i\rangle, |t\rangle, |r\rangle$ from $\mathcal{H}_{prop}$. The plane polarisation modes parallel (resp. perpendicular) to the plane of incidence are represented by normalised vector states $|h\rangle, |v\rangle \in \mathcal{H}_{pol}$. Then an incident photon undergoes the following evolution:

$$\varphi \otimes |i\rangle = \left[a_h|h\rangle + a_v|v\rangle\right] \otimes |i\rangle \mapsto$$
$$= \left[t_h a_h|h\rangle + t_v a_v|v\rangle\right] \otimes |t\rangle + \left[r_h a_h|h\rangle + r_v a_v|v\rangle\right] \otimes |r\rangle \qquad (2.3)$$
$$\equiv (\mathcal{T}\varphi) \otimes |t\rangle + (\mathcal{R}\varphi) \otimes |r\rangle \equiv \Psi_f$$

We shall identify the h and v modes with vectors $\mathbf{e}_1, -\mathbf{e}_1$ in the Poincaré sphere and write $T_{\mathbf{e}_1} = T_h$, $T_{-\mathbf{e}_1} = T_v$. The polarisation analysers C, D are assumed to be calibrated so as to measure the polarisation properties $T_{\pm\mathbf{c}}$ and $T_{\pm\mathbf{d}}$, respectively ($\mathbf{c}, \mathbf{d} \in S^2$). Let P_t, P_r denote the projections onto the vectors $|t\rangle, |r\rangle$. Then the detection process constitutes a measurement of a sharp observable on $\mathcal{H}$ given by the set of pairwise-orthogonal projections on $\mathcal{H}$, $T_{\pm\mathbf{c}} \otimes P_t, T_{\pm\mathbf{d}} \otimes P_r$. The measured polarisation observable is thus obtained from the relations

$$\langle \varphi \mid C_{\pm}\varphi \rangle := \langle \Psi_f \mid T_{\pm\mathbf{c}} \otimes P_t \Psi_f \rangle = \langle \varphi \mid \mathcal{T}^* T_{\pm\mathbf{c}} \mathcal{T} \varphi \rangle$$
$$\langle \varphi \mid D_{\pm}\varphi \rangle := \langle \Psi_f \mid T_{\pm\mathbf{d}} \otimes P_r \Psi_f \rangle = \langle \varphi \mid \mathcal{R}^* T_{\pm\mathbf{d}} \mathcal{R} \varphi \rangle \qquad (2.4)$$

with $C_{\pm}$ and $D_{\pm}$ denoting the resulting measured polarisation effects.

For further analysis one may use the Poincaré sphere representations of these operators. First of all

$$
\begin{aligned}
\mathcal{T} &= t_h T_h + t_v T_v = \tfrac{1}{2} t \left(I + \vartheta \mathbf{e}_1 \cdot \sigma \right) \\
\mathcal{R} &= r_h T_h + r_v T_v = \tfrac{1}{2} r \left(I + \rho \mathbf{e}_1 \cdot \sigma \right) \\
t &= t_h + t_v, \quad \vartheta = (t_h - t_v)/t, \quad r = r_h + r_v, \quad \rho = (r_h - r_v)/r
\end{aligned}
\tag{2.5}
$$

The operators $C_\pm$ and $D_\pm$ are multiples of projections; thus they have the form $c_\pm T_{\gamma\pm}, d_\pm T_{\delta\pm}$. In fact a straightforward calculation yields

$$
\begin{aligned}
C_\pm &= \tfrac{1}{4} |t|^2 \left[1 \pm 2 Re(\vartheta) \mathbf{c} \cdot \mathbf{e}_1 + |\vartheta|^2 \right] T_{\gamma\pm} \\
D_\pm &= \tfrac{1}{4} |r|^2 \left[1 \pm 2 Re(\rho) \mathbf{d} \cdot \mathbf{e}_1 + |\rho|^2 \right] T_{\delta\pm}
\end{aligned}
\tag{2.6}
$$

Choosing $\mathbf{c} = \mathbf{e}_2$, $\mathbf{d} = \mathbf{e}_3$ and making use of the reality of the t's and r's, one gets

$$
\begin{aligned}
C_\pm &= \tfrac{1}{4} t^2 \left[1 + \vartheta^2 \right] T_{\gamma\pm}, \quad \gamma_\pm = (1 + \vartheta^2)^{-1} \left[2\vartheta \mathbf{e}_1 \pm (1 - \vartheta^2) \mathbf{e}_2 \right] \\
D_\pm &= \tfrac{1}{4} r^2 \left[1 + \rho^2 \right] T_{\delta\pm}, \quad \delta_\pm = (1 + \rho^2)^{-1} \left[2\rho \mathbf{e}_1 \pm (1 - \rho^2) \mathbf{e}_3 \right]
\end{aligned}
\tag{2.7}
$$

Taking into account the normalisation conditions for the t's and r's, one easily verifies that the four positive operators $C_\pm, D_\pm$ add up to the unit operator of $\mathcal{H}_{pol}$. To comply with the formal requirements of a POV measure, one may introduce any set of four elements, e.g., $\{1,2,3,4\}$, representing the readings corresponding to the counts at $C_+, C_-, D_+,$ and D_-. The mappings

$$
1 \mapsto C_+, \; 2 \mapsto C_-, \; 3 \mapsto D_+, \; 4 \mapsto D_-
\tag{2.8}
$$

define then uniquely an $\mathcal{L}(\mathcal{H}_{pol})$-valued POV measure on the set $\mathcal{F}$ of subsets of $\{1,2,3,4\}$.

The above four-valued observable entails three two-valued coarse-grainings of it, corresponding to the three different subalgebras of $\mathcal{F}$:

$$
\begin{aligned}
\{1,3\} &\mapsto E_+^{(1)} = C_+ + D_+ = \tfrac{1}{2}\big(I + [t_h t_v \mathbf{e}_2 + r_h r_v \mathbf{e}_3] \cdot \sigma \big) \\
\{2,4\} &\mapsto E_-^{(1)} = C_- + D_- = \tfrac{1}{2}\big(I - [t_h t_v \mathbf{e}_2 + r_h r_v \mathbf{e}_3] \cdot \sigma \big) \\
\{1,4\} &\mapsto E_+^{(2)} = C_+ + D_- = \tfrac{1}{2}\big(I + [t_h t_v \mathbf{e}_2 - r_h r_v \mathbf{e}_3] \cdot \sigma \big) \\
\{2,3\} &\mapsto E_-^{(2)} = C_- + D_+ = \tfrac{1}{2}\big(I - [t_h t_v \mathbf{e}_2 - r_h r_v \mathbf{e}_3] \cdot \sigma \big) \\
\{1,2\} &\mapsto E_+^{(3)} = C_+ + C_- = \tfrac{1}{2}[t_h^2 + t_v^2]\big(I + \tfrac{4\vartheta}{1+\vartheta^2} \mathbf{e}_1 \cdot \sigma \big) \\
\{3,4\} &\mapsto E_-^{(3)} = D_+ + D_- = \tfrac{1}{2}[r_h^2 + r_v^2]\big(I + \tfrac{4\rho}{1+\rho^2} \mathbf{e}_1 \cdot \sigma \big)
\end{aligned}
\tag{2.9}
$$

By construction the resulting simple observables are coexistent. Therefore the present experiment constitutes a joint measurement of three unsharp polarisation observables. It should be noted that the effects $E_{\pm}^{(k)}$ are indeed unsharp polarisation properties.

An important feature of the joint observable constituted by the effects $C_+, C_-,$ D_+, D_- is its informational completeness. In fact a direct calculation shows that the set $\{C_+, C_-, D_+, D_-\}$ is informationally complete if and only if the corresponding set of vectors $\{\gamma_+, \gamma_-, \delta_+, \delta_-\}$ contains three linearly independent vectors (cf. Sec. V.2). We calculate

$$\delta_+ \cdot (\gamma_+ \times \gamma_-) \;=\; -8\vartheta\rho\left(1 - \vartheta^2\right)\left(1 + \vartheta^2\right)^{-2}\left(1 + \rho^2\right)^{-2} \tag{2.10}$$

which is nonzero provided that $\vartheta \neq 0$, $\rho \neq 0$, and $\vartheta^2 \neq 1$, or equivalently, $t_h \neq t_v$, $r_h \neq r_v$, and $t_h \neq 0 \neq t_v$, which can be easily satisfied. In this case the informational completeness of the four-valued observable constituted by the present polarisation experiment is guaranteed.

Coexistence and informational completeness are features of physical observables which are familiar in classical physics but rare in quantum physics. Our example provides an illustration of the price to be paid for obtaining these classical properties: a set of noncommuting observables can be coexistent only if these observables are unsharp, and an informationally complete observable is necessarily an unsharp one.

VII.3 Measurement schemes involving photons

Recent advances in experimental quantum physics have made it possible to apply the basic photon-photon and photon-atom interactions as measurement couplings in studying the properties of individual photons or atoms. In this section we investigate several measurement schemes involving such interactions. We determine, in each case, the actually measured observable, which typically turns out to be a smeared version of the observable intended to be measured. The study of the induced state transformers allows one to estimate repeatability features and the degree of nonideality of the measurements.

VII.3.1 Photon counting and a lossless beam splitter.
The number of counts $\mathcal{N}$ (per a given counting time) in a photodetector for a single-mode radiation field is a random variable. For a detector with unity quantum efficiency, $\mathcal{N}$ comprises a measurement of the number observable $N = a^*a = \sum n|n\rangle\langle n|$ of the mode a in the sense that if T is its state, then

$$\mathrm{prob}\,(\mathcal{N} = n) \;=\; p_T^N(n) \tag{3.1}$$

is the probability of detecting n photons. On the other hand if the quantum efficiency ε of the detector is less that one, the counting statistics is known to be of the form

$$\mathrm{prob}\,(\mathcal{N} = n) \;=\; \sum_{m=n}^{\infty} \binom{m}{n} \varepsilon^n (1 - \varepsilon)^{m-n}\, p_T^N(m) \tag{3.2}$$

showing that such a photodetector constitutes a measurement of an unsharp number observable

$$n \mapsto E_n^\varepsilon \ := \ \sum_{m=n}^{\infty} \binom{m}{n} \varepsilon^n (1-\varepsilon)^{m-n} |m\rangle\langle m| \tag{3.3}$$

It is equally well known that a photodetector with quantum efficiency ε can be modelled by a lossless beam splitter with transparency ε (or just by a two-mode mixer with mixing strength ε) followed by a photodetector of unity quantum efficiency (see Figure 1.3). We shall demonstrate that this scheme leads to a definition of the unsharp number observable (3.3) by means of the experimentally confirmed counting statistics (3.2).

A two-mode mixer models an interaction between any pair of modes of an electromagnetic field (with the same frequency) that preserves the total number of photons in the mode pair but not in each mode separately. Denoting the annihilation operators as a and b, the two-mode mixing operator is then

$$U_\alpha \ = \ \exp(\overline{\alpha} a \otimes b^* - \alpha a^* \otimes b) \tag{3.4}$$

with $\alpha = |\alpha| e^{i\vartheta}$, $\cos|\alpha| = \sqrt{\varepsilon}$, $0 \le |\alpha| \le \frac{\pi}{2}$, $-\frac{\pi}{2} < \vartheta \le \frac{\pi}{2}$. It is unitary, $U_\alpha^{-1} = U_\alpha^* = U_{-\alpha}$, and transforms the annihilation operators of the two modes into each other as follows

$$\begin{aligned}
\tilde{a} \ &:= \ U_\alpha^* (a \otimes I) U_\alpha \ = \ \sqrt{\varepsilon}\,(a \otimes I) + e^{i\vartheta}\sqrt{1-\varepsilon}\,(I \otimes b) \\
\tilde{b} \ &:= \ U_\alpha^* (I \otimes b) U_\alpha \ = \ -e^{-i\vartheta}\sqrt{1-\varepsilon}\,(a \otimes I) + \sqrt{\varepsilon}\,(I \otimes b)
\end{aligned} \tag{3.5}$$

U_α preserves the bosonic commutation relations of the modes a and b; in particular, $[\tilde{a}, \tilde{a}^*] = [\tilde{b}, \tilde{b}^*] = I$. The most general transformation of the form (3.5) preserving these commutation relations is given by the unitary operator

$$C(\varepsilon, \vartheta_t, \vartheta_r) \ = \ \exp\left[i\vartheta_t(a^*a \otimes I - I \otimes b^*b)\right] U_\alpha \tag{3.6}$$

where $\alpha = \arccos\left(\sqrt{\varepsilon}\right) e^{i(\vartheta_t - \vartheta_r)}$. It describes a lossless beam splitter with transparency ε producing the phase shifts ϑ_t and ϑ_r for the transmitted and the reflected signals, respectively.

A two-mode mixer or a beam splitter changes the field state S according to $S \mapsto U_\alpha S U_\alpha^*$, or $S \mapsto C S C^*$, respectively. In particular if the two modes are initially independent, with the states T and T', then

$$W \ := \ U_\alpha (T \otimes T') U_\alpha^* \tag{3.7}$$

is the state of the field after the mixing. Therefore the probability of detecting n photons in the detector D (of unity quantum efficiency) is

$$p_W^{N \otimes I}(n, 1) \ = \ \mathrm{tr}\left[W |n\rangle\langle n| \otimes I\right] \tag{3.8}$$

When these statistics are interpreted with respect to the incoming a-mode field, then, for each initial state T' of the b-mode, one obtains an observable $E \equiv E^{T';\varepsilon}$ of the a-mode such that for every T and all n,

$$p_T^E(n) := p_W^{N\otimes I}(n, 1) = \mathrm{tr}\big[T \otimes T' U_\alpha^* (|n\rangle\langle n| \otimes I) U_\alpha\big] \tag{3.9}$$

A lossless beam splitter followed by an ideal photodetector defines exactly the same observable as a two-mode mixer in this configuration.

We are now ready to determine the observable measured by a photodetector with quantum efficiency ε. In this case the b-mode is taken to be in the vacuum state $T' = |0\rangle\langle 0|$. The observable $E^\varepsilon \equiv E^{0;\varepsilon}$ is obtained from the detection statistics (3.9) by putting $T' = |0\rangle\langle 0|$. In order to derive the explicit form of the effects E_n^ε we use the identity [7.2]

$$U_\alpha = \exp(xa \otimes b^*) \exp\left(-\tfrac{1}{2}y\left(a^*a \otimes I - I \otimes b^*b\right)\right) \exp(-\overline{x}a^* \otimes b), \tag{3.10}$$

where $x = (\tan|\alpha|)\, e^{-i\vartheta}$, $y = -2\ln(\cos|\alpha|)$. A straightforward computation gives

$$E_n^\varepsilon = \sum_{m=n}^{\infty} \binom{m}{n} \varepsilon^n (1-\varepsilon)^{m-n} |m\rangle\langle m| \tag{3.11}$$

The measurement outcome statistics of this observable is the Bernoulli distribution (3.2). Moreover the first moment of E^ε is readily found to be

$$N_\varepsilon^0 := \sum_{n=0}^{\infty} n E_n^\varepsilon = \varepsilon N \tag{3.12}$$

so that, in particular, $\langle N_\varepsilon^0 \rangle := \sum n\, p_T^{E^\varepsilon}(n) = \varepsilon \sum n\, p_T^N(n) = \varepsilon\langle N \rangle = \langle \tilde{a}^* \tilde{a} \rangle$. We conclude that the observable $n \mapsto E_n^\varepsilon$ describes the photon counting of a single-mode field with quantum efficiency ε. The effects E_n^ε are projection operators exactly when $\varepsilon = 1$, in which case $E_n^\varepsilon = |n\rangle\langle n|$.

There is another interesting special case of the observables $E^{T';\varepsilon}$, which arises from mixing the input signal with the local oscillator (the b-mode) prepared in a coherent state $T' = |z\rangle\langle z|$. As this observable is relevant in the analysis of homodyne detection, Sec. VII.3.6, we present it already here. Evaluating Eq. (3.9) for this case leads to the observable $E^{z;\varepsilon}$ [7.3]

$$E_n^{z;\varepsilon} = D_{xz}\, E_n^\varepsilon\, D_{xz}{}^* \tag{3.13}$$

where D_{xz} is the displacement operator associated with the mode a according to Eq. (III.5.29). The first moment of this unsharp number observable is found to be:

$$N_\varepsilon^z := \sum_{n=0}^{\infty} n\, E_n^{z;\varepsilon} = \varepsilon N + (1-\varepsilon)|z|^2 I + \sqrt{\varepsilon(1-\varepsilon)}\,(\overline{z}a + za^*). \tag{3.14}$$

The last term describes the interference of the two signals entangled in the detector or in the beam splitter. We note that if z is real, then N_ε^z is proportional to the field quadrature component a^q, whereas for purely imaginary z it is related to the conjugate quadrature a^p.

As a final example we consider the case where the signal mode is mixed with a Gaussian optical field. Such a field is known to describe, for instance, the effect of thermal noise in photocounting [7.4]. The state T' is then the Gibbs state $T_\beta = Z^{-1} \exp(-\beta b^* b)$, where β is the Boltzmann inverse temperature and $Z = \sum e^{-\beta n}$ is the normalisation constant. Writing this state in the P-representation

$$T_\beta = \frac{1}{\pi} \int P(z) |z\rangle\langle z| \, d^2 z \tag{3.15}$$

with $P(z) = \langle b^* b\rangle_\beta^{-1} \exp(-|z|^2 / \langle b^* b\rangle_\beta)$ the observable $E^{\beta;\varepsilon}$ determined by the detection statistics (3.9) obtains the form

$$E_n^{\beta;\varepsilon} = \frac{1}{\pi} \int P(z) \, E_n^{z;\varepsilon} \, d^2 z \tag{3.16}$$

Note that in the limit of zero temperature this approaches the vacuum case observable $E^{0,\varepsilon}$. The first moment is again related to the number observable N of the signal mode,

$$N_\varepsilon^\beta := \sum n E_n^{\beta;\varepsilon} = \varepsilon N + (1 - \varepsilon)\langle b^* b\rangle_\beta \tag{3.17}$$

VII.3.2 First kind measurements of the photon number. We discuss next measurement schemes involving an interaction between two field modes of the form

$$U = e^{i\chi(N_1 \otimes N_2)} \tag{3.18}$$

Here $N_1 = a_1^* a_1$ and $N_2 = a_2^* a_2$ are the number observables of the modes and χ is a coupling constant (including the counting time). This is the standard form of an interaction employed for measuring observable N_1, with the conjugate quantity of N_2 suggesting itself as the pointer observable. Indeed U effects a phase shift in the second ('probe') mode proportional to the number of photons in the first mode. The map (3.18) is used in quantum optics to model the coupling of the two modes by means of an optical Kerr medium or via four-wave mixing [7.5, 7.6].

A straightforward understanding of this scheme as an N_1-measurement has been somewhat obscured by the ambiguity in the choice of the conjugate quantity of N_2. For instance the quadrature component $a_2^q = \frac{1}{\sqrt{2}}(a_2 + a_2^*)$ of mode 2 as well as the sine-phase $S_2 = \frac{i}{2}(V_2^* - V_2)$ of N_2 have been used as pointer observables in [7.5] and [7.6], respectively. However, in view of the results of Sec. III.5.1, it is natural to take the phase observable M_2 (III.5.12) for that purpose. We shall pursue some choices of pointers and explore the implications of fixing different probe states. Measurement theory tells us that the actually measured observable depends crucially not only on the interaction but also on the pointer observable as well as the initial probe state. Therefore we start with a general description of the present scheme.

Let T_2 be the initial state of the second mode and let F_2 denote the pointer observable (including a possible pointer function). The observable E of the first mode which is measured by the scheme U, T_2 and F_2 is then determined through the condition

$$\text{tr}\big[TE(X)\big] \; := \; \text{tr}\big[UT \otimes T_2 U^* \, I \otimes F_2(X)\big] \tag{3.19}$$

which is to hold for all the initial states T of the first mode and for all value sets X of the pointer observable F_2. One obtains

$$E(X) \; = \; \sum_{n=0}^{\infty} \text{tr}\Big[T_2 \, e^{-i\chi n N_2} F_2(X) e^{i\chi n N_2}\Big] \, |n\rangle\langle n| \tag{3.20}$$

This POV measure is a smeared version of the number observable N_1, associated with the confidence measure $p(X,n) = \text{tr}\big[T_2 \, e^{-i\chi n N_2} \, F_2(X) \, e^{i\chi n N_2}\big]$. The corresponding state transformer is readily determined:

$$\mathcal{I}_X(T) \; = \; \sum_{n,m} \text{tr}\Big[T_2 \, e^{-i\chi n N_2} F_2(X) e^{i\chi m N_2}\Big] \, |n\rangle\langle n| \, T \, |m\rangle\langle m| \tag{3.21}$$

The state of the first mode after the measurement (with no reading of the result) is therefore

$$\mathcal{I}_\Omega(T) \; \equiv \; T^f \; = \; \sum_{n,m} \text{tr}\Big[T_2 \, e^{-i\chi(n-m)N_2}\Big] \, |n\rangle\langle n| \, T \, |m\rangle\langle m| \tag{3.22}$$

The dependence of the actually measured smeared number observable E on the pointer observable F_2 and the initial probe state T_2 is apparent from Eq. (3.20). Moreover (3.22) shows that the final object state T^f can be influenced to some extent by choosing different preparations T_2. Finally one finds that independently of the choice of F_2 and T_2, one always has

$$\text{tr}\big[TE(X)\big] \; = \; \text{tr}\big[T^f E(X)\big] \tag{3.23}$$

In other words the E-measurement in question is of the first kind. From the general theory one can say, in addition, that this measurement could be repeatable only if F_2 were discrete, and this would not be sufficient, in general.

The present scheme has been considered as a realisation of a quantum non-demolition measurement for the photon number. It does indeed fulfill some such requirements: the number eigenstates are left unchanged, that is, if $T = |n\rangle\langle n|$

then $T^f = T = |n\rangle\langle n|$; also the observable intended to be measured, the number, commutes with the free evolution Hamiltonian as well as with the measurement interaction. Yet the actually measured observable is not N_1 but a smeared version of it, given in (3.20). It is only on the level of expectation values that one may speak of a measurement of (a truncated version of) the number. Indeed the first moment of the observable E is

$$
\begin{aligned}
E^{(1)} &= \int x\, E(dx) = \sum_n \int x\, p(dx, n)\, |n\rangle\langle n| \\
&= \sum_n \mathrm{tr}\left[T_2\, e^{-i\chi n N_2} F_2^{(1)} e^{i\chi n N_2} \right] |n\rangle\langle n| =: f(N_1)
\end{aligned}
\tag{3.24}
$$

which in certain circumstances (e.g., restricted set of input states) can be approximated by a linear function of N_1.

The limited use of self-adjoint operators for the analysis of an experiment becomes strikingly evident at this point: while the POV measure (3.20) comprises the totality of all moments of the outcome statistics, this cannot be said about the self-adjoint operator (3.24); in fact denoting the k-th moment operator of E as $E^{(k)}$, one must realise that $E^{(k)} \neq f(N^k)$ in general. Thus one would need an infinity of self-adjoint operators in order to represent the moments of all distributions as expectation values (cf. Sec. II.2.5).

Next we specify the above formulas with reference to the pointer observable being either a quadrature component a_2^q or the phase observable M_2. Other pointers, such as the sine-phase S_2, may be considered, but they do not yield any essentially different features. Note that taking T_2 to be a number state leads to the trivial (constant) observable and no measurement at all. It will turn out that coherent states with not too small amplitudes do lead to satisfactory results.

Considering first the quadrature component a_2^q, we let F_2 be its spectral measure E_2^q. Assume that the initial state T_2 of the second mode is a coherent state $|z_0\rangle\langle z_0|$ such that $\langle z_0 | a_2^q | z_0 \rangle = 0$ and $\langle z_0 | a_2^p | z_0 \rangle = \sqrt{2}p_0$; then the measured observable (3.20) assumes the form

$$
E(X) = \sum_n \langle e^{i\chi n} i p_0 | E_2^q(X) | e^{i\chi n} i p_0 \rangle\, |n\rangle\langle n|
\tag{3.25}
$$

We are facing here an unsharp measurement of a discrete quantity by means of a continuous-scale readout observable. In fact the actually measured observable is a continuously smeared version of the number observable. Its first moment is

$$
E^{(1)} = p_0\, \sin(\chi N_1)
\tag{3.26}
$$

Since the measurement is of the first kind we find again that the average of the first moment is the same in both the initial as well as the final state of mode 1; hence

$\langle \sin(\chi N_1) \rangle_T = \langle \sin(\chi N_1) \rangle_{T'} = \langle E^{(1)} \rangle_T / p_0$. It therefore appears as if the measured observable were the sine of the scaled photon number N_1. But this is only true on the level of expectation values. The higher moments are not given as $\sin^k(\chi N_1)$.

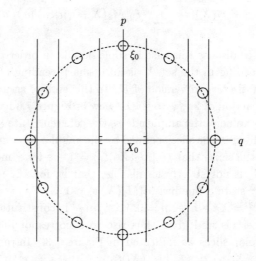

Figure 7.2. Choices of a partition for the pointer E_2^q and of the coherent state amplitude p_0 depending on the coupling constant χ

Nevertheless one may consider the following approximation. Suppose that one is interested only in states φ of mode 1 which are superpositions of the first 1000 number states, say. If the parameter χ is taken to be of the order 10^{-5}, say, then to a good approximation one has $\langle \varphi | E^{(1)} | \varphi \rangle \simeq p_0 \chi \langle \varphi | N_1 \varphi \rangle$.

In order to visualise the phase sensitive measurement afforded by the pointer a_2^q, we consider the phase space picture shown in Figure 7.2. Taking a small value of χ, e.g., $\chi = \frac{2\pi}{m}$, with m being a positive integer greater than unity, one must choose a sufficiently large value of p_0 so that the coherent states $\xi_n := | e^{i\chi n} i p_0 \rangle$ have negligible overlaps. In order to get a high confidence for the inference from the pointer readings on the measured observable, one then should choose a partitioning (X_k) of the value space of E_2^q such that every ξ_n is essentially localised in one of the corresponding phase space slices. It is evident that due to the 2π-periodicity and the reflection symmetry with respect to the q-axis, there will be m distinct groups of infinitely many contributions to the sum in (3.25) associated with the m states ξ_n ($n = 0, 1, ..., m - 1$). Accordingly there are m groups of number states of mode 1 such that only members from different groups can be distinguished by their measurement statistics.

Turning to the second example we take the pointer F_2 to be the phase observable M_2 of Eq. (III.5.12). The state T_2 shall be chosen as above, that is, as the coherent state $P[\xi_0] := |ip_0\rangle\langle ip_0|$. Then Eq. (3.20) gives

$$E(X) = \sum_{n=0}^{\infty} \langle \xi_0 | M_2(X - \chi n)\xi_0 \rangle |n\rangle\langle n| \tag{3.27}$$

It is instructive to discuss this case in some detail. In order to illustrate the interpretation of a reading in the set X as an unsharp reading of the photon number, let us consider a discretised version of E. To this end we assume again that the parameter χ is a fraction of 2π, $\chi = \frac{2\pi}{m}$. In view of Figure 7.3 it is plausible that for a sufficiently large value of the amplitude p_0 the coherent state $\xi_0 = |ip_0\rangle$ can be arbitrarily well localised in the slice spanned by the angular interval $X_o = [\pi - \frac{\pi}{m}, \pi + \frac{\pi}{m})$ of length $\frac{2\pi}{m}$ in the sense that $\langle \xi_0 | M_2(X_o)\xi_0 \rangle = 1 - \varepsilon$ for some small ε. (It may be recalled that M_2 is not strictly localisable, that is, for any set X of measure less than 2π and any state ϕ one has $\langle \phi | M_2(X)\phi \rangle < 1$.)

The sets $X_n = X_o + n\frac{2\pi}{m}$, $n = 0, 1, \cdots, m-1$, constitute a partition of $[0, 2\pi)$ into disjoint intervals and gives thus rise to a corresponding partition of 'phase space' into angular slices as indicated in Figure 7.3. Introducing the stochastic matrix $p_{ln} := p(X_l, n) = \langle \xi_n | M_2(X_l)\xi_n \rangle$, with $\xi_n = e^{i\chi n N_2}\xi_0 = |e^{i\chi n}ip_0\rangle$, one obtains a discrete POV measure with the generating effects

$$E_l := E(X_l) = \sum_{n=0}^{\infty} p_{ln} |n\rangle\langle n| \tag{3.28}$$

Due to the periodicity property $X_l + 2\pi k = X_l$ one has $p_{l,n+km} = p_{ln}$ and therefore

$$E_l = \sum_{n=0}^{m-1} p_{ln} \sum_{k=0}^{\infty} |n + km\rangle\langle n + km| \tag{3.29}$$

In view of the unsharp localisation condition on ξ_0, $p_{ll} = 1 - \varepsilon$ and $p_{ln} < \varepsilon$ whenever $l \neq n$. Letting $\varepsilon \to 0$, we therefore see that in the first sum of (3.29) only the term p_{ll} survives so that the effects E_l become approximately projections,

$$E_l \simeq P_l := \sum_{k=0}^{\infty} |l + km\rangle\langle l + km| \tag{3.30}$$

These projections generate a PV measure associated with a periodic function of N_1.

It is intuitively clear that a finer partition does not increase the amount of information offered by this measurement scheme. The phase spread of the pointer state ξ_0 determines the resolution in separating the different groups of number states labeled by n in (3.29). In fact if the first mode is prepared in some number state $|\,n_0\rangle$, then an outcome l would allow one to conclude almost with certainty that n_0 was one of the numbers $l + km$.

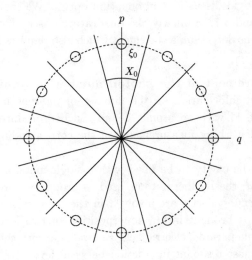

Figure 7.3. Phase space picture of the choice of phase partitioning and pointer states ξ_n depending on the coupling strength χ

The first moment $E^{(1)}$ of the measured observable E is directly obtained from (3.24), or also from (3.27):

$$E^{(1)} = \langle M_2^{(1)}\rangle_{\xi_0} I_1 + \chi N_1 - \sum_n 2\pi \langle M_2(2\pi - \chi n, 2\pi)\rangle_{\xi_0} |n\rangle\langle n| \qquad (3.31)$$

The last term is a periodic function of N_1 if $\frac{\chi}{2\pi}$ is rational.

We are now in a position to determine conditions under which the measured observable (3.29) (thus considering again the periodic case) can approach the sharp number observable. This happens when one is measuring field states φ that are superpositions of the first m_0 number states, $\varphi = \sum_{n=0}^{m_0} c_n |\,n\rangle$, where $m_0 \ll m$. In fact for such states one finds $\langle\varphi\,|\,E_l\varphi\rangle = \sum_{n=0}^{m_0} p_{ln}\,|\langle n\,|\,\varphi\rangle|^2 \simeq |\langle l\,|\,\varphi\rangle|^2$. In the last step we have used the approximation (3.30).

Finally it is interesting to note that for irrational values of $\frac{\chi}{2\pi}$ the set of points $n\chi \;[\mathrm{mod}\; 2\pi]$ is dense in the interval $[0, 2\pi)$. This implies that the periodicity of E and $E^{(1)}$ is lost, and the points $n\chi$ will be randomly distributed over the sections of arbitrary partitions. Consequently one can obtain more and more refined functions of N_1 by taking increasingly finer partitions and larger amplitudes p_0 (assuring an appropriate resolution that keeps $p_{ll} = 1 - \varepsilon$). We may pictorially describe this situation by saying that 'a chaotic system entails an infinite amount of information'.

The choice of the pointer observables a_2^q and M_2 in the the present measurement scheme raises the important question whether there exist concrete measurement devices associated with these observables. The problem of measuring the phase is still an issue of vivid discussions in quantum optics. We shall return to this topic in Sections 3.4 and 3.7. Similarly the quadrature observables are usually said to describe homodyne detection, a statement which also requires further investigations (Sec. 3.6).

VII.3.3 Measurement of the photon number in a microwave cavity.
It is possible to get information on the number of photons in a microwave cavity by probing the field with an atomic beam. Such procedures have been claimed to provide non-disturbing measurements of the corresponding number observable [7.7–10].

We assume the cavity to be filled with a single-mode field, with number observable $N = a^*a$; the Rydberg atoms used as probes are prepared in such a way that only three energy levels are involved in the interaction. Before entering the cavity the atom is brought into a superposition of two levels $|1\rangle$ and $|2\rangle$ (by means of a suitable periodic electric field E_1 resonant with this transition). Within the cavity an almost resonant interaction between levels $|2\rangle$ and $|3\rangle$ establishes a correlation between the atom and the field, without allowing an absorption of a photon. After leaving the cavity the atom passes again an intense electric field E_2 and is then analysed in an ionisation counter in order to determine whether the atom is in state $|1\rangle$ or $|2\rangle$ (Figure 7.4). The effects of the electric fields on the probe atom are described by unitary operators $R_k(\theta_k) = e^{-i\theta_k J_k^{12}}$, $k = 1, 2$, where $J_1^{12} = \frac{1}{2}(|2\rangle\langle 1| + |1\rangle\langle 2|)$ and $J_2^{12} = -\frac{i}{2}(|2\rangle\langle 1| - |1\rangle\langle 2|)$ are the inversion operators for the energy levels $|1\rangle$ and $|2\rangle$. The cavity-atom interaction can be modelled as

$$U = \exp\left[-i\lambda N \otimes J_3^{23}\right] \tag{3.32}$$

with the polarisation operator $J_3^{23} = \frac{1}{2}(|3\rangle\langle 3| - |2\rangle\langle 2|)$.

If T_A is the initial state of the probe atom, then the state of the field-atom system transforms under the interaction according to

$$T \otimes T_A \mapsto W := I \otimes R_2(\theta_2)\, U\left[T \otimes R_1(\theta_1)T_A R_1(\theta_1)^*\right] U^* I \otimes R_2(\theta_2)^* \tag{3.33}$$

$$= \sum_{n,m=0}^{\infty} |n\rangle\langle n|T|m\rangle\langle m| \otimes R_2(\theta_2)e^{-i\lambda n J_3^{12}} R_1(\theta_1)T_A R_1(\theta_1)^* e^{i\lambda m J_3^{12}} R_2(\theta_2)^*$$

Therefore if F is the atomic observable to be used as the pointer, then the measured field observable E is determined from the formula

$$\text{tr}\left[TE(X)\right] := \text{tr}\left[WI \otimes F(X)\right] \tag{3.34}$$

One finds

$$E(X) = \tag{3.35}$$
$$\sum_{n=0}^{\infty} \text{tr}\Big[R_2(\theta_2)e^{-i\lambda n J_3^{12}} R_1(\theta_1)T_A R_1(\theta_1)^* e^{i\lambda n J_3^{12}} R_2(\theta_2)^* F(X)\Big] |n\rangle\langle n|$$

Hence E is a smeared number observable. The state transformer determined by this measurement is also easily obtained:

$$\mathcal{I}_X(T) = \sum_{n,m} \text{tr}\Big[R_2(\theta_2)e^{-i\lambda n J_3^{12}} R_1(\theta_1)T_A R_1(\theta_1)^* e^{i\lambda m J_3^{12}} R_2(\theta_2)^* F(X)\Big] \cdot$$
$$|n\rangle\langle n|T|m\rangle\langle m| \tag{3.36}$$

In particular this gives

$$T_\Omega \equiv \mathcal{I}_\Omega(T) = \sum_{n,m} \text{tr}\Big[e^{-i\lambda(n-m)J_3^{12}} R_1(\theta_1)T_A R_1(\theta_1)^*\Big] |n\rangle\langle n|T|m\rangle\langle m| \tag{3.37}$$

which shows that

$$\text{tr}[TE(X)] = \text{tr}[T_\Omega E(X)] \tag{3.38}$$

for all T and X. In other words one is facing a first kind measurement of the field observable E, the 'photon statistics' remains undisturbed in the present scheme.

(a)

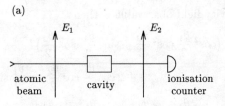

(b)

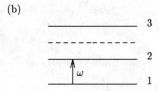

Figure 7.4. Measurement of the number of photons in a microwave cavity: (a) the setup; (b) atomic level scheme

The first moment of E is

$$E^{(1)} = \int x E(dx) = \sum_n \langle F^{(1)} \rangle_{T_A^n} |n\rangle\langle n| \tag{3.39}$$

where $\langle F^{(1)} \rangle_{T_A^n}$ denotes the mean value of the pointer F in the final n-conditioned state of the atom $T_A^n := R_2(\theta_2) e^{-i\lambda n J_3^{12}} R_1(\theta_1) T_A R_1(\theta_1)^* e^{i\lambda n J_3^{12}} R_2(\theta_2)^*$. Clearly $E^{(1)}$ is a function of N.

The important question then is to choose the pointer observable F as well as the initial state T_A of the atom such that the measured observable E gives optimal information on the photon statistics in the cavity field. Without trying to solve the optimisation problem in its full generality, we take, first of all, a physically realisable initial state for the probe atom: $T_A = |1\rangle\langle 1|$. Then

$$E(X) = \sum_n \langle \psi_n | F(X)\psi_n \rangle |n\rangle\langle n| \tag{3.40}$$

with

$$\begin{aligned}
\psi_n &= \left(\cos\frac{\theta_1}{2}\cos\frac{\theta_2}{2} - i\sin\frac{\theta_1}{2}\sin\frac{\theta_2}{2}e^{in\lambda/2}\right) |1\rangle \\
&\quad - \left(\cos\frac{\theta_1}{2}\sin\frac{\theta_2}{2} + i\sin\frac{\theta_1}{2}\cos\frac{\theta_2}{2}e^{in\lambda/2}\right) |2\rangle
\end{aligned} \tag{3.41}$$

This shows that a natural minimal choice of the pointer observable F is one that distinguishes between the involved energy levels $|1\rangle$ and $|2\rangle$ of the probe atom; the simplest realisation being a three valued pointer

$$F_1 = |1\rangle\langle 1|, \quad F_2 = |2\rangle\langle 2|, \quad F_3 = I - (|1\rangle\langle 1| + |2\rangle\langle 2|) \tag{3.42}$$

The measured cavity field observable is then

$$\begin{aligned}
E_1 &= \left(\cos^2\frac{\theta_1}{2}\cos^2\frac{\theta_2}{2} + \sin^2\frac{\theta_1}{2}\sin^2\frac{\theta_2}{2}\right)I \\
&\quad + 2\cos\frac{\theta_1}{2}\cos\frac{\theta_2}{2}\sin\frac{\theta_1}{2}\sin\frac{\theta_2}{2}\sin\left(\frac{\lambda}{2}N\right) \\
E_2 &= \left(\cos^2\frac{\theta_1}{2}\sin^2\frac{\theta_2}{2} + \sin^2\frac{\theta_1}{2}\cos^2\frac{\theta_2}{2}\right)I \\
&\quad - 2\cos\frac{\theta_1}{2}\cos\frac{\theta_2}{2}\sin\frac{\theta_1}{2}\sin\frac{\theta_2}{2}\sin\left(\frac{\lambda}{2}N\right) \\
E_3 &= O
\end{aligned} \tag{3.43}$$

Fixing the phases $\theta_1 = \theta_2 = \frac{\pi}{2}$ one finds simply

$$\begin{aligned}
E_1 &= \frac{1}{2}\left(I + \sin\frac{\lambda}{2}N\right) \\
E_2 &= \frac{1}{2}\left(I - \sin\frac{\lambda}{2}N\right) \\
E_3 &= O
\end{aligned} \tag{3.44}$$

The state transformer of the measurement is also simplified,

$$\mathcal{I}_1(T) = \tfrac{1}{4}\left(I + ie^{i\frac{\lambda}{2}N}\right) T \left(I - ie^{-i\frac{\lambda}{2}N}\right)$$
$$\mathcal{I}_2(T) = \tfrac{1}{4}\left(I - ie^{i\frac{\lambda}{2}N}\right) T \left(I + ie^{-i\frac{\lambda}{2}N}\right)$$
$$\mathcal{I}_3(T) = O$$
$$\mathcal{I}_\Omega(T) = \tfrac{1}{2}\left(T + e^{-i\frac{\lambda}{2}N}\, T\, e^{i\frac{\lambda}{2}N}\right)$$
(3.45)

which shows that the measurement remains nonrepeatable even with this choice of the initial atomic state and the pointer observable.

Specifying still further the pointer observable to be $Z = |1\rangle\langle 1| - |2\rangle\langle 2|$, say, one may determine the first moment of E:

$$E^{(1)} = E_1 - E_2 = \sin(\tfrac{\lambda}{2}N)$$
(3.46)

Thus from the point of view of expectation values the measured observable looks like the sine of the number observable

$$\langle E^{(1)}\rangle_T = \langle \sin(\tfrac{\lambda}{2}N)\rangle_T = \langle Z\rangle_{T_A^f} = \langle E^{(1)}\rangle_{T_\Omega}$$
(3.47)

where T_A^f is the final state of the atom. The last two equalities follow from the condition (3.34) and the first kind property (3.38) of the measurement.

To summarise, in probing a single-mode cavity field with a Rydberg atom initially in a state $|1\rangle$ and detecting in which of the two energy states $|1\rangle$ or $|2\rangle$ the atom finally is, one finds a two-valued field observable constituted by the effects E_1, E_2 which are diagonal in the number state basis. Depending on the parameters one may distinguish different number states by means of the averages of E_1, E_2. For example, if $\lambda < \frac{\pi}{1000}$, then the first 1000 number states can be uniquely identified due to the one-to-one correspondence between n and $\sin\left(\frac{n\lambda}{2}\right)$ for $n \leq 1000$. If the effects are periodic functions of the sine argument, which occurs for rational values of $\frac{\lambda}{2\pi}$, then no unique statistical distinction of the photon number states is possible. On the other hand for irrational values of $\frac{\lambda}{2\pi}$, arbitrary pairs of number states can be distinguished according to the expectation values of E_1. Returning to the level of first moments, Eqs. (3.47), we note that for the above choice of small parameters the sine can be linearised, so that one obtains a truncated number observable which is 'proportional' to the pointer observable.

VII.3.4 Phase distributions from number statistics. Phase observables of a single-mode electromagnetic field were introduced in Sec. III.5.1 as observables that are covariant under the shifts generated by the number observable. We have investigated various examples and in Sec. VII.3.2 two of them were found to be quite useful pointer observables in measuring the photon number of another mode. Moreover states with fairly well defined phase, coherent and squeezed states, can be prepared. Hence the quest for experimental procedures determining the phase is getting more and more urgent and has been a subject of recent investigations [3.17]. Up to now no concrete measurement setup has been proposed that would define a phase observable according to the probability reproducibility rule. That phase statistics can be obtained as marginal distributions from joint measurements of the field quadratures will be demonstrated in Sec. 3.7. For the time being we describe a method of obtaining phase information from the combined statistics of other measurements.

Consider the photodetection of a signal mode mixed with a local oscillator prepared in a coherent state $|z\rangle$. If the quantum efficiency of the detector is ε, then the measured observable $E^{z;\varepsilon}$ is the smeared number observable $n \mapsto E_n^{z;\varepsilon} = D_{xz} E_n^{0;\varepsilon} D_{xz}{}^*$ of Eq. (3.13). The expectation value of this observable in the input state T of the signal depends, in particular, on the strength $|z|^2$ of the local oscillator, and it contains an interference term $\sqrt{\varepsilon(1-\varepsilon)}\,\mathrm{tr}\big[T\,(\bar{z}a + za^*)\big]$, see Eq. (3.14). This suggests that the interference pattern could be used to gain information on the phase of the input signal. To achieve high resolution the strength of the local field should be strong. However, in the limit $|z| \to \infty$, all the effects $E_n^{z;\varepsilon}$ tend to the null operator weakly so that one has to consider another limit allowed by the apparatus parameter ε. Indeed letting $\varepsilon \to 1$ together with $|z| \to \infty$ such that $xz = u$ is a fixed complex number, one has $(1-\varepsilon)|z|^2 = \varepsilon|x|^2|z|^2 = \varepsilon|u|^2$ and

$$E_n^{z;\varepsilon} \;\to\; D_u\,|n\rangle\langle n|\,D_u{}^* \;=:\; E_n^u \tag{3.48}$$

Thus a photodetection with an almost ideal photocounter of a single-mode field mixed in the active part of the detector with a strong local oscillator defines a signal observable $n \mapsto E_n^u$, where $|u|^2 = |x|^2|z|^2$ describes the percentage of energy which the signal gains from the coherent pulse $|z\rangle$. The first moment of this limiting observable is

$$N^u \;:=\; \sum n E_n^u \;=\; D_u\,N\,D_u{}^* \;=\; N + |u|^2 I + (ua + \bar{u}a^*) \tag{3.49}$$

showing still a similar interference pattern as Eq. (3.14). In order to obtain a reliable phase information on the signal we consider many of such photodetection schemes with the ensuing number statistics $p_T^{E^u}(n) = \mathrm{tr}\big[T E_n^u\big]$, labelled with $u = xz \in \mathbf{C}$. Now for any fixed n one may collect all the measurement outcome probabilities $p_T^{E^u}(n)$, $u \in \mathbf{C}$, into a single probability distribution on 'phase space'

$$Z \;\mapsto\; \frac{1}{\pi} \int_Z p_T^{E^u}(n)\, d^2u, \quad Z \in \mathcal{B}(\mathbf{C}) \tag{3.50}$$

These probability measures define the (normalised) POV measure

$$A^{|n\rangle} : Z \mapsto A^{|n\rangle}(Z) \;:=\; \frac{1}{\pi} \int_Z D_u\,|n\rangle\langle n|\,D_u{}^*\, d^2u \tag{3.51}$$

studied in Sec. III.5.2. This is an unsharp joint observable for the quadrature components of the signal mode. Unsharp quadrature observables as well as unsharp number and phase observables were found as its marginals with reference to the Cartesian and polar coordinates, respectively. In particular, $A_{ph}^{|n\rangle}$ contains information on the phase of the input signal, see Sec. III.5.2. We conclude that by collecting the number statistics from different photodetection schemes, with the parameters ε, z, one obtains the phase distribution $X \mapsto \text{tr}\big[T A_{ph}^{|n\rangle}(X)\big]$ for each possible number outcome n.

It should be emphasised that the probability measures (3.50), as they are constituted here, do not correspond to a measurement of the observable (3.51). They should rather be seen as a pattern that is encrypted in the totality of statistics collected in the manifold of measurements labelled with the parameter u.

VII.3.5 Nondegenerate amplification and two-mode squeezing. Besides the two-mode interaction $H_\alpha = i\alpha a^* \otimes b - i\bar{\alpha} a \otimes b^*$ modelling the beam splitter coupling (VII.3.1), there is another basic two-mode coupling:

$$H_\kappa = i\bar{\kappa} a \otimes b - i\kappa a^* \otimes b^* \qquad (3.52)$$

Unlike H_α, this interaction does not conserve the total photon number $N_a \otimes I + I \otimes N_b$ of the two modes but it conserves the difference of the photon number in these modes, $[H_\kappa, N_a \otimes I - I \otimes N_b] = O$. Writing $\kappa = re^{i\varphi}$, with $0 \leq r < \infty$, $-\pi < \varphi \leq \pi$, we have

$$U \equiv U_{r,\varphi} := e^{-iH_\kappa} = \exp\big[r(e^{-i\varphi} a \otimes b - e^{i\varphi} a^* \otimes b^*)\big] \qquad (3.53)$$

which is the two-mode squeeze operator. It transforms the annihilation operators of the modes as

$$\begin{aligned}
\tilde{a} &:= U^*(a \otimes I)U = a \otimes I \cosh r - I \otimes b^* \, e^{i\varphi} \sinh r \\
\tilde{b} &:= U^*(I \otimes b)U = I \otimes b \cosh r - a^* \otimes I \, e^{i\varphi} \sinh r
\end{aligned} \qquad (3.54)$$

The two-mode squeeze operator (3.53) is a basic tool in the theory of squeezed states and it models, for instance, a nondegenerate parametric amplifier and a four-wave mixer [7.11]. We consider the implications of the coupling (3.53) within the POV measure description of some detection models, this time putting emphasis on the amplification process.

Suppose that the a-mode, the signal, is subjected to a two-mode squeezer, getting thereby coupled with an idle b-mode. The state T of the signal is then 'amplified',

$$T \mapsto \Phi(T) \qquad (3.55)$$

the state $\Phi(T)$ being defined by

$$\text{tr}\big[\Phi(T)A\big] := \text{tr}\big[U(T \otimes |0\rangle\langle 0|)U^* A \otimes I\big] \qquad (3.56)$$

where A is any observable of the signal. Factorising [7.12] the operator U as

$$U = \exp(-xa^* \otimes b^*) \exp(-y(N_a \otimes I + I \otimes N_b + I)) \exp(\overline{x}a \otimes b) \quad (3.57)$$

with $x = e^{i\varphi} \tanh r$, $y = \ln \cosh r$, the state $\Phi(T)$ assumes the form

$$\Phi(T) = \sum_{k=0}^{\infty} \frac{1}{k!} (\tanh r)^{2k} (a^*)^k e^{-y(N_a+1)} T e^{-y(N_a+1)} a^k \quad (3.58)$$

As an illustration, if the signal is in a number state $T = |n\rangle\langle n|$, then the 'squeezer' amplifies the signal state to a mixture of the number states $|n+k\rangle\langle n+k|$, $k = 0, 1, 2, \cdots$, all with nonzero weights. We consider the dual process of amplifying number states; that is, we determine the signal observable resulting from detecting (with an ideal detector) the photon number of the signal after it has passed the squeezer. If T is the input state of the signal, then $\mathrm{tr}\left[U(T \otimes |0\rangle\langle 0|)U^* |n\rangle\langle n| \otimes I\right]$ is the probability of detecting n photons in the output port. Therefore the measured signal observable $E^{0;\kappa} : n \mapsto E_n^{0;\kappa}$ is again the one satifying the relation

$$p_T^{E^{0;\kappa}}(n) = p_W^{N \otimes I}(n) \quad (3.59)$$

for all T and n, with $W := U(T \otimes |0\rangle\langle 0|)U^*$. Due to the duality of states and observables we have

$$E_n^{0;\kappa} = \Phi^*(|n\rangle\langle n|) \quad (3.60)$$

where Φ^* is the dual map of the amplification operation Φ. Therefore

$$E_n^{0;\kappa} = \sum_{k=0}^{\infty} \frac{1}{k!} (\tanh r)^{2k} e^{-y(N+1)} a^k |n\rangle\langle n| (a^*)^k e^{-y(N+1)} \quad (3.61)$$

The effects $E_n^{0;\kappa}$ constitute an unsharp number observable, which reduces to the sharp number observable if the squeeze factor r is 0, in which case no squeezing takes place.

The first moment of the observable (3.61) is determined to be

$$N_\kappa^0 := \sum n E_n^{0;\kappa} = \cosh^2(r(N+I)) \quad (3.62)$$

which shows that the average number of photons in the signal mode is increased in the amplification:

$$\langle N \rangle_{\Phi(T)} = \langle N_\kappa^0 \rangle_T \geq \langle N \rangle_T \quad (3.63)$$

VII.3.6 Homodyne detection. The homodyne detector is a fundamental device suited for phase-sensitive (single-mode) field measurements. Indeed the measured observables have been commonly regarded to be the quadrature components of the field [7.13, 7.14]. In such a detector, sketched in Figure 7.5, the signal field is coupled, via a beam splitter, with a strong local oscillator, a single-mode field in a coherent state, oscillating at the same frequency as the signal. The photocurrent from the detector is then filtered to select the appropriate frequency component.

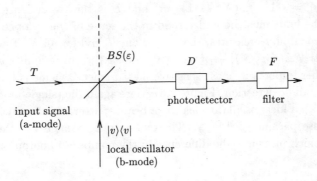

Figure 7.5. Homodyne detector

In this configuration, the beam splitter, with transparency ε, followed by an ideal photo(emissive) detector, defines the signal observable $E^{z;\varepsilon}$ with the effects (3.13):

$$E_n^{z;\varepsilon} \;=\; D_{xz}\, E_n^{0;\varepsilon}\, D_{xz}{}^{*} \tag{3.64}$$

The first moment of this observable is

$$N_\varepsilon^z \;:=\; \sum_{n=0}^{\infty} n E_n^{z;\varepsilon} \;=\; \varepsilon N + (1-\varepsilon)|z|^2 + \sqrt{\varepsilon(1-\varepsilon)}(\bar z a + z a^{*}) \tag{3.65}$$

Thus one shows that

$$\frac{1}{\sqrt{2}}\,\frac{1}{z\sqrt{1-\varepsilon}}\,\operatorname{tr}\left[T\left(N_\varepsilon^z - (1-\varepsilon)z^2\right)\right] \;\longrightarrow\; \operatorname{tr}\left[Ta^q\right] \tag{3.66}$$

$$\text{as } z \to \infty,\ \varepsilon \to 1 \text{ such that } z\sqrt{1-\varepsilon} \to \infty$$

where z is taken to be real. Removing the bias term $(1-\varepsilon)z^2$, normalising with the factor $z\sqrt{1-\varepsilon}$, and letting the local oscillator strength grow to infinity amounts to modelling the filtering process [7.13, 7.14]. Hence the homodyne detector constitutes, in the sense of expectation values, a measurement of the quadrature a^q. That this scheme does yield, in an appropriate limit, a full measurement of the observable a^q (in the sense of all moments) was shown in [7.14] using the method of characteristic functions. The Lévy-Cramér theorem [7.15] allows one to obtain the same result also for the POV measures. Indeed one can prove [7.16] that

$$E^{z,\varepsilon}(X^{z,\varepsilon}) \longrightarrow E(X)$$

$$\text{as } z \to \infty, \ \varepsilon \to 1 \text{ such that } z\sqrt{1-\varepsilon} \to \infty, \ z(1-\varepsilon)^{\frac{3}{2}} \to 0 \qquad (3.63)$$

where $X^{z,\varepsilon} = \sqrt{1-\varepsilon}z\left(X + \sqrt{1-\varepsilon}z\right)$ and E is the PV measure associated with $a + a^*$. The limit must be understood in the sense of the weak convergence of the ensuing probability measures, that is, for any T and for all X of nonzero Lebesgue measure, $\text{tr}\left[TE^{z,\varepsilon}(X^{z,\varepsilon})\right] \to \text{tr}\left[TE(X)\right]$ under the above limit conditions.

Another important phase-sensitive measurement scheme is furnished by a balanced homodyne detection (Figure 7.6). Here again the (single-mode) signal field is coupled with a local oscillator field by a beam splitter, this time with transparency $\frac{1}{2}$ and phase parameter $\vartheta = \frac{\pi}{2}$. The two output channels are then directed to a detector which measures the difference in the number of photons counted.

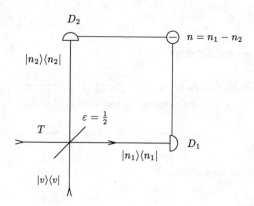

Figure 7.6. Balanced homodyne detector.

The detection observable is the difference of the photon numbers in the two arms of the apparatus:

$$N_1 \otimes I - I \otimes N_2 = \sum_{n_1} n_1 |n_1\rangle\langle n_1| \otimes I - \sum_{n_2} n_2 I \otimes |n_2\rangle\langle n_2|$$

$$= \sum_{n_1,n_2} (n_1 - n_2)|n_1\rangle\langle n_1| \otimes |n_2\rangle\langle n_2| =: \sum_{k=-\infty}^{+\infty} k\, A_k \qquad (3.68)$$

where we have defined

$$A_k := \sum_{n_2 \geq max\{0,-k\}} |k+n_2\rangle\langle k+n_2| \otimes |n_2\rangle\langle n_2| \qquad (3.69)$$

If $W = U_\alpha \big(T \otimes |z\rangle\langle z| \big) U_\alpha^*$, with $\alpha = i\frac{\pi}{4}$, then the detection statistics $p_W^A(k)$ defines a signal observable $E : k \mapsto E_k$ such that

$$p_T^E(k) := p_W^A(k) \tag{3.70}$$

for all values k and for all initial states T of the signal. Using the decomposition (3.10) of $U_{i\frac{\pi}{4}}$ the observable E can be computed from (3.70):

$$E_k = e^{-|z|^2} e^{i\bar{z}a} 2^{-\frac{1}{2}N_a} \cdot$$

$$\sum_{n_2 = max\{0,-k\}}^{\infty} \frac{1}{n_2!} \big(\sqrt{2}\bar{z} + ia^* \big)^{n_2} \ |k + n_2\rangle\langle k + n_2| \ \big(\sqrt{2}z - ia \big)^{n_2} \cdot \tag{3.71}$$

$$2^{-\frac{1}{2}N_a} e^{-iza^*}$$

The first moment of this observable is

$$E^{(1)} := \sum_{k=-\infty}^{+\infty} k \, E_k = i(\bar{z}a - za^*) \tag{3.72}$$

which is proportional to a^p or a^q, for real or purely imaginary z, respectively. Therefore, in the sense of expectation values, the measurement of the observable E is equivalent to that of a^p. In [7.17] it is argued that the difference statistics can be smoothed into continuous distributions which in the limit $z \to \infty$ approach the probability distributions of a^p. It can be shown that the POV measure E does indeed tend to E^p in this limit [7.18].

VII.3.7 Joint measurement of the quadrature components.
The beam splitter can be applied to perform a joint measurement of the two field quadrature components a^q and a^p of a one-mode signal. To see this we write the two-mode coupling U_α in terms of these operators so that for $\alpha = r \in (0, \frac{\pi}{2})$ (a beam splitter with transmission $\varepsilon = \cos^2 r$ and without phase shift), we obtain

$$U_r = \exp \big[ir(a^q \otimes b^p - a^p \otimes b^q) \big] \tag{3.73}$$

The signal field and local oscillator, prepared in the states T and T', are coupled in the beam splitter and enter then two detectors measuring the quadrature components a^q and b^p, respectively (Figure 7.7). It turns out that the ensuing joint outcome statistics is that of a phase space observable in the input state T. The operator in the exponent of (3.73) is formally identical to the angular momentum component L_3. This observation makes it straightforward to evaluate the defining condition for the measured observable $\tilde{G}$,

$$p_T^{\tilde{G}}(X \times Y) := p_W^{E^q \otimes E^p}(X \times Y) = \text{tr}\big[U_\alpha \, (T \otimes T') \, U_\alpha^* \, E^q(X) \otimes E^p(Y) \big] \tag{3.74}$$

where E^q and E^p are the spectral measures of a^q and a^p. For simplicity we shall consider the case of pure input states $T = P[\varphi]$ and $T' = P[\psi]$. One obtains

$$
\begin{aligned}
\left\langle \varphi \,\middle|\, \tilde{G}(X \times Y)\varphi \right\rangle &= \left\langle \varphi \otimes \psi \,\middle|\, U_r^* \, E^q(X) \otimes E^p(Y) \, U_r \, \varphi \otimes \psi \right\rangle \\
&= \int_{X \times Y} \left| U_F^{(b)} \, U_r \varphi \otimes \psi(q,p) \right|^2 dq \, dp \\
&= \frac{1}{2\pi} \int_{X_r \times Y_r} dq \, dp \, \langle \varphi \,|\, \xi_{qp} \rangle \langle \xi_{qp} \,|\, \varphi \rangle \\
&=: \langle \varphi \,|\, G_\xi(X_r \times Y_r)\varphi \rangle
\end{aligned}
\tag{3.75}
$$

Here $U_F^{(b)}$ denotes the Fourier-Plancherel operator with respect to the second degree of freedom, and we have introduced the scaled sets $X_r = X/\cos r$ and $Y_r = -Y/\sin r$. Furthermore, ξ_{qp} denotes the phase space translate of the (normalised) state function

$$
\xi(y) := \frac{1}{\sqrt{\tan r}} \, \overline{\psi\left(-\frac{y}{\tan r}\right)}
\tag{3.76}
$$

The accordingly rescaled observable is thus

$$
Z \mapsto G_\xi(Z) = \frac{1}{\pi} \int_Z d^2 z \, D_z \, |\xi\rangle\langle\xi| \, D_z^*
\tag{3.77}
$$

which is a phase space observable with unsharp quadrature components as its Cartesian marginals (cf. Secs. III.2.4, III.5.2).

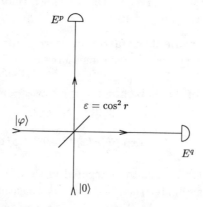

Figure 7.7. Joint measurement scheme for the quadrature components

For the case of a vacuum input in the b-mode, one obtains [7.19] the Q-distribution associated with the observable

$$
Z \mapsto G_0(Z) = \frac{1}{\pi} \int_Z |z\rangle\langle z| \, d^2 z
\tag{3.78}
$$

It is important to note that one can choose ψ so as to have ξ be a number state. One has thus found a class of measurement schemes yielding the statistics for all the phase observables described in III.5.2.

VII.3.8 Atomic position measurement. When an atom passes through a standing light wave its position may get correlated with the phase of the field. With a phase sensitive field measurement one can therefore obtain information on the atom's localisation along the cavity axis (Figure 7.8). In order to corroborate this expectation with determining the measured observable, we follow the ideas presented in [7.20].

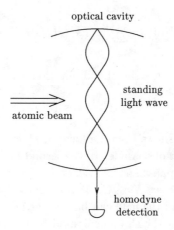

optical cavity

standing
light wave

atomic beam

homodyne
detection

Figure 7.8. Localisation of an atom in a cavity field

According to the most simplified picture, the atom is modelled by a two-level system with one spatial degree of freedom, its Hilbert space thus being $L^2(\mathbf{R}) \otimes \mathbf{C}^2$. Furthermore only a single-mode field, with the number observable $N = a^*a$, is taken into account. In the rotating-wave approximation, with a large detuning $\delta := \omega_o - \omega_a$ of the atomic transition frequency ω_o from the cavity mode frequency ω_a, the effective coupling between atom and cavity field can be given as

$$U \;=\; \exp\!\big(-i\alpha \cos^2(kQ + \zeta) \otimes \sigma_3 \otimes N\big) \tag{3.79}$$

Here Q is the (center of mass) position of the atom (along the cavity axis), σ_3 the Pauli spin operator, $\alpha = 2|g|^2/\delta$ and g is the coupling constant (the single photon Rabi frequency). The field is assumed to be initially in a coherent state $|\,z\rangle$.

In order to obtain information on the position of the atom relative to the nodes and antinodes of the standing cavity wave, a phase-sensitive observable F is to be applied as the pointer. To this end one may take any phase-shifted quadrature observable $e^{i\varphi N} a^q e^{-i\varphi N}$. For illustrative purposes we base our analysis on the joint observable $F = G_0$, Eq. (3.78). The field leaks out of the cavity through the end mirror to produce a signal that can be measured. If T_A is the initial state of the atom, then the condition

$$\text{tr}\left[T_A E(X)\right] := \text{tr}\left[U\left(T_A \otimes |z\rangle\langle z|\right) U^* I \otimes I \otimes F(X)\right] \tag{3.80}$$

determines the atomic observable E measured with this arrangement. A straightforward calculation gives

$$\begin{aligned} E(X) &= \int dE^Q(q) \otimes \sum_{s=\pm 1} |s\rangle\langle s| \langle e^{is\phi_q} z| F(X) |e^{is\phi_q} z\rangle \\ &= \int dE^Q(q) \otimes \sum_{s=\pm 1} |s\rangle\langle s| \langle z| G_0(e^{is\phi_q} X)|z\rangle \end{aligned} \tag{3.81}$$

Here we have introduced the shorthand notation

$$\phi_q := \alpha \cos^2(kq + \zeta) \tag{3.82}$$

The first line of (3.81) shows that the form of E does not significantly depend on the choice of the pointer observable. In the second line we have exploited the phase shift covariance of the phase space observable $F = G_0$, which is represented by a rotation of the complex set $X \subset \mathbf{C}$. It is evident that a correlation between the position of the atom and the field can be obtained only if the detection observable F is not invariant under the shifts generated by the number observable; hence G_0 is an appropriate choice. The phase change induced by the atom depends on the vacuum light shift $\frac{\alpha}{2}\delta \cos^2(kq + \zeta)$ and is independent of the number of photons in the cavity. If the atom is initially in its ground state, $T_A = T \otimes |-1\rangle\langle -1|$, it also remains so under the coupling. In this case the observable (3.81) reduces to

$$E(X) = \int dE^Q(q) \langle z| G_0(e^{i\phi_q} X)|z\rangle =: \int p(X, q)\, dE^Q(q) \tag{3.83}$$

It follows that E is in fact a smeared (though non-covariant) position observable. The degree of smearing and the resolution with which a phase change can be detected depend crucially on the pointer and the amplitude of the field $|z|^2$.

We determine next the state transformer associated with this measurement. In the position representation it assumes the form

$$\mathcal{I}_X(T) = \int_{\mathbf{R}} dq \int_{\mathbf{R}} dq'\, \langle z| G_0(e^{i\phi_q} X)|z\rangle \langle q|T|q'\rangle |q\rangle\langle q'| \tag{3.84}$$

which still applies to any phase space observable A. One can immediately confirm that for each T and X, $\text{tr}\left[\mathcal{I}_X \mathcal{I}_\Omega(T)\right] = \text{tr}\left[\mathcal{I}_X(T)\right]$, that is, the measurement is of the first kind: the probability for a particular reading is the same before and after the measurement. Nevertheless the measurement is not repeatable.

Next making use of the specific form of the observable G_0, that is, $G_0(X) = \frac{1}{\pi} \int_X |u\rangle\langle u| \, d^2u$, allows one to write the state transformer (3.84) as

$$\mathcal{I}_X(T) = \frac{1}{\pi} \int_X K_u T K_u^* \, d^2u \qquad (3.85)$$

where

$$K_u = \int_{\mathbf{R}} dq \, \langle u \,|\, e^{-i\phi_q} z \rangle \, |q\rangle\langle q| \qquad (3.86)$$

To get an idea how this measurement scheme operates, we consider the questions of calibration and inference. First assume the state T is so well localised in an interval around the point q_0 that in (3.83) and (3.84) one may replace ϕ_q with ϕ_{q_0} (as a crude approximation). Then one obtains

$$\begin{aligned}
\mathrm{tr}\big[T\,E(X)\big] &\simeq \frac{1}{\pi} \int dq \, \langle q\,|T|q\rangle \int_X d^2u \, \big|\langle u|\, e^{i\phi_{q_0}} z\rangle\big|^2 \\
&= \frac{1}{\pi} \int_X d^2u \, \exp\big(-|u - e^{i\phi_{q_0}} z|^2\big)
\end{aligned} \qquad (3.87)$$

This probability will approach unity if X is an angular slice centered around the ray containing the point $e^{i\phi_{q_0}} z$ and of width sufficiently large so as to contain the bulk of the phase space distribution associated with the coherent state $|z\rangle$. It is clear that the resolution can be increased indefinitely by choosing large amplitudes $|z|^2$. With the same crude approximation one finds from (3.85) that the state T remains nearly unchanged (modulo normalisation), that is, $\mathcal{I}_X(T) \simeq T \, \mathrm{tr}\big[T\,E(X)\big]$. Thus localised states can be used for calibrating this measurement procedure.

On the other hand given only the information that the state was localised in some small interval, it is not possible to determine uniquely the location of this interval from the measurement statistics. This is due to the periodic dependence of the readout observable E on the position parameter via the phase ϕ_q.

VII.4 Wave-particle duality of photons

The *which path* experiments for photons and other quantum objects have remained an issue of intensive experimental and theoretical investigations throughout the history of quantum mechanics. The famous two-slit arrangement, well known as a source of interference phenomena in classical light optics, was quickly recognised as an excellent illustration of the nonobjectivity of quantum observables of individual systems. The more recent quantum optical split-beam analogue provided by the Mach-Zehnder interferometer offered the possibility of actually realising the thought experiments that were invented by Bohr and Einstein in their attempts to

demonstrate the idea of complementarity or to circumvent the measurement limitations due to the uncertainty relations. The wave-particle duality for single photons
has been strikingly confirmed in a series of modern experiments [7.21–23]. We shall
rederive the mutual exclusiveness of the particle and the wave behaviour, reflected
in the two options of path determination and interference measurements. Moreover
both aspects can be reconciled with each other if one does not require absolute
certainty with respect to the path nor optimal interference contrast.

To put the subsequent formulation of these features into its proper perspective,
some general remarks are in order. The mathematical description of the interference experiments is sufficiently simple so as to admit a fairly exhaustive account
of the physical situation. It turns out that there is not just one description but
a variety of them, each yielding the same experimental figures but nevertheless
leading to totally different (though equivalent) mathematical representations and
physical interpretations. In particular we encounter illustrations of an instrumen-
talist account, a phenomenalistic description and a realistic picture of physical
experiments. In the first case one is only concerned with computing the counting
frequencies, treating the whole experimental setup as a 'black box' with a variety of
control parameters. The second type of account acknowledges that there is an input
system influencing the black box, the measuring apparatus, and one may interpret
the counting statistics with respect to this input system. In both cases varying
the control parameters amounts to specifying another measurement. It is only in
the third, realistic, account that the mathematical language utilised matches the
wordings used by the experimenters in devising the setup, carrying out the preparations and measurements, and interpreting the outcomes in terms of the prepared
system. In effect, the other two approaches also must base their way of computing
on a certain interpretation; they do introduce a splitting of the whole process into
an observing and an observed part (measurement performed after a preparation);
but the *cut* is placed at different locations, and this decision determines the ensuing mathematical picture. In this way different degrees of reality are ascribed to
the phenomena, ranging from mere measurement outcomes over highly contextual
entities to something that may be considered as a kind of quantum objects.

VII.4.1 Photon split-beam experiments. Using the tools developed in the
preceding sections we may now reanalyse the photon split-beam experiments performed with a Mach-Zehnder interferometer. Figure 7.9 shows the scheme of such
a device consisting of two beam splitters $BS_1(\varepsilon_1)$ and $BS_2(\varepsilon_2)$, with transparencies ε_1 and ε_2, reflecting mirrors M_1 and M_2, and a phase shifter $PS(\delta)$, allowing
for the variation of the path difference between the two arms of the interferom-
eter. The detectors D_1 and D_2 are assumed to register the number of photons
$N_1 = \sum n_1 |n_1\rangle\langle n_1|$ and $N_2 = \sum n_2 |n_2\rangle\langle n_2|$ emerging from the second beam splitter when a photon pulse is impinging on the first one. We shall assume a single-mode
input field in a state T, with $N_a = a^*a$ denoting its number operator. The first

beam splitter $BS_1(\varepsilon_1)$ effects a coupling of this mode with an idle single-mode field, with $N_b = b^*b$. It will be useful to consider first arbitrary states T' of the b-mode, and only later fix it to be idle, $T' = |0\rangle\langle0|$.

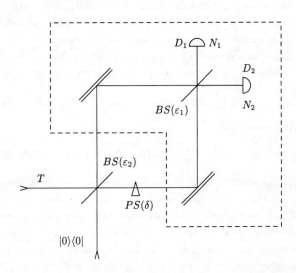

Figure 7.9. Scheme of a Mach-Zehnder interferometer. The dotted box indicates the measuring device according to the 'realistic' cut.

The action of a beam splitter $BS(\varepsilon)$ is given by a coupling of the form (3.6),

$$U_\alpha = \exp\left[i\vartheta(N_a \otimes I - I \otimes N_b)\right] \exp(\overline{\alpha}a \otimes b^* - \alpha a^* \otimes b) \qquad (4.1)$$

with $\alpha = |\alpha|e^{i\vartheta}$, $\cos|\alpha| = \sqrt{\varepsilon}$. The phase shifter $PS(\delta)$ acts according to

$$V_\delta = e^{i\delta N_a} \otimes I \qquad (4.2)$$

Therefore if $T \otimes T'$ is the initial state of the two-mode field entering the interferometer, then the state of the field emerging from the second beam splitter is

$$W := U_\beta V_\delta U_\alpha(T \otimes T')U_\alpha^* V_\delta^* U_\beta^* \qquad (4.3)$$

where $\alpha = |\alpha|e^{i\vartheta_1}$, $\cos|\alpha| = \sqrt{\varepsilon_1}$, and $\beta = |\beta|e^{i\vartheta_2}$, $\cos|\beta| = \sqrt{\varepsilon_2}$. The probability of detecting n_1 photons in the detector D_1 and n_2 photons in the detector D_2 is thus

$$p_W^{N_1 \otimes N_2}(n_1, n_2) = \langle n_1, n_2 | W | n_1, n_2 \rangle \qquad (4.4)$$

In this reading the measured observable is the two-mode number observable $N_1 \otimes N_2$, and the measured system is the output field emerging from the interferometer.

The field's passage through the interferometer is treated as an indivisible part of the preparation of the phenomenon to be observed, and no attempt is made at analysing the various stages of this process. All adjustable variables, the two-mode input state, the transparencies, and the phase shift, are treated on equal footing as control parameters of the black box determining the preparation of the output state W which is subjected to a counting measurement.

The counting probability (4.4) can equivalently be written as a measurement outcome probability with respect to the input state $T \otimes T'$ of the two-mode field,

$$p_{T \otimes T'}^E(n_1, n_2) \; := \; p_W^{N_1 \otimes N_2}(n_1, n_2) \tag{4.5}$$

the observable being now

$$E(n_1, n_2) \; := \; U_\alpha^* V_\delta^* U_\beta^*\big(|n_1\rangle\langle n_1| \otimes |n_2\rangle\langle n_2|\big) U_\beta V_\delta U_\alpha \tag{4.6}$$

Here the interferometer is taken as a part of the measuring device, which now serves to yield information on the input state. Accordingly, the set of variables mentioned above is split into two parts, the input state representing the preparation, and the interferometer parameters belonging to the measurement. In order to realise wave or particle phenomena, one must take into account, in this view, all the details of the experiment, including the measuring system parameters. The 'particle' or 'wave' cannot be described as an entity existing independently of the constituting measurement context. In fact the input state alone does not determine whether the field passing the interferometer behaves like a particle or a wave.

In order to find the explicit form of the POV measure (4.6), we note first that $V_\delta U_\alpha = U_{\alpha'} V_\delta$, with $\alpha' = e^{i\delta}\alpha$. Next observe that the operators $a \otimes b^*$, $a^* \otimes b$, and $\frac{1}{2}(N_a \otimes I - I \otimes N_b)$ satisfy the standard commutation relations of the generators of the group $SU(2)$. Therefore for any α and β, there is a γ such that

$$U_\alpha U_\beta \; = \; e^{\frac{i}{2} f(\alpha, \beta)(N_a \otimes I - I \otimes N_b)} \, U_\gamma \tag{4.7}$$

[7.2]. This allows one to write the observable E as

$$E(n_1, n_2) \; = \; E^\varepsilon(n_1, n_1) \; := \; U_\gamma^*\big(|n_1\rangle\langle n_1| \otimes |n_2\rangle\langle n_2|\big) U_\gamma \tag{4.8}$$

for an appropriate γ. It follows that the Mach-Zehnder interferometer acts like a single beam splitter $BS(\varepsilon)$, where $\sqrt{\varepsilon} = \cos|\gamma|$, and $\gamma = \gamma(\alpha, \beta, \delta)$. The explicit form of the effective transparency ε shall be determined subsequently.

If the state T' of the second mode is kept fixed, one may view the a-mode alone as the input system. This step is necessary if one wants to represent the idea that a light pulse, or even one photon, coming from one source is subjected to an interferometric measurement. In that view the counting statistics (4.4) define an observable $F^{T', \varepsilon}$ of this mode such that for any T and for all n_1, n_2 one has

$$\mathrm{tr}\big[T F^{T'; \varepsilon}(n_1, n_2)\big] \; := \; \mathrm{tr}\big[T \otimes T' E^\varepsilon(n_1, n_2)\big] \tag{4.9}$$

If T' is a vector state, then this observable is just the Neumark projection of E^ε with the projection $I \otimes T'$:

$$F^{T';\varepsilon} \equiv I \otimes T' E^\varepsilon I \otimes T' \tag{4.10}$$

Finally taking the second mode to be in the vacuum state, the explicit form of this observable is obtained by a simple computation:

$$F^{0;\varepsilon}(n_1, n_2) = \frac{(n_1 + n_2)!}{n_1! n_2!} \varepsilon^{n_1} (1 - \varepsilon)^{n_2} |n_1 + n_2 \rangle \langle n_1 + n_2| \tag{4.11}$$

The counting statistics of a single-mode light field sent through an interferometer define thus an observable of this field. This observable is a PV measure exactly when $\varepsilon = 0$, or $\varepsilon = 1$. Any choice of the parameter ε refers to a different experimental arrangement consisting of $BS_1(\varepsilon_1)$, $BS_2(\varepsilon_2)$, and $PS(\delta)$. As a rule they all give rise to different observables $F^{0;\varepsilon}$, which are thus mutually exclusive in the trivial sense that any choice of ε_1, ε_2, and δ excludes another one. However, these observables all are mutually commuting. Nevertheless they reflect *on a phenomenological level* the wave-particle duality of a single photon, as will become clear below. We call this description phenomenological as the production of the wave or the particle phenomena is described in terms of the instrumental tools, without making reference to the behaviour of the observed entity.

The marginals $F_i^{0;\varepsilon}$, $i = 1, 2$, of the observable $F^{0;\varepsilon}$ are found to be of the form (3.11), the first marginal associated with the parameter ε, the second with the parameter $1 - \varepsilon$. They are mutually commuting observables as well, representing unsharp versions of the number observable N_a.

We consider next the case that the incoming a-mode is prepared in a number state $T = |n \rangle \langle n|$. The probabilities (4.9) obtain then the simple form

$$\begin{aligned} p_{|n\rangle}^{F^{0;\varepsilon}}(n_1, n_2) &= \langle n | F^{0;\varepsilon}(n_1, n_2)|n \rangle \\ &= \frac{(n_1 + n_2)!}{n_1! n_2!} \varepsilon^{n_1} (1 - \varepsilon)^{n_2} \delta_{n, n_1 + n_2} \end{aligned} \tag{4.12}$$

From these probabilities the conservation of photon number (energy) in the interferometer is manifest: n input photons give rise to a total of $n = n_1 + n_2$ counts in the two detectors.

We are now ready to discuss the wave-particle duality for a single photon input $T = |1 \rangle \langle 1|$. Formula (4.12) gives

$$\langle 1 | F^{0;\varepsilon}(1, 0)|1 \rangle = \varepsilon \tag{4.13a}$$
$$\langle 1 | F^{0;\varepsilon}(0, 1)|1 \rangle = 1 - \varepsilon \tag{4.13b}$$
$$\langle 1 | F^{0;\varepsilon}(n_1, n_2)|1 \rangle = 0 \quad \text{if } n_1 + n_2 \neq 1 \tag{4.13c}$$

This case offers a particularly simple way of determining the dependence of ε on the parameters of the interferometer:

$$\varepsilon = \varepsilon_1 \varepsilon_2 + (1 - \varepsilon_1)(1 - \varepsilon_2) + 2\sqrt{\varepsilon_1(1 - \varepsilon_1)\varepsilon_2(1 - \varepsilon_2)} \, \cos(\vartheta_2 - \vartheta_1 - \delta) \quad (4.14)$$

There are three cases of special interest: ε_1 variable, $\varepsilon_2 = 1$; $\varepsilon_1 = \varepsilon_2 = \frac{1}{2}$; and $\varepsilon_1 = \frac{1}{2}$, ε_2 variable. The first choice gives $\varepsilon = \varepsilon_1$, the second $\varepsilon = \cos^2\left(\frac{1}{2}(\vartheta_2 - \vartheta_1 + \delta)\right)$, and the third $\varepsilon = \frac{1}{2}\left[1 + 2\sqrt{\varepsilon_2(1 - \varepsilon_2)} \cos(\vartheta_2 - \vartheta_1 - \delta)\right]$ $= \frac{1}{2}\left[(1 + 2\sqrt{\varepsilon_2(1 - \varepsilon_2)}) \cos\delta\right]$, where the last equality is under the assumption that $\vartheta_1 = \vartheta_2$. Some of these cases have been investigated experimentally, confirming thus the predicted quantum mechanical probabilities (4.13) [7.21–23].

The experiment with $\varepsilon_2 = 1$ would allow one to decide from a single count event whether ε_1 was 1 or 0 if one of these values was given. This is interpreted as the calibration for a path measurement. If ε_1 differs from these values then, of course, the notion of a path taken by the photon is meaningless. But the very ability of the device to detect the path (if it was fixed) destroys any interference. Next the statistics obtained in the case $\varepsilon_2 = \frac{1}{2}$ reproduce the expected interference pattern resulting from many runs of this single-photon experiment. More precisely the interference disappears if ε_1 is 0 or 1, which corresponds to the situation where the photon is forced to take exactly one path. In this sense a precise fixing of the path destroys again the interference. Maximal path indeterminacy, $\varepsilon_1 = \frac{1}{2}$, gives rise to optimal interference, while there is no way to get any information on the path when $\varepsilon_2 = \frac{1}{2}$. In this way we recover the wave-particle duality in Bohr's complementarity interpretation. There are mutually exclusive options for both, the *preparation*, as well as the *registration*, of path or wave behaviour.

In [7.24] a modified Mach-Zehnder interferometer, with $\varepsilon_1 = \frac{1}{2}$, and variable ε_2, was introduced in order to test the detection probability $\frac{1}{2}\left[1 + 2\sqrt{\varepsilon_2(1 - \varepsilon_2)} \cos\delta\right]$. This experiment was interpreted as providing simultaneous information on the two complementary properties of a photon. Indeed letting ε_2 vary from $\frac{1}{2}$ to 1, one recognises that the interference fades away gradually from the pattern with optimal contrast, $\cos^2\left(\frac{1}{2}(\vartheta_2 - \vartheta_1 + \delta)\right)$, to no interference at all, $\frac{1}{2}$. The experimentally realised case $\varepsilon_2 = 0,994$ still leads to a recognisable interference pattern ($\varepsilon = \frac{1}{2}(1 + 0,154\cos\delta)$) even though there is, loosely speaking, already a high (84%) confidence on the path of the photon. In a suitable measure this situation was characterised by ascribing 98,2% particle nature and 1,8% wave nature to the photon. Unfortunately in the present experiment [7.24] the incoming light pulses originated from a laser so that no genuine single photon situation was guaranteed. That is, the intensity was low enough to ensure, with high probability, the presence of only one photon in the interferometer but the detection was not sensitive to single counts.

The analysis carried out so far rephrases the common view that the detection statistics of single-photon Mach-Zehnder interferometry exhibit both the wave-particle duality as well as the unsharp wave-particle behaviour for single photons.

Note, however, that the language used here goes beyond the formal description that could be given in terms of the observables $F^{0;\varepsilon}$. An account based solely on the latter is phenomenological in the sense that the relevant photon observables $F^{0,\varepsilon}$, which pertain to the object under investigation, are mutually commutative; hence on the object level there is no complementarity. Only the various statistics for single photon input states show the 'complementary' behaviour in question. We shall change now our point of view to show that the same statistics can also be interpreted on the basis of complementary observables.

To this end we redefine again the cut to be placed in the experimental setup of Figure 7.10. Instead of taking the whole interferometer together with the detectors as the registration device, we consider the first beam splitter $BS_1(\varepsilon_1)$ and the phase shifter $PS(\delta)$ as parts of the preparation device. The object system is therefore the two-mode field prepared in a state

$$S := V_\delta U_\alpha (T \otimes T') U_\alpha^* V_\delta^* \tag{4.15}$$

The detection statistics (4.4) can then be written as

$$p_S^{E^\beta}(n_1, n_2) = p_W^{N_1 \otimes N_2}(n_1, n_2) \tag{4.16}$$

for the observable E^β,

$$E^\beta(n_1, n_2) := U_\beta^* \left(|n_1\rangle\langle n_1| \otimes |n_2\rangle\langle n_2| \right) U_\beta \tag{4.17}$$

We restrict our considerations to the single photon case, $T = |1\rangle\langle 1|$, $T' = |0\rangle\langle 0|$, so that the possible initial states of the two-mode field are the vector states

$$\psi_{\alpha,\delta} := V_\delta U_\alpha |10\rangle = \sqrt{\varepsilon_1} |10\rangle + e^{-i(\vartheta_1 + \delta)}\sqrt{1 - \varepsilon_1} |01\rangle \tag{4.18}$$

where, for instance, $|10\rangle = |1\rangle \otimes |0\rangle$. Let P_{10} and P_{01} denote the one dimensional projections $|1\rangle\langle 1| \otimes |0\rangle\langle 0|$ and $|0\rangle\langle 0| \otimes |1\rangle\langle 1|$ of the two-mode Fock space. Then $P_{10} + P_{01}$ projects onto the two-dimensional subspace of the vectors (4.18) which we take to represent the object system to be investigated. Due to the number conservation under the unitary map U_β, the projection operators (4.17) commute with $P_{10} + P_{01}$, so that the following operators define a PV measure on the state space of the object system:

$$F^{1,0;\varepsilon_2}(n_1, n_2) := \left(P_{10} + P_{01} \right) E^\beta(n_1, n_2) \left(P_{10} + P_{01} \right) \tag{4.19}$$

For the states (4.18) the observables E^β and $F^{1,0;\varepsilon_2}$ have the same expectations,

$$\langle \psi_{\alpha,\delta} | F^{1,0;\varepsilon_2}(n_1, n_2)\psi_{\alpha,\delta} \rangle = \langle \psi_{\alpha,\delta} | E^\beta(n_1, n_2)\psi_{\alpha,\delta} \rangle \tag{4.20}$$

for all n_1, n_2 and for each α and δ. In particular this means that the observable $F^{1,0;\varepsilon_2}$ is, in fact, determined by the detection statistics.

The effects of Eq. (4.19) read as follows:

$$F^{1,0;\varepsilon_2}(1,0) \; = \; E^\beta(1,0) \; = \; P\left[U_\beta^* \mid 10\right] \tag{4.21a}$$
$$= \; P\left[\sqrt{\varepsilon_2} \mid 10\rangle + e^{-i\vartheta_2}\sqrt{1-\varepsilon_2} \mid 01\rangle\right]$$
$$F^{1,0;\varepsilon_2}(0,1) \; = \; E^\beta(0,1) \; = \; P\left[U_\beta^* \mid 01\right] \tag{4.21b}$$
$$= \; P\left[\sqrt{1-\varepsilon_2} \mid 10\rangle - e^{-i\vartheta_2}\sqrt{\varepsilon_2} \mid 01\rangle\right]$$
$$F^{1,0;\varepsilon_2}(n_1,n_2) \; = \; O \;\; \text{if } n_1 + n_2 \neq 1 \tag{4.21c}$$

Also for any α and δ,

$$\langle \psi_{\alpha,\delta} \mid F^{1,0;\varepsilon_2}(1,0)\psi_{\alpha,\delta} \rangle \; = \; \varepsilon \tag{4.22a}$$
$$\langle \psi_{\alpha,\delta} \mid F^{1,0;\varepsilon_2}(0,1)\psi_{\alpha,\delta} \rangle \; = \; 1 - \varepsilon \tag{4.22b}$$
$$\langle \psi_{\alpha,\delta} \mid F^{1,0;\varepsilon_2}(n_1,n_2)\psi_{\alpha,\delta} \rangle \; = \; 0 \;\; \text{if } n_1 + n_2 \neq 1 \tag{4.22c}$$

with ε given by Eq. (4.14).

On the level of a statistical description the change of the viewpoint has brought nothing new; the measurement outcome probabilities (4.22) are, as they should, the same as those of Eq. (4.13), or just the detection statistics (4.4). There is, however, an essential new aspect in the description. All the photon observables $F^{1,0;\varepsilon_2}$ are mutually complementary in the strong measurement theoretical sense: these observables cannot be measured or even tested together. In particular the observables $F^{1,0;1}$ and $F^{1,0;\frac{1}{2}}$ associated with the extreme choices $\varepsilon_2 = 1$ and $\varepsilon_2 = \frac{1}{2}$ are complementary path and interference observables,

$$F^{1,0;1}(1,0) \; = \; P_{10} \tag{4.23a}$$
$$F^{1,0;\frac{1}{2}}(1,0) \; = \; P\left[\tfrac{1}{\sqrt{2}}\big(\mid 10\rangle + e^{-i\vartheta_2} \mid 01\rangle\big)\right] \tag{4.23b}$$

In [7.24] a measurement of $F^{1,0;\varepsilon}$ was interpreted as a joint unsharp determination of the complementary path and interference observables $F^{1,0;1}$ and $F^{1,0;\frac{1}{2}}$. While the observable $F^{1,0;\varepsilon}$ does entail probabilistic information on the two complementary observables, it is a sharp observable, and cannot therefore be considered to represent an unsharp joint measurement in the sense of the general point of view followed in this text. In the next subsection two proposals are reviewed which do lead to such joint measurements as can be read off from the ensuing POV measures.

VII.4.2 Joint path-interference measurements for single photons.
We describe next an experimental scheme in which a Mach-Zehnder interferometer is again used for measuring an interference observable, but with an additional component introduced that is capable of carrying out a nondemolishing path determination. The experimental setup is sketched out in Figure 7.10.

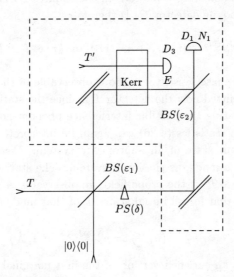

Figure 7.10. Mach-Zehnder interferometer with a Kerr medium

The new component is a Kerr medium placed in the second arm of the interferometer, which will couple the b-mode field with a single-mode probe field according to the interaction

$$U_K = I_1 \otimes e^{-i\lambda(N_2 \otimes N_3)} \tag{4.24}$$

where $N_3 = c^*c$ is the number observable of the probe field. This coupling will not change the number of photons of the interferometer field but it will affect the phase of the probe field. Therefore analysing the latter with a phase sensitive detector D_3 yields information on the number of photons in the second arm of the interferometer. At the same time the detectors D_1 and D_2 register the number of photons $N_1 = a^*a$ and $N_2 = b^*b$ emerging from the second beam splitter, exhibiting thereby the possible interference pattern. Such a scheme was proposed in [7.25], where the homodyne (quadrature) observable $\frac{1}{2}(c^* + c)$ was used as the readout observable for the probe mode. In fact any phase observable conjugate to N_3 would do as well.

Let T, $|0\rangle\langle0|$, and T' be the input states of the incoming photon pulse, the b-mode, and the probe field. One is again facing the task of splitting the whole experiment into an observed and an observing part, or into a preparation and a registration. Taken as a whole the state of the total field will change in the Mach-Zehnder-Kerr apparatus according to

$$T \otimes |0\rangle\langle0| \otimes T' \rightarrow W := U_\beta U_K V_\delta U_\alpha \left(T \otimes |0\rangle\langle0| \otimes T'\right) U_\alpha^* V_\delta^* U_K^* U_\beta^* \tag{4.25}$$

The probability of detecting n photons in the counter D_1 and reading a value in a set X in the homodyne detector D_3 is therefore

$$p_W^{N_1 \otimes I_2 \otimes E}(n, 1, X) = \text{tr}\big[W \,|n\rangle\langle n| \otimes I_2 \otimes E(X)\big] \tag{4.26}$$

where E is some (phase sensitive) readout observable of the third mode. For simplicity we have omitted now the detector D_2, since the statistics of D_1 are already sufficient for indicating the possible interference phenomenon.

The detection statistics (4.26) can again be interpreted in various ways with respect to the input state of an appropriate system. We consider the incoming photon pulse, the a-mode, as the input system. The statistics (4.26) then define, for each initial state T' of the probe field, an observable $A^{0,T'}$ associated with the a-mode field such that for all initial states T of that mode and for all D_1-values n and D_3-values X

$$p_T^{A^{0,T'}}(n, X) := p_W^{N_1 \otimes I_2 \otimes X}(n, 1, X) \tag{4.27}$$

This observable is an enriched version of the first marginal $F_1^{0;\varepsilon}$ of the observable (4.11). It depends on the whole variety of the 'apparatus parameters' $|0\rangle\langle 0|$, T', α, β, δ, and λ. Unlike $F_1^{0;\varepsilon}$, this observable contains direct information on the dual aspects of a single photon. To see this we determine this observable specifying the beam splitters to be semitransparent ($\varepsilon_1 = \varepsilon_2 = \frac{1}{2}$) with $\vartheta_1 = \vartheta_2 = \frac{\pi}{2}$, the case where one expects optimal interference. The observable $A^{0,T'}$ is now found by direct computation:

$$A^{0,T'}(n, X) = \sum_{m=0}^{\infty} \frac{(m+n)!}{m!n!} \,|m+n\rangle\langle m+n|$$
$$\text{tr}\Big[T' \sin^n\big(\tfrac{\delta}{2}I_3 - \tfrac{\lambda}{2}N_3\big)\, \cos^m\big(\tfrac{\delta}{2}I_3 - \tfrac{\lambda}{2}N_3\big)\, e^{-i(m+n)\frac{\lambda}{2}N_3}$$
$$E(X)\, e^{i(m+n)\frac{\lambda}{2}N_3} \sin^n\big(\tfrac{\delta}{2}I_3 - \tfrac{\lambda}{2}N_3\big)\, \cos^m\big(\tfrac{\delta}{2}I_3 - \tfrac{\lambda}{2}N_3\big)\Big] \tag{4.28}$$

One may go on to determine the marginal observables of $A^{0,T'}$. The first one is obtained by putting $X = \mathbf{R}$:

$$A_1^{0,T'}(n) = \sum_{m=0}^{\infty} \frac{(m+n)!}{m!n!} |m+n\rangle\langle m+n|$$
$$\text{tr}\Big[T' \sin^{2n}\big(\tfrac{\delta}{2}I_3 - \tfrac{\lambda}{2}N_3\big) \cos^{2m}\big(\tfrac{\delta}{2}I_3 - \tfrac{\lambda}{2}N_3\big)\Big] \tag{4.29}$$

showing that $A_1^{0,T'}$ is an unsharp number observable. The second marginal observable $X \mapsto A_2^{0,T'}(X) := \sum_n A^{0,T'}(n, X)$ is also directly obtained from (4.28) but it is less straightforward to exhibit a simplifying expression for it. However, this is not needed here since our primary interest is in the case of the single photon input state $T = |1\rangle\langle 1|$. For this state all the relevant probabilities are easily computed:

$$p_{|1\rangle}^{A^{0,T'}}(0, X) = \text{tr}\left[T'\cos\left(\tfrac{\delta}{2}I_3 - \tfrac{\lambda}{2}N_3\right)e^{-i\frac{\lambda}{2}N_3}E(X)e^{i\frac{\lambda}{2}N_3}\cos\left(\tfrac{\delta}{2}I_3 - \tfrac{\lambda}{2}N_3\right)\right] \quad (4.30a)$$

$$p_{|1\rangle}^{A^{0,T'}}(1, X) = \text{tr}\left[T'\sin\left(\tfrac{\delta}{2}I_3 - \tfrac{\lambda}{2}N_3\right)e^{-i\frac{\lambda}{2}N_3}E(X)e^{i\frac{\lambda}{2}N_3}\sin\left(\tfrac{\delta}{2}I_3 - \tfrac{\lambda}{2}N_3\right)\right] \quad (4.30b)$$

$$p_{|1\rangle}^{A^{0,T'}}(0, \mathbf{R}) = p_{|1\rangle}^{A^{0,T'}}(0) = \text{tr}\left[T'\cos^2\left(\tfrac{\delta}{2}I_3 - \tfrac{\lambda}{2}N_3\right)\right] \quad (4.31a)$$

$$p_{|1\rangle}^{A^{0,T'}}(1, \mathbf{R}) = p_{|1\rangle}^{A^{0,T'}}(1) = \text{tr}\left[T'\sin^2\left(\tfrac{\delta}{2}I_3 - \tfrac{\lambda}{2}N_3\right)\right] \quad (4.31b)$$

$$p_{|1\rangle}^{A^{0,T'}}(\mathbf{N}, X) = p_{|1\rangle}^{A^{0,T'}}(0, X) + p_{|1\rangle}^{A^{0,T'}}(1, X) = p_{|1\rangle}^{A_2^{0,T'}}(X) \quad (4.32)$$

$$= \tfrac{1}{2}\left(\text{tr}\left[T'E(X)\right] + \text{tr}\left[T'e^{-i\lambda N_3}E(X)e^{i\lambda N_3}\right]\right)$$

As a consistency check one may first observe that for $\lambda = 0$ the probabilities (4.31) are just the single photon counting probabilities in a Mach-Zehnder interferometer with $\varepsilon_1 = \varepsilon_2 = \tfrac{1}{2}$ and $\vartheta_1 = \vartheta_2 = \tfrac{\pi}{2}$. The introduction of the Kerr medium ($\lambda \neq 0$) in the second arm of the interferometer affects these probabilities with a T'-dependent phase shift. Moreover the detector D_3 allows one to collect the additional single-photon statistics (4.32) which contain information on the path of the photon. It must be emphasised that the a-mode observable $A^{0,T'}$ and the ensuing single photon probability measures $p_{|1\rangle}^{A^{0,T'}}$ do depend on the state of the probe field T' as well as on the path-indicating observable E. Up to this point no properties of T' or E are used, and the problem is to choose these 'control parameters' in such a way that the probabilities (4.30–32) would provide a good interference pattern together with a reliable path determination. Clearly if T' is a number state or E is compatible with the number observable N_3 there will be no path information available from (4.32). On the other hand if E is any phase observable conjugate to the number observable N_3, we have $e^{-i\lambda N_3}E(X)e^{i\lambda N_3} = E(X + \lambda)$, which shows that it is possible to obtain information about the path; a photon traversing through the second arm of the interferometer leaves a track on the statistics of the second marginal $A_2^{0,T'}$ collected at the detector D_3. In [7.25] E was chosen to be the homodyne observable $\tfrac{1}{2}(c + c^*)$ and T' a coherent or squeezed state.

We may again describe the whole experiment on the basis of the 'realistic' cut introduced in the preceding subsection, where the first beam splitter and the phase shifter belong to the preparation device. We assume from the outset that the b-mode is idle so that the prepared state S of the (a,b)-mode field is

$$S := V_\delta U_\alpha\left(T \otimes |0\rangle\langle 0|\right)U_\alpha^* V_\delta^* \quad (4.33)$$

From the counting statistics (4.26) one then obtains a unique observable $E^{T';\varepsilon_2}$ of the (a,b)-mode field for any fixed input state T' of the probe mode:

$$p_S^{E^{T';\varepsilon_2}}(n, X) = \text{tr}\left[S \otimes T' U_K^* U_\beta^*\left(|n\rangle\langle n| \otimes I_2 \otimes E(X)\right)U_\beta U_K\right] \quad (4.34)$$

For the explicit determination of this observable we shall only consider a single photon input state $T = |1\rangle\langle 1|$ so that the object system is given again by the subspace of states $\psi_{\alpha,\delta}$ from Eq. (4.18). On that state space the observable $E^{T';\varepsilon_2}$ reduces to the following one:

$$F^{T';\varepsilon_2}(n, X) := \left(P_{10} + P_{01}\right) E^{T';\varepsilon_2}(n, X) \left(P_{10} + P_{01}\right) \tag{4.35}$$

Evaluation of Eq. (4.34) yields

$$E^{T';\varepsilon_2}(n, X) = \sum_{n_1, m_1, n_2, m_2} |n_1, n_2\rangle\langle m_1, m_2| \tag{4.36}$$

$$\langle n_1, n_2 | U_\beta^* \left(|n\rangle\langle n| \otimes I_2\right) U_\beta |m_1, m_2\rangle \, \mathrm{tr}\left[T' e^{i\lambda m_1 N_3} E(X) e^{-i\lambda m_2 N_3}\right]$$

The observable (4.35) is obtained simply by carrying out the above sum under the constraint $n_1 + n_2 = m_1 + m_2 = 1$:

$$
\begin{aligned}
F^{T';\varepsilon_2}(n, X) = \ & |10\rangle\langle 10| \left[\varepsilon_2 \delta_{n1} + (1-\varepsilon_2)\delta_{n0}\right] \mathrm{tr}\left[T' E(X)\right] \\
& + |01\rangle\langle 01| \left[(1-\varepsilon_2)\delta_{n1} + \varepsilon_2 \delta_{n0}\right] \mathrm{tr}\left[T' e^{i\lambda N_3} E(X) e^{-i\lambda N_3}\right] \\
& + |10\rangle\langle 01| \sqrt{\varepsilon_2(1-\varepsilon_2)}\, e^{i\vartheta_2} \left(\delta_{n1} - \delta_{n0}\right) \mathrm{tr}\left[T' E(X) e^{-i\lambda N_3}\right] \\
& + |01\rangle\langle 10| \sqrt{\varepsilon_2(1-\varepsilon_2)}\, e^{-i\vartheta_2} \left(\delta_{n1} - \delta_{n0}\right) \mathrm{tr}\left[T' e^{i\lambda N_3} E(X)\right]
\end{aligned}
\tag{4.37}
$$

It is instructive to determine the marginal observables $F_1^{T';\varepsilon_2}(n) := F^{T';\varepsilon_2}(n, \mathbf{R})$ and $F_2^{T';\varepsilon_2}(X) := \sum_n F^{T';\varepsilon_2}(n, X)$:

$$
\begin{aligned}
F_1^{T';\varepsilon_2}(1) = \ & |10\rangle\langle 10| \varepsilon_2 + |01\rangle\langle 01| (1-\varepsilon_2) \\
& + |10\rangle\langle 01| \sqrt{\varepsilon_2(1-\varepsilon_2)}\, e^{i\vartheta_2} \, \mathrm{tr}\left[T' e^{-i\lambda N_3}\right] \\
& + |01\rangle\langle 10| \sqrt{\varepsilon_2(1-\varepsilon_2)}\, e^{-i\vartheta_2} \, \mathrm{tr}\left[T' e^{i\lambda N_3}\right]
\end{aligned}
\tag{4.38a}
$$

$$
\begin{aligned}
F_1^{T';\varepsilon_2}(0) = \ & |10\rangle\langle 10| (1-\varepsilon_2) + |01\rangle\langle 01| \varepsilon_2 \\
& + |10\rangle\langle 01| \sqrt{\varepsilon_2(1-\varepsilon_2)}\, e^{i\vartheta_2} \, \mathrm{tr}\left[T' e^{-i\lambda N_3}\right] \\
& + |01\rangle\langle 10| \sqrt{\varepsilon_2(1-\varepsilon_2)}\, e^{-i\vartheta_2} \, \mathrm{tr}\left[T' e^{i\lambda N_3}\right]
\end{aligned}
\tag{4.38b}
$$

$$F_2^{T';\varepsilon_2}(X) = |10\rangle\langle 10| \, \mathrm{tr}\left[T' E(X)\right] + |01\rangle\langle 01| \, \mathrm{tr}\left[T' e^{i\lambda N_3} E(X) e^{-i\lambda N_3}\right] \tag{4.39}$$

It is obvious that irrespective of the choice of T' and the observable E, the first marginal is a smeared interference observable, while the second one is a smeared path observable. This shows that the device serves its purpose to establish a joint measurement of these complementary properties. With reference to the latter marginal it should be noted that neither T' nor E may commute with N_3 since otherwise the phase sensitivity is lost. One may now proceed to analyse Eqs. (4.38–39), showing that high path confidence will lead to low interference contrast, and vice versa. Indeed an optimal interference would be obtained if $\varepsilon_2 = \frac{1}{2}$ and if the numbers $\mathrm{tr}\left[e^{\pm i\lambda N_3} T'\right]$ were equal to unity. But this requires T' to be an eigenstate of N_3, which destroys the path measurement. On the other hand imagine that E is a phase observable, $X = [0, \frac{\pi}{2}]$, and T' is chosen such that the number $\mathrm{tr}\left[T' E(X)\right]$ is close to unity. This can be achieved, e.g., for a coherent state with a large amplitude. Such states have a slowly varying number distribution so that the modulus of the complex numbers $\mathrm{tr}\left[e^{\pm i\lambda N_3} T'\right]$ is small compared to unity. But this is to say that the first marginal observable is also 'close to' a path observable, thus yielding only low interference contrast.

There exists yet another scheme of a joint measurement of complementary path and interference observables that is based solely on mirrors, beam splitters and phase shifters, as they were used in the original Mach-Zehnder interferometer (Figure 7.11).

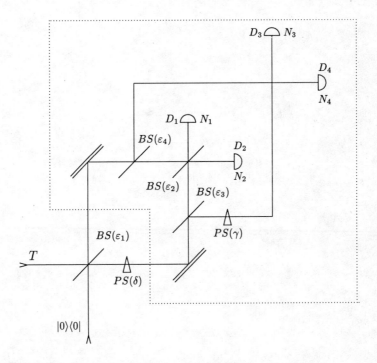

Figure 7.11. Expanded Mach-Zehnder interferometer

Regarding again the first beam splitter and the first phase shifter as parts of the preparation device, the residual elements constitute the measuring instrument which now has four detectors. An explicit analysis of this experiment with respect to a single-photon input has been carried out in [7.26] and is found to be formally very similar to the example of Section 2; therefore we may restrict ourselves to a nontechnical summary. The detection statistics give rise again to a unique observable of the object which now has four outcomes. One may combine pairs of these outcomes to add them up to three two-valued marginal observables. By a suitable adjustment of the system parameters it is possible to ensure that two of these observables are path and interference observables; moreover one may achieve conditions such that the full detection statistics uniquely determine the prepared state; that is, the measurement can be managed to be informationally complete. In particular the statistics would allow one to find out the values of α and δ. From the mathematical point of view it is interesting that in this experiment (as well as in the polarisation experiment of Section 2) the 'positivistic' output description corresponds to the minimal Neumark extension of the measured object observable obtained in the realistic description, so that we are facing here an experimental realisation of this formal construction.

Epilogue

One may view the world with the p-eye and one may view it with the q-eye but if one opens both eyes simultaneously then one gets crazy.

Wolfgang Pauli in a letter to Werner Heisenberg, 19 October 1926

We hope to have demonstrated that one can safely open a pair of complementary 'eyes' simultaneously. He who does so may even 'see more' than with one eye only. The means of observation being part of the physical world, Nature Herself protects him from seeing too much and at the same time protects Herself from being questioned too closely: quantum reality, as it emerges under physical observation, is intrinsically unsharp. It can be forced to assume sharp contours – real properties – by performing repeatable measurements. But sometimes unsharp measurements will be both, less invasive and more informative.

References

Chapter I.

1.1 P. BUSCH, P. LAHTI, P. MITTELSTAEDT. *The Quantum Theory of Measurement*, Springer-Verlag, Berlin, 1991.

1.2 O. STERN. Ein Weg zur experimentellen Prüfung der Richtungsquantelung im Magnetfeld. *Zeitschrift für Physik 7*, 249–253, 1921.
W. GERLACH, O. STERN. Der experimentelle Nachweis der Richtungsquantelung im Magnetfeld. *Zeitschrift für Physik 9*, 349–352, 1922.

1.3 G. LUDWIG. *Foundations of Quantum Mechanics, Vol I*. Springer-Verlag, Berlin, 1983.

1.4 K. KRAUS. *States, Effects and Operations*. Springer-Verlag, Berlin, 1983.

1.5 R. HAAG, D. KASTLER. An algebraic approach to quantum field theory, *Journal of Mathematical Physics 5*, 846–861, 1964.

1.6 B. MIELNIK. Theory of filters, *Communications in Mathematical Physics 15*, 1–46, 1969.

1.7 E.B. DAVIES, J.T. LEWIS. An operational approach to quantum probability. *Communications in Mathematical Physics 17*, 239–259, 1970.

1.8 E.B. DAVIES. *Quantum Theory of Open Systems*. Academic Press, London, 1976.

1.9 G. LINDBLAD. *Non-Equilibrium Entropy and Irreversibility*. Reidel, Dordrecht, 1983.

1.10 R. ALICKI, K. LENDI. *Quantum Dynamical Semigroups and Applications*. Springer-Verlag, Berlin, 1987.

1.11 C.W. HELSTROM. *Quantum Detection and Estimation Theory*. Academic Press, New York, 1976.

1.12 A.S. HOLEVO. *Probabilistic and Statistical Aspects of Quantum Theory*. North Holland Publishing Corporation, Amsterdam, 1982.

1.13 M. OHYA, D. PETZ. *Quantum Entropy and Its Use*. TMP, Springer-Verlag, Berlin, 1993.

1.14 S.T. ALI. Stochastic Localization, Quantum Mechanics on Phase Space and Quantum Space-Time. *La Rivista del Nuovo Cimento 8* (11), 1–128, 1985.

1.15 E. PRUGOVEČKI. *Stochastic Quantum Mechanics and Quantum Space Time*. D. Reidel Publishing Corporation, Dordrecht, 2nd edition, 1986.

1.16 E. PRUGOVEČKI. *Quantum Geometry*. Kluwer Academic Publishers, Dordrecht, 1992.

Chapter II.

2.1 E. BELTRAMETTI and G. CASSINELLI. *The Logic of Quantum Mechanics*. Addison-Wesley, Reading, Massachusetts, 1981.

2.2 J.M. JAUCH. *Foundations of Quantum Mechanics*. Addison-Wesley, Reading, Massachusetts, 1968.

2.3 J. VON NEUMANN. *Mathematische Grundlagen der Quantenmechanik*. Springer-Verlag, Berlin, 1932. English translation: *Mathematical Foundations of Quantum Mechanics*, Princeton University Press, Princeton, 1955.

2.4 M. REED, B. SIMON. *Methods of Modern Mathematical Physics*. I: Functional Analysis, Academic Press, New York, 1972.

2.5 G.P. CATTANEO, G. NISTICO. Axiomatic foundations of quantum physics. Critiques and misunderstandings, I. Piron's question-proposition system. *International Journal of Theoretical Physics 30*, 1217–1227, 1991.
M.L. DALLA CHIARA, R. GIUNTINI. Partial and unsharp quantum logics. *Foundations of Physics 24*, 1161–1177, 1994.

R. Giuntini, H. Greuling. Towards a formal language for effects and unsharp properties. *Foundations of Physics 19*, 931–945, 1989.

R. Greechie, D. Foulis. Transition to effect algebras. *International Journal of Theoretical Physics*, 1995.

2.6   V.S. Varadarajan. *Geometry of Quantum Theory*, 2nd edition, Springer-Verlag, New York, 1985.

2.7   P. Kruszyński, W.M. de Muynck. Compatibility of observables represented by positive operator-valued measures. *Journal of Mathematical Physics 28*, 1761–1763, 1987.

2.8   K. Ylinen. On a theorem of Gudder on joint distributions of observables. In: *Symposium on the Foundations of Modern Physics 1985*. Eds P. Lahti and P. Mittelstaedt, World Scientific, Singapore, pp. 691–694, 1985.

2.9   F. Riesz, B. Sz.-Nagy. *Functional Analysis*. Dover Publications, Inc., New York, 1990.

2.10 N.I. Akhiezer, I.M. Glazman. *Theory of Linear Operators in Hilbert Space*. Fredrik Ungar, New York, 1963.

2.11 W. Mlak. *Hilbert Spaces and Operator Theory*. PWN-Polish Scientific Publishers, Warsaw, 1991.

2.12 H. Dehmelt. Less is more: experiments with an individual atomic particle at rest in free space, *American Journal of Physics 58*, 17–27, 1990.

2.13 E. Beltrametti, G. Cassinelli, P. Lahti. Unitary Measurements of Discrete Quantities in Quantum Mechanics, *Journal of Mathematical Physics 31*, 91–98, 1990.

2.14 P. Busch, G. Cassinelli, P. Lahti. On the quantum theory of sequential measurements, *Foundations of Physics 20*, 757–775, 1990.

2.15 M. Ozawa. Quantum measuring processes of continuous observables. *Journal of Mathematical Physics 25*, 79–87, 1984.

2.16 A. Łuczak. Instruments on von Neumann algebras, Institute of Mathematics, Łódź University, Poland, 1986.

Chapter III.

3.1   H. Weyl. *The Theory of Groups and Quantum Mechanics*. Dover Publications, Inc., New York, 1950. German original 1931.

3.2   E.P. Wigner. Unitary representations of the inhomogeneous Lorentz group. *Annals of Mathematics 40*, 149–204, 1939.

3.3   G.W. Mackey. *The Theory of Unitary Group Representations in Physics, Probability, and Number Theory*. Benjamin/Cummings, Reading, Massachusetts, 1978.

3.4   A.S. Wightman. On the localizability of quantum mechanical systems. *Reviews of Modern Physics 34*, 845–872, 1962.

3.5   J.-M. Lévy-Leblond. Galilei group and Galilean invariance. In: *Group Theory and Its Applications*. Ed. E.M. Loebl, Academic Press, New York, pp. 221–299, 1971.

3.6   B. Simon. Quantum dynamics: from automorphism to Hamiltonian. In: *Studies in Mathematical Physics. Essays in Honor of Valentine Bargman*. Eds. E.H. Lieb, B. Simon and A.S. Wightman, Princeton Series in Physics, Princeton University Press, pp. 327–349, 1976.

3.7   S.T. Ali. A geometrical property of pov-measures and systems of covariance. In: *Differential Geometric Methods in Mathematical Physics*. Eds. H.-D. Doebner, S.I. Anderson and H.R. Petri, Lecture Notes in Mathematics, Vol. 905, Springer-Verlag, Berlin, pp. 207–228, 1982.

3.8   S.T. Ali, E. Prugovečki. Systems of imprimitivity and representations of quantum mechanics in fuzzy phase spaces. *Journal of Mathematical Physics 18*, 219–228, 1977.

3.9 E.B. DAVIES. On repeated measurements of continuous observables. *Journal of Functional Analysis 6*, 318–346, 1970.

3.10 R. WERNER. Screen observables in relativistic and nonrelativistic quantum mechanics. *Journal of Mathematical Physics 27*, 793–803, 1986.

3.11 Y. AHARONOV, D. BOHM. Time in the quantum theory and the uncertainty relation for time and energy. *Physical Review 122*, 1649–1658, 1961.

3.12 A. PERES. Measurement of time by quantum clocks. *American Journal of Physics 48*, 552–557, 1980.

3.13 E.P. WIGNER. On the time–energy uncertainty relation. In: *Aspects of Quantum Theory*. Eds. A. Salam and E.P. Wigner, Cambridge University Press, Cambridge, pp. 237–247, 1972.

3.14 A. BARCHIELLI. Direct and heterodyne detection and other applications of quantum stochastic calculus to quantum optics. *Quantum Optics 2*, 423–441, 1990.

V.P. BELAVKIN. A continuous counting observation and posterior quantum dynamics. *Journal of Physics A 22*, 1109–1114, 1989.

3.15 R.L. HUDSON, K.R. PARTHASARATHY. Quantum Ito's formula and stochastic evolutions. *Communications in Mathematical Physics 93*, 301–323, 1984.

3.16 K.R. PARTHASARATHY. *An Introduction to Quantum Stochastic Calculus*. Monographs in Mathematics, Vol. 85, Birkhäuser Verlag, Basel, 1992.

3.17 W. SCHLEICH, S.M. BARNETT (EDS.). *Quantum Phase and Phase Dependent Measurements*. Special Issue, *Physica Scripta T48*, 1993.

3.18 R.G. NEWTON. Quantum action-variables for harmonic oscillators. *Annals of Physics 124*, 327–346, 1980.

3.19 A. GALINDO. Phase and number. *Letters in Mathematical Physics 8*, 495–500, 1984.

J.C. GARRISON and J. WONG. Canonically conjugate pairs, uncertainty relations, and phase operators. *Journal of Mathematical Physics 11*, 2242–2249, 1970.

3.20 P. CARRUTHERS and M.M. NIETO. Phase and angle variables in quantum mechanics. *Reviews of Modern Physics 40*, 411–440, 1968.

3.21 D.T. PEGG, S.M. BARNETT. Phase properties of the quantized single-mode electromagnetic field. *Physical Review A 39*, 1665–1675, 1989.

S.M. BARNETT and D.T. PEGG. Quantum theory of rotation angles. *Physical Review A 41*, 3427–3435, 1990.

3.22 V.N. POPOV and V.S. YARUNIN. Photon phase operator. *Theoretical and Mathematical Physics 89*, 1292–1297, 1992.

3.23 R. GLAUBER. The quantum theory of optical coherence. *Physical Review 130*, 2529–2539, 1963.

Coherent and incoherent states of the radiation field. *Physical Review 131*, 2766–2788, 1963.

3.24 M. GRABOWSKI. New observables in quantum optics and entropy. In: *Symposium on the Foundations of Modern Physics 1993*. Eds P. Busch, P. Lahti, and P. Mittelstaedt, World Scientific, pp 182–191.

3.25 P. BUSCH, M. GRABOWSKI, P. LAHTI. Who is afraid of POV measures? Unified approach to quantum phase observables. *Annals of Physics 237*, 1–11, 1995.

3.26 T.D. NEWTON, E.P. WIGNER. Localized states for elementary systems. *Reviews in Modern Physics 21*, 400–406, 1949.

3.27 K. KRAUS. Position observable of the photon. In: *The Uncertainty Principle and Foundations of Quantum Mechanics*. Eds. W.C. Price and S.S. Chissick, John Wiley & Sons, New York, pp. 293–320, 1976.

S.T. ALI, G.G. EMCH. Fuzzy observables in quantum mechanics. *Journal of Mathematical Physics 15*, 176–182, 1974.

3.28 J.A. BROOKE, F.E. SCHROECK. Localization of the photon on phase space. *Preprint*, 1994.

Further Reading

S.T. ALI. Survey of quantization methods. In: *Classical and Quantum Systems – Foundations and Symmetries*. Eds. H.D. Doebner, W. Scherer, and F.E. Schroeck, Jr., World Scientific, Singapore, 1993.

Chapter IV.

4.1   H. RAUCH. Some aspects of neutron quantum optics. In: *Symposium on the Foundations of Modern Physics 1990*. Eds. P. Lahti and P. Mittelstaedt, World Scientific, pp. 347–363, 1991.

Steps towards neutron quantum optics. In: *Symposium on the Foundations of Modern Physics 1993*. Eds. P. Busch, P. Lahti, and P. Mittelstaedt, World Scientific, pp. 341–360, 1993.

4.2   A. TONOMURA *et al.* Electron-interferometric observation of magnetic flux quanta using the Aharonov-Bohm effect. In: *Foundations of Quantum Mechanics in the Light of New Technology 1989*. Eds. S. Kobayashi *et al*, The Physical Society of Japan, pp. 15–24, 1990.

4.3   M. SCULLY , B.-G. ENGLERT, H. WALTHER. Quantum optical tests of complementarity. *Nature 351*, 111–116, 1991.

4.4   B.C. VAN FRAASSEN. *Quantum Mechanics. An Empiricist View*, Clarendon Press, Oxford, 1991.

4.5   P. BUSCH, M. GRABOWSKI, P. LAHTI. Repeatable measurements in quantum theory: their role and feasibilty, *Preprint*, 1994.

4.6   G. CASSINELLI, P. LAHTI. Conditional probability in the quantum theory of measurement. *Il Nuovo Cimento 108*, 45–56, 1993.

4.7   P. BUSCH, P. LAHTI. Some remarks on unsharp quantum measurements, quantum non-demolition, and all that. *Annalen der Physik 47*, 369–382, 1990.

4.8   S. SAKAI. *Operator Algebras in Dynamical Systems*, Cambridge University Press, Cambridge, 1991.

4.9   P. BUSCH, T.P. SCHONBEK, F.E. SCHROECK. Quantum observables: Compatibility versus commutativity and maximal information. *Journal of Mathematical Physics 28*, 2866–2872, 1987.

K. YLINEN. Commuting functions of the position and momentum observables on locally compact Abelian groups. *Journal of Mathematical Analysis and Applications 137*, 185–192, 1989.

4.10 K. KRAUS. Complementary observables and uncertainty relations. *Physical Review D 35*, 3070–3075, 1987.

4.11 A. LENARD. The numerical range of a pair of projections. *Journal of Functional Analysis 10*, 410–423, 1972.

4.12 P. LAHTI, K. YLINEN. On total noncommutativity in quantum mechanics. *Journal of Mathematical Physics 28*, 2614–2617, 1987.

4.13 P. BUSCH, P. LAHTI. To what extent do position and momentum commute? *Physics Letters A 115*, 259–264, 1986.

4.14 E. WIGNER. Die Messung quantenmechanischer Operatoren. *Zeitschrift für Physik 133*, 101–108, 1952.

4.15 H. ARAKI, M. YANASE. Measurements of quantum mechanical operators. *Physical Review 120*, 622–626, 1960.

4.16 H. STEIN, A. SHIMONY. Limitations on measurements. In: *Foundations of Quantum Mechanics*. Ed. B. d'Espagnat, Academic Press, New York, pp. 56–76, 1971.

A. SHIMONY, H. STEIN. A problem in Hilbert space theory arising from the quantum theory of measurement. *American Math. Monthly 86*, 292–293, 1979.

4.17 G.C. GHIRARDI, F. MIGLIETTA, A. RIMINI, T. WEBER. Limitations on quantum measurements. I. Determination of the minimal amount of nonideality and identification of the optimal measuring apparatuses. *Physical Review D24*, 347–352, 1981. II. Analysis of a model example. *Physical Review D24*, 353–358, 1981.

G.C. GHIRARDI, A. RIMINI, T. WEBER. Quantum evolution in the presence of additive conservation laws and the quantum theory of measurement. *Journal of Mathematical Physics 23*, 1792–1796, 1982.

4.18 P. BUSCH. Momentum conservation forbids sharp localisation. *Journal of Physics A 18*, 3351–3354, 1985. Can quantum mechanical reality be considered sharp? In: *Recent Development in Quantum Logic*. Eds. P. Mittelstaedt and E.-W. Stachow, B.I.-Wissenschaftsverlag, Mannheim, pp. 81–101, 1985.

4.19 M. OZAWA. Quantum Limits of Measurements and Uncertainty Principle. In *Quantum Aspects of Optical Communications*, Eds. C. Bendjaballah, O. Hirota, S. Reynaud, *Lecture Notes in Physics 378*, Springer-Verlag, Berlin, pp. 3–17, 1991.

4.20 S. KUDAKA, K. KAKAZU. The Wigner-Araki-Yanase theorem and its extension in quantum measurement with generalized coherent states. *Progress of Theoretical Physics 87*, 61–76, 1992.

Chapter V.

5.1 S. BRAUNSTEIN. Quantum limits on precision measurement of phase. *Physical Review Letters 69*, 3598–3601.

5.2 A. PERES. *Quantum Theory: Concepts and Methods*. Kluwer, Dordrecht, 1994.

5.3 G. CASSINELLI, P. LAHTI. Spectral properties of observables and convex mappings in quantum mechanics. *Journal of Mathematical Physics 34*, 5468–5475, 1993.

5.4 R. QUADT. *Doctoral Thesis, University of Cologne, 1992*. The part relevant here is published in: R. QUADT, P. BUSCH. Coarse graining and the quantum-classical connection. *Open Systems and Information Dynamics 2* , 129–155, 1994.

P. BUSCH, R. QUADT. Concepts of coarse graining in quantum mechanics. *International Journal of Theoretical Physics 32*, 2261–2269, 1993.

5.5 P. BUSCH, G. CASSINELLI, P. LAHTI. Probability structures for quantum state spaces. *Reviews in Mathematical Physics*, 1995, to appear.

5.6 P. BUSCH, P. LAHTI. The determination of the past and the future of a physical system in quantum mechanics. *Foundations of Physics 19*, 633–678, 1989.

5.7 M. PAVIČIC. Complex Gaussians and the Pauli non-uniqueness, *Physics Letters A 122*, 280–282, 1987.

5.8 S.T. ALI, H.D. DOEBNER. On the equivalence of nonrelativistic quantum mechanics based upon sharp and fuzzy measurements. *Journal of Mathematical Physics 17*, 1105–1111, 1976.

5.9 S.T. ALI, E. PRUGOVECKI. Classical and quantum statistical mechanics in a common Liouville space. *Physica 89A*, 501–521, 1977.

5.10 I. PITOWSKI. *Quantum Probability – Quantum Logic*. Springer-Verlag, Berlin, 1989.

E. BELTRAMETTI, M.J. MACZYNSKI. On a Characterization of Classical and Nonclassical Probabilities. *Journal of Mathematical Physics 32*, 1280–1286, 1991.

5.11 A. SHIMONY. *Search for a Naturalistic World View. Volume II. Natural Science and Metaphysics*. Cambridge University Press, Cambridge, 1993.

5.12 G. KAR, S. ROY. Unsharp spin-$\frac{1}{2}$ observable and CHSH inequalities. *Physics Letters A*, 1995.

5.13 I.D. IVANOVIC. How to differentiate between non-orthogonal states. *Physics Letters A 123*, 257–259, 1987.

D. DIEKS. Overlap and distinguishability of quantum states. *Physics Letters A 126*, 303–306, 1988.

A. PERES. How to differentiate between non-orthogonal states. *Physics Letters A 128*, 19, 1988.

M.G. ALFORD, S. COLEMAN, J. MARCH-RUSSELL. Disentangling nonabelian discrete quantum hair. *Nuclear Physics B351*, 735–748, 1991.

G. JAEGER, A. SHIMONY. Optimal distinction between two non-orthogonal quantum states. *Physics Letters A 197*, 83–87, 1995.

5.14 A. ROYER. Reversible quantum measurements on a spin 1/2 and measuring the state of a single system. *Physical Review Letters 73*, 913–917, 1994. Erratum: ibid., 1995.

5.15 G. CASSINELLI, G. OLIVIERA. The statistics of unbounded observables in Hilbert space quantum mechanics, *Il Nuovo Cimento 84B*, 43–52, 1984.

5.16 J.M. LÉVY-LEBLOND. Who is afraid of nonhermitian operators? A quantum description of angle and phase. *Annals of Physics 101*, 319–341, 1976.

5.17 H.J. LANDAU, H.O. POLLAK. Prolate spheroidal wave functions, Fourier analysis and uncertainty - II, III. *The Bell System Technical Journal 40*, 65–84, 1962, *41*, 1295–1336, 1962.

5.18 J. HILGEVOORD, J. UFFINK. The mathematical expression of the uncertainty principle. In: *Proceedings of the International Conference on Microphysical Reality and Quantum Description*. Eds. F. Selleri, A. van der Merwe, G. Tarozzi, Reidel, Dordrecht, pp. 91–114, 1988.

J. UFFINK. Measures of uncertainty and the uncertainty principle. Thesis, Utrecht, 1990.

5.19 G. LINDBLAD. Entropy, information and quantum measurements *Communications in Mathematical Physics 33*, 305–322, 1973.

5.20 M. GRABOWSKI. New observables in quantum optics and entropy. In: *Symposium on the Foundations of Modern Physics 1993*. Eds. P. Busch, P. Lahti, P. Mittelstaedt, World Scientific, Singapore, pp. 182–191, 1993. *A*-entropy for spectrally absolutely continuous observables. *Reports on Mathematical Physics 14*, 377–384, 1978.

5.21 D. DEUTSCH. Uncertainty in quantum mechanics, *Physical Review Letters 50*, 631–633, 1983.

M.H. PARTOVI. Entropic formulation of uncertainty for quantum measurement, *Physical Review Letters 50*, 1883–1885, 1983.

P. BUSCH, P. LAHTI. Minimal uncertainty and maximal information for quantum position and momentum, *Journal of Physics A: Math. Gen. 20*, 899–906, 1987.

F.E. SCHROECK. On the entropic formulation of uncertainty for quantum measurements. *Journal of Mathematical Physics 30*, 2078–2082, 1989.

5.22 M. GRABOWSKI. Entropic uncertainty relations for "phase-number of quanta" and "time-energy", *Physics Letters A 124*, 19–21, 1987.

5.23 E. LIEB. Proof of an entropy conjecture of Wehrl, *Communications in Mathematical Physics 62*, 35–41, 1978.

5.24 M. GRABOWSKI. Wehrl-Lieb's inequality for entropy and the uncertainty relation, *Reports on Mathematical Physics 20*, 153–155, 1984.

Further Reading

H. MARTENS, W.M. DE MUYNCK. Nonideal quantum measurements. *Foundations of Physics 20*, 255–281, 1990. The inaccuracy principle. *Foundations of Physics 20*, 357–380, 1990. Towards a new uncertainty principle: quantum measurement noise. *Physics Letters A 157*, 441–448, 1991.

W.M. DE MUYNCK. Information in neutron interference experiments. *Physics Letters A 182*, 201–206, 1993.

F.E. SCHROECK. Unsharpness in measurement yields informational completeness. In: *Symposium on the Foundations of Modern Physics 1990*. Eds P. Lahti, P. Mittelstaedt, World Scientific, Singapore, pp. 375–389, 1991.

D.M. HEALY, F.E. SCHROECK, JR. On informational completeness and covariant localization observables and Wigner coefficients. *Journal of Mathematical Physics 36*, 453–507, 1995.

H.P. YUEN, M. OZAWA. Ultimate information carrying limit of quantum systems, *Physical Review Letters 70*, 363–366, 1993.

Chapter VI.

6.1 E.P. WIGNER. On the quantum corrections for thermodynamic equilibrium. *Physical Review 40*, 749–759, 1932.
6.2 J.E. MOYAL. Quantum mechanics as a statistical theory. *Proceedings of the Cambridge Philosophical Society 45*, 99–124, 1949.
6.3 J.C.T. POOL. Mathematical aspects of the Weyl correspondence. *Journal of Mathematical Physics 7*, 66–76, 1966.
6.4 K. HUSIMI. Some formal properties of the density matrix. *Proceedings of the Physico-Mathematical Society of Japan 22*, 264–314, 1940.
6.5 J.R. KLAUDER, E.C.G. SUDARSHAN. *Fundamentals of Quantum Optics*, W.A. Benjamin, New York, 1968.
6.6 E. ARTHURS, J.L. KELLY. On the simultaneous measurements of a pair of conjugate observables. *Bell System Technical Journal 44* 725–729, 1965.
6.7 P. BUSCH. *Doctoral Thesis, University of Cologne, 1982*. English translation: Indeterminacy relations and simultaneous measurements in quantum theory. *International Journal of Theoretical Physics 24*, 63–92, 1985.
6.8 H. SCHERER. *Doctoral Thesis, University of Cologne, 1994*. The part relevant here is published in: H. SCHERER, P. BUSCH. Weakly disturbing phase space measurements in quantum mechanics. In: *Quantum Communication and Measurement*. Eds. V.P. Belavkin, O. Hirota, R.L. Hudson, Plenum Press, New York, 1995.

Further Reading

E.G. BELTRAMETTI, S. BUGAJSKI. Decomposability of mixed states into pure states and related properties. *International Journal of Theoretical Physics 32*, 2235–2244, 1993.

S. BUGAJSKI. Nonlinear quantum mechanics is a classical theory. *International Journal of Theoretical Physics 30*, 961–971, 1991.

P. BUSCH, P. LAHTI, P. MITTELSTAEDT (EDS.). *Symposium on the Foundations of Modern Physics 1993*. World Scientific, Singapore, 1993.

Therein:

L. LANZ, O. MELSHEIMER. Quantum mechanics and trajectories, pp. 233–241.

G. LUDWIG. The minimal interpretation of quantum mechanics and the objective description of macrosystems, pp. 242–250.

H. NEUMANN. Macroscopic properties of photon quantum fields, pp. 303–308.

R. OMNÈS. ¿From Hilbert space to common sense. *Annals of Physics 201*, 354–447, 1990

M. SINGER, W. STULPE. Phase-space representations of general statistical physical theories. *Journal of Mathematical Physics 33*, 131–142, 1992.

Chapter VII.

7.1 G. LUDWIG. *Foundations of Quantum Mechanics. Part II*. Springer-Verlag, Berlin, 1986.
7.2 A.M. PERELOMOV. *Generalized Coherent States and their Applications*. Springer-Verlag, Berlin, 1986.
7.3 M. GRABOWSKI. Quantum measurement scheme and new examples of generalized observables. In: *Symposium on the Foundations of Modern Physics 1990*. Eds. P. Lahti, P. Mittelstaedt, World Scientific, Singapore, pp. 124–137, 1991.

7.4   R. LOUDON. *The Quantum Theory of Light.* Clarendon Press, Oxford, 1983.

7.5   G.J. MILBURN, D.F. WALLS. Quantum nondemolition measurements via quadratic coupling. *Physical Review A 28*, 2065–2070, 1983.

7.6   N. IMOTO, H.A. HAUS, Y. YAMAMOTO. Quantum nondemolition measurement of the photon number via the optical Kerr effect. *Physical Review A 32*, 2287–2292, 1985.

7.7   M. BRUNE, S. HAROCHE, V. LEFEVRE, J.M. RAIMOND, N. ZAGURY. Quantum nondemolition measurement of small photon numbers by Rydberg-atom phase-sensitive detection. *Physical Review Letters 65*, 976–979, 1990.

7.8   H. PAUL. Phase destruction in quantum non-demolition measurement of the photon number. *Quantum Optics 3* 169–178, 1991.

7.9   M.J. HOLLAND, D.F. WALLS, P. ZOLLER. Quantum nondemolition measurements of photon number by atomic-beam deflection. *Physical Review Letters 67*, 1716–1719, 1991.

7.10 G.J. MILBURN, M.J. GAGEN. Rydberg-atom phase-sensitive detection and the quantum Zeno effect. *Physical Review A 46*, 1578 –1585, 1992.

7.11 B.L. SCHUMAKER. Quantum mechanical pure states with Gaussian wave functions. *Physics Reports 135*, 317–408, 1986.

7.12 B.L. SCHUMAKER, C.M. CAVES. New formalism for two-photon quantum optics. II. Mathematical foundation and compact notation. *Physical Review A 31*, 3093–3111, 1985.

7.13 H.P. YUEN, J.H. SHAPIRO. Quantum statistics of homodyne and heterodyne detection. In: *Coherence and Quantum Optics IV.* Eds L. Mandel, E. Wolf, Plenum Press, New York, pp. 719–727, 1978.

7.14 H.P. YUEN, J.H. SHAPIRO. Optical communication with two-photon coherent states - part III: quantum measurements realizable with photoemissive detectors. *IEEE Transactions in Information Theory IT-26*, 78–92, 1980.

7.15 K.R. PARTASARATHY. *Introduction to Probability and Measure*, Academic Press, New York, 1980.

7.16 M. GRABOWSKI. Homodyne detection and positive operator-valued measures. *Open Systems & Information Dynamics 1*, 307–312, 1993.

7.17 W. VOGEL, W. SCHLEICH. Phase distribution of a quantum state without using phase states. *Physical Review A 44*, 7642–7646, 1991.

7.18 E. VOGEL. Unpublished results, University of Cologne, 1994.

7.19 U. LEONHARDT, H. PAUL. Phase measurement and Q function. *Physical Review A 47*, R2460–R2463, 1993.

7.20 P. STOREY, M. COLLET, D. WALLS. Measurement-induced diffraction and interference of atoms. *Physical Review Letters 68*, 472–475, 1992. Atomic–position resolution by quadrature–field measurement. *Physical Review A 47*, 405–418, 1993.

7.21 PH. GRANGIER, G. ROGER, A. ASPECT. Experimental evidence for a photon anticorrelation effect in a beam splitter: a new light on single-photon interferences. *Europhysics Letters 1*, 173–179, 1986.

7.22 A. ASPECT, PH. GRANGIER. Wave-particle duality for single photons. *Hyperfine Interactions 37*, 3–18, 1987.

7.23 A. ASPECT, PH. GRANGIER. Wave-particle duality: a case study. In: *Sixty-Two Years of Uncertainty: Historical, Philosophical, and Physical Inquiries into the Foundations of Quantum Mechanics.* Ed. A.I. Miller, Plenum, New York, pp. 45–59, 1990.

7.24 P. MITTELSTAEDT, A. PRIEUR, R. SCHIEDER. Unsharp particle-wave duality in a photon split beam experiment. *Foundations of Physics 17*, 893–903, 1987.

7.25 B.C. SANDERS, G.J. MILBURN. Complementarity in a quantum nondemolition measurement. *Physical Review A 39*, 694–702, 1989.

7.26 P. BUSCH. Some realizable joint measurements of complementary observables. *Foundations of Physics 17*, 905–937, 1987.

Further Reading

H. MARTENS, W.M. DE MUYNCK. Neutron Interferometry and the Joint Measurement of Incompatible Observables, *Physical Review A 42*, 5079–5085, 1990.

H. MARTENS, W.M. DE MUYNCK. Compensation of self-phase modulation in a QND measurement of photon number based on the optical Kerr effect, *Quantum Optics 4*, 303–316, 1992.

H. MARTENS, W.M. DE MUYNCK. Single and joint spin measurements with a Stern-Gerlach device, *Journal of Physics A 26*, 2001–2010, 1993.

D.F. WALLS, G.J. MILBURN. *Quantum Optics*, Springer-Verlag, Berlin, 1994.

Notations

Subject Index

Boldface numbers show the pages where definitions are given.

Lecture Notes in Physics

For information about Vols. 1–408
please contact your bookseller or Springer-Verlag

New Series m: Monographs